National Center for Construction Education and Research

Pipeline Mechanical Level Three

Annotated Instructor's Guide

Prentice Hall
Upper Saddle River, NJ
Columbus, OH

contren™ Learning Series

American Petroleum Institute
Pipeliner Training and Assessment Program (PTAP)

This information is general in nature and intended for training purposes only. Actual performance of activities described in this manual requires compliance with all applicable operating, service, maintenance, and safety procedures under the direction of qualified personnel. References in this manual to patented or proprietary devices do not constitute a recommendation of their use.

10 9 8 7 6
ISBN 0-13-101081-6

Preface

This volume was developed by the National Center for Construction Education and Research (NCCER) in response to the training needs of the construction, maintenance, and pipeline industries. It is one of many in NCCER's *Contren*™ *Learning Series*. The program, covering training for close to 40 construction and maintenance areas, and including skills assessments, safety training, and management education, was developed over a period of years by industry and education specialists.

NCCER also maintains a National Registry that provides transcripts, certificates, and wallet cards to individuals who have successfully completed modules of NCCER's *Contren*™ *Learning Series*, when the training program is delivered by an NCCER Accredited Training Sponsor.

The NCCER is a not-for-profit 501(c)(3) education foundation established in 1995 by the world's largest and most progressive construction companies and national construction associations. It was founded to address the severe workforce shortage facing the industry and to develop a standardized training process and curricula. Today, NCCER is supported by hundreds of leading construction and maintenance companies, manufacturers, and national associations, including the following partnering organizations:

PARTNERING ASSOCIATIONS

- American Fire Sprinkler Association
- American Petroleum Institute
- American Society for Training & Development
- American Welding Society
- Associated Builders & Contractors, Inc.
- Associated General Contractors of America
- Association for Career and Technical Education
- Carolinas AGC, Inc.
- Carolinas Electrical Contractors Association
- Citizens Democracy Corps
- Construction Industry Institute
- Construction Users Roundtable
- Design-Build Institute of America
- Merit Contractors Association of Canada
- Metal Building Manufacturers Association
- National Association of Minority Contractors
- National Association of State Supervisors for Trade and Industrial Education
- National Association of Women in Construction
- National Insulation Association
- National Ready Mixed Concrete Association
- National Utility Contractors Association
- National Vocational Technical Honor Society
- North American Crane Bureau
- Painting & Decorating Contractors of America
- Portland Cement Association
- SkillsUSA-VICA
- Steel Erectors Association of America
- Texas Gulf Coast Chapter ABC
- U.S. Army Corps of Engineers
- University of Florida
- Women Construction Owners & Executives, USA

Some features of NCCER's *Contren*™ *Learning Series* are:

- An industry-proven record of success
- Curricula developed by the industry for the industry
- National standardization providing portability of learned job skills and educational credits
- Credentials for individuals through NCCER's National Registry
- Compliance with Apprenticeship, Training, Employer, and Labor Services (ATELS) requirements for related classroom training (CFR 29:29)
- Well-illustrated, up-to-date, and practical information

Acknowledgments

This curriculum was revised as a result of the farsightedness and leadership of the following sponsors:

Alyeska Pipeline Service Company
Colonial Pipeline Company
Koch Industries
Sun Pipeline Company
Marathon Ashland Pipeline LLC
Minnesota Pipeline
Plains All American Pipeline
Plantation Pipe Line Company

This curriculum would not exist were it not for the dedication and unselfish energy of those volunteers who served on the Authoring Team. A sincere thanks is extended to:

Bill Coulter
Ron Gunter
Debra Haifliegh
Jordan Janak
Scott Neujahr
Rich OBeid
Steve Sanders
Dave Yeager

A final note: This book is the result of a collaborative effort involving the production, editorial, and development staff at Prentice-Hall, Inc., and the National Center for Construction Education and Research. Thanks to all of the dedicated people involved in the many stages of this project.

Contents

Module 63301-02

Installing Rotating Equipment

Installing Rotating Equipment
Annotated Instructor's Guide

Module 63301-02

MODULE OVERVIEW

This module discusses methods of inspecting and preparing rotating equipment for installation. Trainees will learn how to install rotating equipment and to relieve pipe stress from rotating equipment.

PREREQUISITES

Please refer to the Course Map in the Trainee Module. Prior to training with this module, it is recommended that the trainee shall have successfully completed the following:

Core Curriculum; Pipeline Core, Modules 66101-02 and 66102-02; Pipeline Mechanical Levels One and Two

OBJECTIVES

Upon completion of this module, the trainee will be able to:

1. Identify inspection requirements for an equipment pad.
2. Describe the requirements for equipment base preparation.
3. Inspect equipment prior to installation.
4. Prepare equipment prior to installation.
5. Describe the installation process for rotating equipment.
6. Describe the process to relieve pipe stress from rotating equipment.

PERFORMANCE TASKS

Under the supervision of the instructor, the trainee should be able to:

1. Identify and describe the inspection requirements for an equipment pad.
2. Describe, inspect, and prepare equipment prior to installation.
3. Describe the installation process for rotating equipment.
4. Describe the process to relieve pipe stress from rotating equipment.

NCCER STANDARDIZED CRAFT TRAINING PROGRAM

The National Center for Construction Education and Research (NCCER) provides a standardized national program of accredited craft training. Key features of the program include instructor certification, competency-based training, and performance testing. The program provides trainees, instructors, and companies with a standard form of recognition through a National Craft Training Registry. The program is described in full in the *Guidelines for Accreditation*, published by the NCCER. For more information on standardized craft training, contact the NCCER by writing us at P.O. Box 141104, Gainesville, FL 32614-1104; calling 352-334-0911; or e-mailing info@nccer.org. More information may be found at our Web site, www.nccer.org.

HOW TO USE THIS ANNOTATED INSTRUCTOR'S GUIDE

Each page presents two sections of information. The larger section displays each page exactly as it appears in the Trainee Module. The narrow column ties suggested trainee and instructor actions to each page and provides icons to call your attention to material, safety, audiovisual, or testing requirements. The bottom of each page includes space for your notes.

Teaching Tip

If you see the Teaching Tip icon, that means there is a teaching tip associated with this section. Also refer to any suggested teaching tips at the end of the module.

SAFETY CONSIDERATIONS

Ensure that the trainees are equipped with appropriate personal protective equipment. Emphasize the importance of following all safety precautions and procedures when working with power tools.

PREPARATION

Before teaching this module, you should review the Module Outline, Objectives, Performance Tasks, and the Materials and Equipment List. Be sure to allow ample time to prepare your own training or lesson plan and gather all required equipment and materials.

MATERIALS AND EQUIPMENT LIST

Transparencies

Markers/chalk

Blank acetate sheets

Transparency pens

Pencils and scratch paper

Calculators

Module Examinations*

Performance Profile Sheets*

Overhead projector and screen

Whiteboard/chalkboard

Appropriate personal protective equipment

Copies of *49 CFR Part 192* (Gas) and/or *49 CFR Part 195* (Liquid)

Copies of your company policy and procedures manual

A variety of base plates

Installation drawings

Sample pumps and/or motors with physical damage

Sample base plates that are warped, damaged, or cracked

Equipment to perform the bearing heating method

Tools and equipment necessary to level base plates, including:
- Jack bolts
- Anchor bolts
- Shims
- Wrenches
- Wedges
- Hammers
- Levels
- Shim packs
- Straightedges
- Feeler gauges

Tools and materials necessary to perform clearance and interference installation, including:
- Brass hammers
- Key and setscrew couplings
- Cleaning solvents
- Shop clothes
- Tapered shafts
- Locking nuts
- Micrometers

Nonshrink and epoxy grout

Forms to practice grouting

Tools to work the grout

* Located in the Test Booklet.

TEACHING TIME FOR THIS MODULE

An outline for use in developing your lesson plan is presented below. Note that each Roman numeral in the outline equates to one session of instruction. Each session has a suggested time period of 2½ hours. This includes 10 minutes at the beginning of each session for administrative tasks and one 10-minute break during the session. Approximately 25 hours are suggested to cover *Installing Rotating Equipment.* You will need to adjust the time required for hands-on activity and testing based on your class size and resources.

Topic	Planned Time
Session I. Checking for Proper Construction of the Equipment Pad	
A. Inspection Process	________
B. Form Preparation	________
Session II. Performing Equipment Base Preparation, Part One	
A. Leveling Using Jack Bolts	________
B. Leveling Using Wedges	________
1. Laboratory	________
Session III. Performing Equipment Base Preparation, Part Two	
A. Leveling Using Shims	________
1. Laboratory	________
B. Precision Leveling Using Shims	________
Session IV. Inspecting Equipment Prior to Installation	
A. Installation Drawings	________
B. Inspection Procedures	________
1. Performance Testing	________
Session V. Preparing Equipment for Installation, Part One	
A. Key and Setscrew Method	________
B. Tapered Shaft and Locking Nut Method	________
Session VI. Preparing Equipment for Installation, Part Two	
A. Heating Couplings	________
1. Laboratory	________
B. Interference-Fit Method	________
1. Performance Testing	________
Session VII. Installing Rotating Equipment, Part One	
A. Setting the Pump	________
B. Setting the Pump Motor	________
Session VIII. Installing Rotating Equipment, Part Two	
A. Site Visit	________
B. Performance Testing	________
Session IX. Stress Relieving and Aligning Rotating Equipment, Part One	
A. Flexibility in Layout	________
B. Installing Expansion Loops	________
C. Cold Springing	________

Session X. Stress Relieving and Aligning Rotating Equipment, Part Two

A. Grouting Baseplates ____________
   1. Laboratory ____________
B. Preliminary Alignment Between Equipment ____________
C. Summary
   1. Summarize module ____________
   2. Answer questions ____________
D. Module Examination ____________
   1. Trainees must score 70% or higher to receive recognition from the NCCER.
   2. Record the testing results on Craft Training Report Form 200 and submit the results to the Training Program Sponsor.
E. Performance Testing ____________
   1. Trainees must perform each task to the satisfaction of the instructor to receive recognition from the NCCER.
   2. Record the testing results on Craft Training Report Form 200 and submit the results to the Training Program Sponsor.

Module 63301-02

Installing Rotating Equipment

Instructor's Notes:

Instructor's Notes:

Introduction to the Pipeline Operator Qualification Rule, API, and NCCER

It could be said that petroleum products keep America running. Petroleum products like oil and natural gas are the principal sources of energy in the United States and are used in many manufacturing and chemical processes. Petroleum products are primarily transported by pipeline. There are more than a million miles of liquid and natural gas pipeline in this country, ranging in diameter from two inches to four feet. These pipelines bring crude oil and natural gas from production fields to processing plants and refineries—and ultimately to the consumer.

Pipeline Operator Qualification

Petroleum products are considered hazardous, so pipeline personnel must take extra precautions and follow specific procedures when working with them. To help ensure the safe handling and transportation of petroleum products, the United States Department of Transportation (DOT) mandates safe practices as federal regulations.

One such regulation, the Operator Qualification (OQ) rule, requires pipeline operators to develop a written qualification plan to evaluate their personnel and their contract personnel (see *49 CFR Part 192 Subpart N—Gas*, and *49 CFR Part 195 Subpart G—Liquid*). The evaluation hinges on the operator's ability to perform certain covered tasks and to be able to recognize and respond to abnormal operating conditions that may be encountered while performing these covered tasks.

Covered Tasks

A pipeline performance task is deemed *covered* if the activity passes the following four-part test:

1. Is it performed on a pipeline facility?
2. Is it an operations or maintenance task?
3. Is it performed as a requirement of *49 CFR Part 195 (Liquid)* or *49 CFR Part 192 (Gas)*?
4. Does it affect the operation or integrity of the pipeline facility?

If the answer to all four questions is *yes*, then the activity is considered *covered* and must figure into the evaluation process of pipeline personnel.

American Petroleum Institute

The OQ rule and its related covered tasks provided the catalyst to develop a unified pipeline training and assessment program—so that pipeline operators and owners have ready access to the assessment and training tools they need. This effort was the direct result of the vision and leadership provided by The American Petroleum Institute (API) and API members.

Efforts to unify the industry are nothing new to API. Indeed, the impetus for forming the Institute in 1919 was the need to standardize engineering specifications for drilling and production equipment. Today, the most pressing issues revolve around public perceptions and government policies toward the industry. Speaking with one voice on these issues has become as essential as having interchangeable parts in the field. API's members determine the petroleum industry's positions on public policy as well as the standards that govern its day-to-day operations worldwide.

Consortium on Operator Qualification

API assembled a focus group whose primary purpose was to identify covered tasks and provide compliance recommendations to pipeline operators.

The Consortium on Operator Qualification (COOQ) became the guiding force for the identification and development of covered tasks for the liquid pipeline industry. Their work product, API publication 1161, *Guidance Document for the Qualification of Liquid Pipeline Personnel*, formed the basis for the resulting Pipeliner Training and Assessment Program.

The covered task list created in API publication 1161 has now evolved to include gas covered tasks and is maintained by NCCER. The most current covered task list can be downloaded at www.nccer.org.

National Center for Construction Education and Research

With the guidance and support of API, its members, and the Consortium on Operator Qualification, the National Center for Construction Education and Research (NCCER) was able to design and implement the Pipeliner Training and Assessment Program.

NCCER is a not-for-profit 501(c)(3) education foundation established in 1995 by the world's largest and most progressive construction and maintenance contractors and national contractor associations. NCCER was founded to address the

severe workforce shortage facing the construction and maintenance industries and to develop a standardized training process and curricula. Today, NCCER is supported by hundreds of leading construction and maintenance companies, manufacturers, and national associations and developed more than 35 construction and maintenance titles.

NCCER also maintains a National Registry, which contains the records of trainees and instructors who have received NCCER training. All completion information for every trainee is sent to the NCCER and filed within the Registry. The Registry can then confirm training and skills for workers as they move from company to company, state to state, or even within their own company.

Standardized Curriculum (NCCER's Contren™ Learning Series)

NCCER curricula are developed using the expertise of industry subject matter experts (SMEs). The use of SMEs as an integral part of every new development or revision ensures that the curriculum is current, industry-driven, and indicative of industry practices. The development and/or revision process is facilitated from start to finish by NCCER project managers, who work closely with the SMEs and a select team of technical writer(s) on the development and presentation of the curriculum.

Pipeline: Five Training Paths

The pipeline curriculum focuses on essential skills and competencies within five skill areas:

- Corrosion Control
- Electrical and Instrumentation
- Maintenance
- Mechanical
- Operations (Field and Control Center)

SMEs from the pipeline industry have determined the skills and competencies needed based on API's covered tasks and the consensus of the group/industry. Specific company information is generally not incorporated into the curriculum unless the SMEs determine that the information is the industry standard.

Training Modules and Covered Tasks

The smallest unit of instruction in the curriculum is a *module*. Each curriculum is composed of modules (the equivalent of stand-alone chapters) that are logically sequenced into complete *levels* of instruction. Every module is task-oriented and requires the successful completion of a written (knowledge-based) examination and a performance (hands-on) test. Modules may be individually selected and/or combined for efficient task training. Modules in the pipeline curricula, to the extent possible, will align with API-developed covered tasks. While not all modules will relate to covered tasks, all covered tasks will relate to at least one module.

Benefits of Training to the Pipeline Marketplace

Applying NCCER's standardized training process allows pipeline owners and contractors in the United States to establish the benchmark for acceptable skill and competency levels within the pipeline industry. This process will assist training organizations in developing and/or improving standardized pipeline training programs that effectively match their resources and operating conditions while responding to regulatory demands and to the needs of the industry. The process will also recognize training organizations that meet the minimum standards of competence, warranting regulatory, public, and professional confidence. This in turn will foster national unity within the pipeline industry for the mutual benefit of individual technicians, owners, contractors, and ultimately pipeline-users and society.

Additional Resources

To review the Pipeliner Training and Assessment Program Implementation and Resource Guide and the latest version of pipeline covered tasks, visit NCCER's Web site at www.nccer.org, or contact:

National Center for Construction Education and Research
PO Box 141104
Gainesville, FL 32614-1104
(888) 622-3720
Web site: www.nccer.org

For more information on API, please contact:

American Petroleum Institute
1220 L Street, Northwest
Washington, D.C. 20005
(202) 682-8000
Web site: www.api.org

To order *API Pub 1161*, contact Global Engineering Documents by calling 1-800-854-7149 or visiting their Web site at www.global.his.com.

Additional sources of information on the pipeline operator qualification regulation include:

DOT – Office of Pipeline Safety
www.ops.dot.gov

Transportation Safety Institute
www.tsi.dot.gov

Instructor's Notes:

COURSE MAP

This course map shows all of the modules in the third level of the Pipeline Mechanical curriculum. The suggested training order begins at the bottom and proceeds up. Skill levels increase as you advance on the course map. The local Training Program Sponsor may adjust the training order.

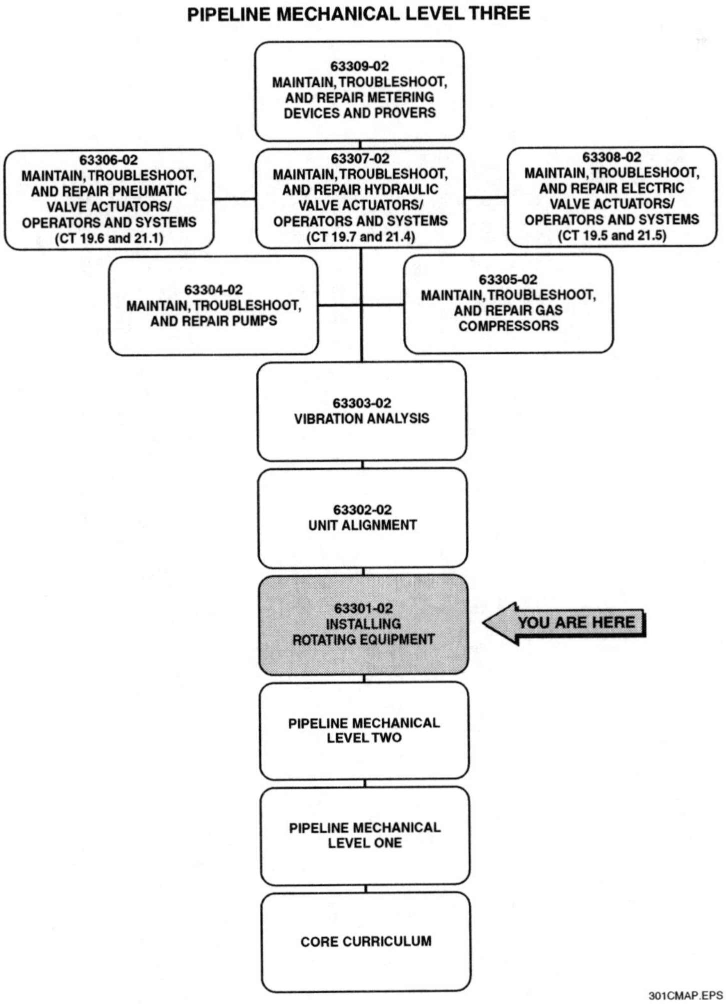

301CMAP.EPS

Homework

Assign reading of Module 63301-02.

Instructor's Notes:

MODULE 63301-02 CONTENTS

Figures

Instructor's Notes:

MODULE 63301-02

Installing Rotating Equipment

Objectives

When you have completed this module, you will be able to do the following:

1. Identify inspection requirements for an equipment pad.
2. Describe the requirements for equipment base preparation.
3. Inspect equipment prior to installation.
4. Prepare equipment prior to installation.
5. Describe the installation process for rotating equipment.
6. Describe the process to relieve pipe stress from rotating equipment.

Prerequisites

Before you begin this module, it is recommended that you successfully complete the following: Core Curriculum; Pipeline Core, Modules 66101-02 and 66102-02; Pipeline Mechanical Level One; Pipeline Mechanical Level Two.

Required Trainee Materials

1. Pencil and paper
2. Appropriate personal protective equipment

1.0.0 ◆ INTRODUCTION

The equipment pad is the key to effective operation of rotating equipment. The equipment pad must conform to the construction specifications that will provide the required support, levelness, and finish for the equipment that is going to be installed on it.

Prealignment is the preliminary, rough alignment of the equipment that will be installed on the equipment pad. The preliminary alignment will provide the starting point for initial setup of the equipment and allow for other work to be accomplished. This module will explain some techniques on how to prealign the pump and motor, install the rotating equipment, and relieve pipe **stress** to the equipment.

2.0.0 ◆ CHECKING FOR PROPER CONSTRUCTION OF THE EQUIPMENT PAD

Normally, the equipment pad is constructed by a contractor at the site that your company requests. Additionally, your company will provide the contractor with specific construction requirements that must be met. These construction requirements are designed to meet the needs of the equipment that will be installed on the pad. It may be your responsibility to perform interval inspections of the construction progress to ensure that the pad meets your company's design criteria. The following section describes some items that will be helpful in your inspection process.

Before concrete is placed in forms, the forms should be checked to ensure that they are accurately set, clean, tight, and adequately braced, and they must have a form surface that will produce the desired appearance after the concrete is cured. All dirt, sawdust, shavings, tie wire, loose nails, and other debris should be removed from the forms. The forms should be coated with a releasing agent such as form oil to prevent adhesion of the concrete and allow for easy removal. The releasing agent should be formulated for the particular usage and material to which it is to be applied.

Materials

Ensure that you have everything required to teach the course. Check the Materials and Equipment List at the front of this module.

Classroom

Explain that terms shown in bold are defined in the Glossary at the back of this module.

Audiovisual

Show Transparency 1, Course Objectives.

Show Transparency 2, Performance Tasks.

Teaching Tip

See the general Teaching Tip at the end of this module.

Classroom

Discuss the role of the equipment pad in rotating equipment. Explain how preliminary alignment provides the starting point for equipment setup.

Discuss considerations for inspecting an equipment pad.

Classroom

Explain how to inspect form preparation. Stress the importance of following specifications when mixing concrete and water.

Ask trainees to write down when inspections of the equipment pad are required. Have trainees review what is involved in each inspection. Answer any questions trainees have.

Homework

Have trainees review Sections 3.0.0–3.2.0 for the next session.

Materials

Ensure that you have everything required for the laboratory during this session.

Classroom

Discuss the methods used to set up equipment base plates.

Audiovisual

Show Transparency 3 (Figure 1). Explain how to level a base plate using jack bolts.

All reinforcing steel must be cleaned. It should be free of mud, oil, loose rust, and mill scale. Loose, flaky, scaly rust that would affect the bond should be removed by wire brushing or other means. If reinforcing steel is sticking out from a former placement of concrete, it should be cleaned of any dried mortar coatings or splashes that might have occurred during the former placement.

Other points to check during form preparation include the following:

- Subgrade firm and on grade with the soil undisturbed or heavily tamped
- Forms properly aligned and braced
- Forms set to proper grade
- **Screeds** set at proper locations for efficient placing and leveling of concrete
- Screeds set to proper grade
- Bulkheads properly braced and set to proper grade
- Reinforcing mesh or rods properly placed and not touching the forms
- Expansion joint material in place and on grade
- Embedded items, such as anchor bolts, traps, pipe, and conduit, held firmly in place using templates or attached to the forms or reinforcing steel

For most pad construction, a formal final inspection of the formwork is required before placement of the concrete can occur. This inspection should include a representative from the contractor, various city or state inspectors (as required), and you or another representative from your company. This inspection should produce a written report of your findings.

The mixture of concrete and the amount of water content that is to be used is critical and should conform to the specifications. During the installation process, the concrete should be tapped down to remove air bubbles from the mixture. After the concrete is poured and is still damp, the surface and edges should be finished to the specifications as outlined in the construction drawings. Adequate time should be allowed for the concrete to properly cure. This will depend on the type of concrete used and the weather conditions.

After the concrete has cured, another inspection of the pad is required. This inspection should follow the written instructions contained in the construction specifications. These instructions will vary depending on the load of the equipment, type of equipment to be installed, and the location. You should always check the surface of the pad to ensure that it will be adequate for the equipment installation.

3.0.0 ◆ PERFORMING EQUIPMENT BASE PREPARATION

The first step in installing rotating equipment is to set up and level the equipment base plate. If the base plate is not level, it is very difficult to align the equipment. The purpose of setting up the base plate is to bring the plate to the proper height and make it perfectly level. There are several ways to set up a base plate, including:

- Leveling using **jack bolts**
- Leveling using **wedges**
- Leveling using **shims**
- Precision leveling using shims

3.1.0 Leveling Using Jack Bolts

Many bases can be quickly and accurately leveled using jack bolts. Jack bolts are large bolts or all-thread rods that screw into threaded holes near each corner of the base plate. Tightening a jack bolt raises that corner of the base plate; loosening the jack bolt lowers the corner. A ¼-inch shim should be placed under each jack bolt to prevent it from digging into the concrete (see *Figure 1*). A small indentation should be drilled in each shim for the jack bolt to fit into to prevent the shim or base plate from moving. Follow these steps to level a base plate using jack bolts:

Step 1 Adjust a jack bolt on one corner of the plate using a wrench to bring that corner to the proper elevation.

Step 2 Place a level on the base plate and check the levelness from the jack bolt adjusted in Step 1 to a second jack bolt.

> **NOTE**
> The base plate must be clean and free of any foreign matter that might keep the level from lying flat on the plate.

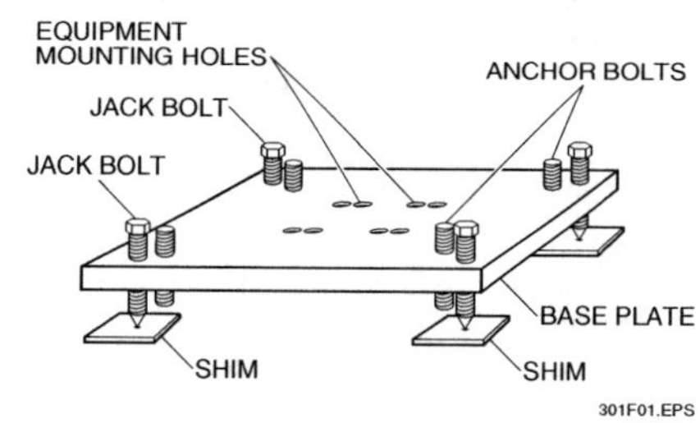

Figure 1 ◆ Base plate with jack bolts.

Instructor's Notes:

Step 3 Adjust the second jack bolt until the bubble in the level is centered.

Step 4 Reposition the level and adjust the jack bolts until the base plate is level in all directions.

Step 5 Install shims as specified by company policies or by the original equipment manufacturer.

Step 6 Repeat Steps 2 through 4 to ensure final set on the base plate.

Step 7 Install the anchor bolt nuts.

Step 8 Tighten the anchor bolts to secure the plate once it is at the correct elevation and is level in all directions.

Step 9 Check the level in all directions.

3.2.0 Leveling Using Wedges

Although leveling with jack bolts is quick and accurate, many base plates do not have jack bolts. Wedges are as easy to work with, and they work with any base plate. Always use wedges in pairs, one on top of the other. A single wedge is less accurate than a double wedge (see *Figure 2*).

One pair of wedges should be used at each corner of the base plate and next to each anchor bolt. The wedges are driven in with a hammer to adjust the height of the plate. Follow these steps to level a base plate using wedges:

Step 1 Place the wedges under the four corners of the base plate.

Step 2 Make a rough adjustment to bring the base plate just below the correct elevation.

Step 3 Using a hammer, adjust the wedges under one corner of the base plate to bring that corner to the correct elevation.

Step 4 Place a level on the base plate and check the levelness from the set corner to a second corner.

Step 5 Adjust the second wedge until the bubble in the level is centered.

Step 6 Reposition the level and adjust the wedges until the base plate is level in all directions.

Step 7 Install the anchor bolt nuts.

Step 8 Tighten the anchor bolts to secure the base plate once it is at the correct elevation and is level in all directions.

Step 9 Check the level in all directions.

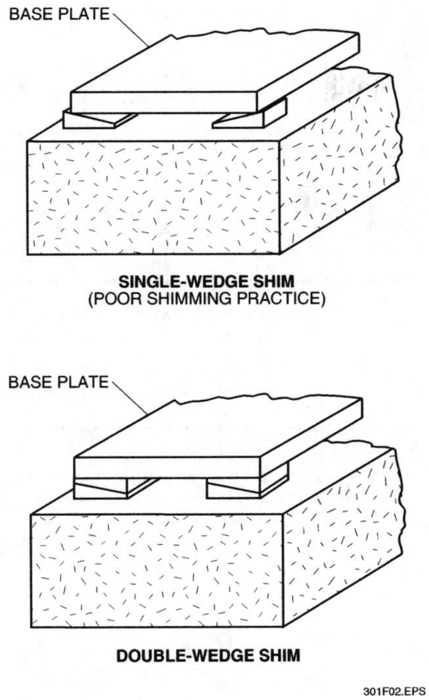

Figure 2 ◆ Single- and double-wedge shims.

3.3.0 Leveling Using Shims

Equipment is often set on hard **shim packs**, although these are more difficult to adjust than wedges or jack bolts. Shim packs are set at the corners of the base plate and around the anchor bolts. Shim packs can be put in place and leveled before the base is set on top of them (see *Figure 3*). After the shim packs are leveled, the base plate is put in place and leveled precisely.

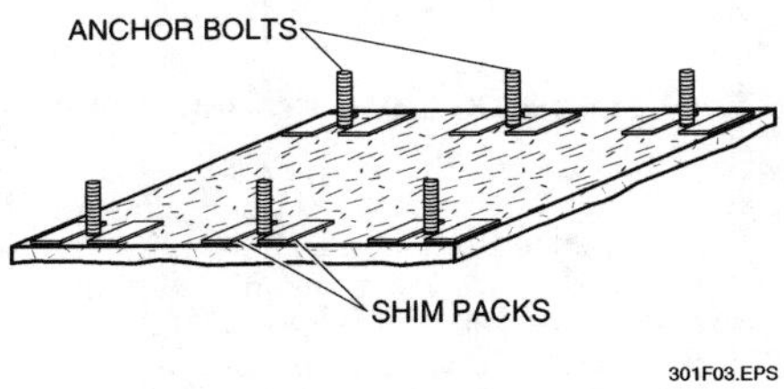

Figure 3 ◆ Shim packs set in place.

Classroom

Refer to Figure 2. Review the steps to level a base plate using wedges. Explain why it is better to use a double-wedge shim than it is to use a single-wedge shim.

Laboratory

Set up stations with jack bolts, wedges, and base plates. Ask trainees to practice leveling the base plates following the procedures for these two methods. Go over any questions trainees have.

Homework

Have trainees review Sections 3.3.0–3.4.0 for the next session.

Materials

Ensure that you have everything required for the laboratory and demonstration during this session.

Audiovisual

Show Transparency 4 (Figure 3). Explain why it is more difficult to adjust shim packs than wedges or jack bolts.

Classroom

Review the procedures to level a base plate using shim packs.

Laboratory

Provide a sample base plate with shim packs and anchor bolts. Ask a trainee to adjust the shim packs until they are at the correct elevation and level. Have trainees discuss what was done correctly and/or incorrectly.

Classroom

Explain how to perform precision leveling on a base plate. Ask whether trainees have any questions.

Audiovisual

Show Transparency 5 (Figure 5). Review the formula to calculate slope.

Demonstration

On the whiteboard/chalkboard, demonstrate how to use slope to determine the total shim thickness needed to make the base plate level.

Teaching Tip

See the Teaching Tip for Section 3.4.0 at the end of this module.

Follow these steps to level a base plate using shim packs:

Step 1 Set a shim pack at one corner of the base plate at the correct elevation.

> **NOTE**
> It is better to use a few thick shims than many thin shims. Thin shims are used in the final, precise leveling. All shims must be cut the same size and neatly packed.

Step 2 Position a level from the first shim pack to the position of a second shim pack.

Step 3 Install the second shim pack and adjust it until the bubble in the level is centered.

> **NOTE**
> If the level does not reach from one shim pack to the other, use a straightedge with the level placed on top of it (see *Figure 4*).

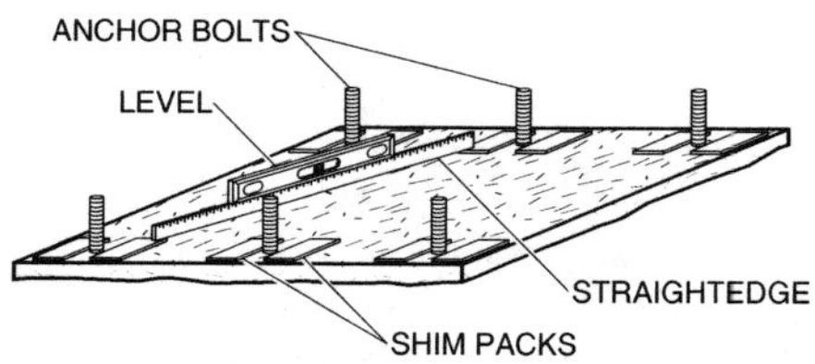

Figure 4 ◆ Leveling shim packs using a level and straightedge.

Step 4 Reposition the level and adjust the shim packs until they are all at the correct elevation and level.

Step 5 Place the base plate on the shim packs.

Step 6 Check the level in all directions.

3.4.0 Precision Leveling Using Shims

Once the base plate has been prealigned, it can be more precisely leveled by further adjusting the shims. Many technicians adjust the shims by trial and error. This works reasonably well with small base plates with shims that are easy to change. On larger base plates, however, each change can take considerable time and effort. Using a level and a feeler gauge, you can calculate the adjustment that is needed.

Follow these steps to perform precision leveling of the base plate:

Step 1 Place the level on the base plate near the center.

> **NOTE**
> The base plate must be clean and free of any foreign matter that might keep the level from lying flat on the base plate.

Step 2 Slip the feeler gauge under the low end of the level, trying different thicknesses until the bubble in the level is centered.

Step 3 Note the thickness of the feeler gauge.

> **NOTE**
> Precision levels have a scale that shows the amount off level per foot. With other levels, you must calculate the amount off level.

Step 4 Divide the feeler gauge thickness by the length of the level to calculate the slope in thousandths of an inch per inch (see *Figure 5*).

Step 5 Multiply the slope determined in Step 4 by the length of the base plate between bearing points to determine the total shim thickness needed to make the base plate level (see *Figure 6*).

$$\frac{.016"}{18"} = .0009 \text{ in./in.}$$

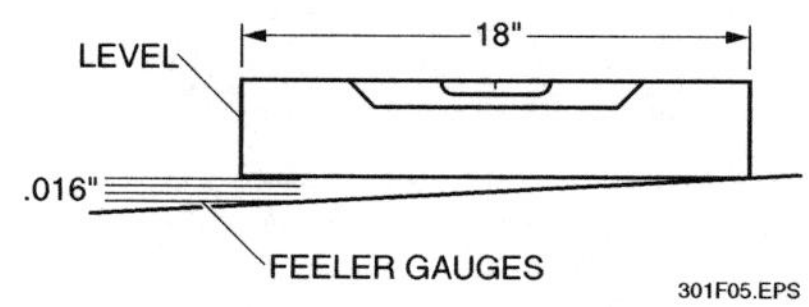

Figure 5 ◆ Calculating slope in inches per inch.

$$.0009 \text{ in./in.} \times 65" = .059"$$

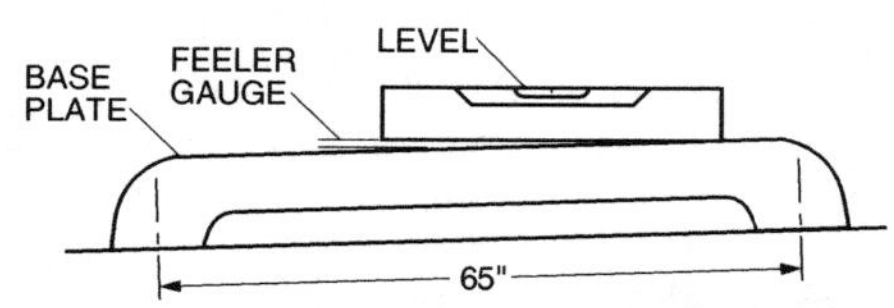

Figure 6 ◆ Calculating total shim thickness.

Instructor's Notes:

Step 6 Add shims to the low end of the base plate or remove shims from the high end to level the base plate.

Step 7 Turn the level 90 degrees on the base plate.

Step 8 Repeat Steps 2 through 6 to level the base plate in the other direction.

Step 9 Tighten all anchor bolts.

Step 10 Check the level in all directions.

4.0.0 ◆ INSPECTING EQUIPMENT PRIOR TO INSTALLATION

Before installing and prealigning a piece of equipment, the equipment must be inspected to ensure that it is suitable for service. Follow these steps to inspect a pump and motor before installation:

Step 1 Check the installation drawings against the equipment delivered for installation.

> **NOTE**
> The equipment delivered must match the drawing requirements to ensure the proper size motor and pump are to be installed in the proper location.

Step 2 Inspect the pump and motor for any signs of physical damage.

> **NOTE**
> If there are signs of physical damage, contact your supervisor about what actions should be taken before the equipment is installed.

Step 3 Check the motor to ensure that it is filled to the proper level with the proper lubricant.

Step 4 Rotate the pump and motor by hand to ensure that they are not locked or binding, that there are no flat places in the bearings, and that the shafts are turning true.

> **NOTE**
> The shafts of the pump and motor should turn smoothly. If there are flat places in the bearings, you can feel a roughness as you rotate the shaft.

Step 5 Check the motor for radial runout according to the manufacturer's specifications.

Step 6 Inspect the equipment base plate to ensure that it is not warped, damaged, or cracked. The base plate should have a smooth, clean surface.

5.0.0 ◆ PREPARING EQUIPMENT FOR INSTALLATION

The first step in setting a pump and motor is to install couplings on the mating shafts. There are two basic types of couplings: rigid and flexible. As the name indicates, a rigid coupling connects the shafts together rigidly and does not allow for any misalignment or movement. Flexible couplings allow for a small amount of misalignment and some torsional and **axial movement**. Two basic methods of coupling installation are clearance installation and interference installation.

5.1.0 Clearance Installation

Clearance installation uses a setscrew or nut to lock the coupling onto the shaft. The two most common methods for clearance installation are the key and setscrew method and the tapered shaft and locking nut method.

5.1.1 Key and Setscrew Method

With the key and setscrew method, the coupling has a key that fits into the shaft. A setscrew then screws through the coupling and presses on the key to lock the coupling onto the shaft (see *Figure 7*).

The coupling should be placed on the shaft by hand or by lightly tapping it with a brass hammer. Do not drive the coupling onto the shaft because this could damage the shaft and the coupling. If the coupling does not go on easily, the shaft or the coupling must be dressed down until the coupling

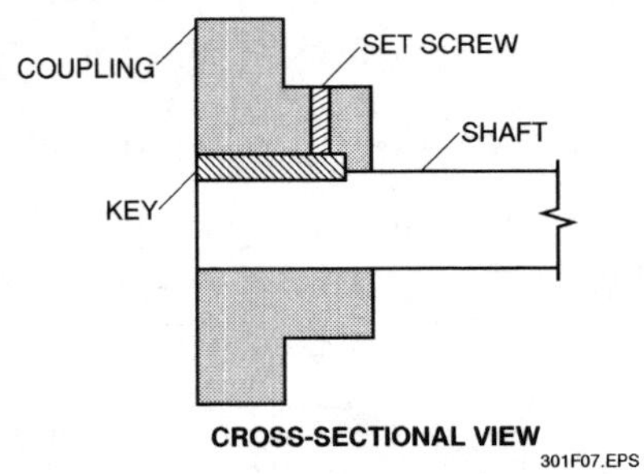

Figure 7 ◆ Key and setscrew coupling.

Homework

Have trainees review Section 4.0.0 for the next session.

Materials

Ensure that you have everything required for testing during this session.

Teaching Tip

See the Teaching Tip for Section 4.0.0 at the end of this module.

Performance Testing

Have each trainee perform Performance Task 1 to your satisfaction. Fill out a Performance Profile Sheet for each trainee.

Homework

Have trainees review Sections 5.0.0–5.1.2 for the next session.

Materials

Ensure that you have everything required to teach this session.

Classroom

Review the two methods used for coupling installation. Discuss the two most common methods for clearance installation.

Refer to Figure 7. Explain how to properly install a key and locking screw coupling. Answer any questions trainees have.

Audiovisual

Show Transparency 6 (Figure 8). Compare the procedures to install a tapered coupling with the key and setscrew method.

Safety

Explain how to prevent damage caused by overtightening the nut.

Homework

Have trainees review Section 5.2.0 for the next session.

Materials

Ensure that you have everything required for the laboratory during this session.

Laboratory

Set up several stations with the tools and materials necessary to preheat couplings. Ask trainees to practice using a bearing heating method to heat a coupling.

does go on easily. When installing a coupling on an old shaft, the shaft can be dressed down to remove burrs or repair damage. Never dress the shaft below its original diameter. Dress down the coupling instead.

Follow these steps to install a key and locking screw coupling:

Step 1 Clean the shaft thoroughly using an approved solvent and a shop cloth.

Step 2 Sand the shaft lightly using an emery cloth to smooth the surface of the shaft.

Step 3 Inspect the shaft to ensure that there are no burrs that would interfere with the coupling fitting onto the shaft.

Step 4 Coat the shaft lightly with oil or grease to make the installation of the coupling easier.

Step 5 Position the key in the shaft **keyway**.

Step 6 Slip the coupling onto the shaft over the key.

Step 7 Align the end of the coupling flush with the end of the shaft.

Step 8 Tighten the setscrew to hold the coupling in place.

5.1.2 Tapered Shaft and Locking Nut Method

The tapered coupling and locking nut use a wedging principle to hold the coupling onto the shaft (see *Figure 8*). The coupling is slipped onto the shaft, and the nut is tightened, causing a wedging action that generates tremendous holding force. This type of coupling also has a key. The combined wedging force and key prevent the coupling from rotating on the shaft.

Follow these steps to install a tapered coupling:

Step 1 Clean the shaft thoroughly.

Step 2 Inspect the shaft threads to ensure that they are clean and not damaged.

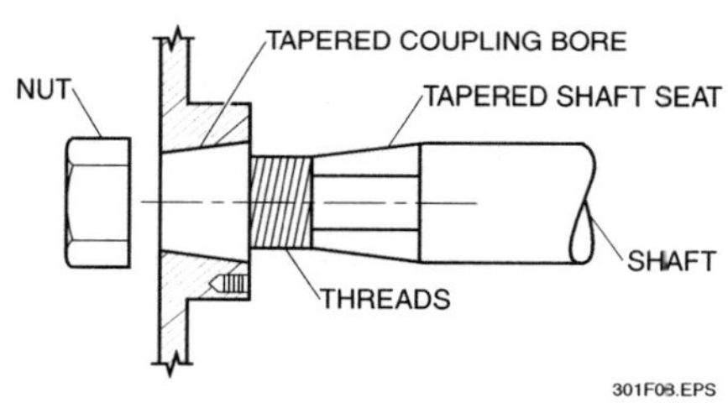

Figure 8 ◆ Tapered coupling and shaft.

Step 3 Install the key in the shaft keyway.

Step 4 Slip the coupling onto the shaft over the key.

Step 5 Screw the nut onto the end of the shaft.

Step 6 Tighten the nut to lock the coupling onto the shaft.

> **CAUTION**
> Do not overtighten the nut because this may damage the nut and the shaft. Use a torque wrench to tighten the nut to the proper torque.

5.2.0 Interference-Fit Method

In the interference-fit method of installation, the coupling has a bore that is slightly smaller than the shaft diameter. The coupling must be **preheated** to expand to the bore before it can be installed on the shaft. As the coupling cools, it shrinks to grip the shaft securely. There are several methods for heating a coupling, and they are the same as those for heating a bearing.

Follow these steps to install a coupling using the interference-fit method:

Step 1 Determine from the manufacturer's installation procedure where the coupling should fit on the shaft.

Step 2 Clean the shaft thoroughly to remove any foreign material that might keep the coupling from sliding onto the shaft.

Step 3 Measure the shaft diameter using a micrometer and record the measurement.

Step 4 Measure the height and width of the key installed on the shaft using an outside micrometer.

Step 5 Measure the inside diameter of the coupling using an inside micrometer to ensure that the coupling is the proper size to fit the shaft.

> **NOTE**
> The inside diameter of the coupling should be smaller than the outside diameter of the shaft by no more than 0.0015 inch per 1 inch of shaft diameter. For example, a 1-inch shaft requires a coupling that is 0.0015 inch smaller, while a 2-inch shaft requires a coupling that is 0.0030 inch smaller.

Instructor's Notes:

Step 6 Install a stop on the shaft that will stop the coupling in its proper position when it is installed.

> **NOTE**
> The stop will help ensure that the coupling is positioned properly and quickly. The stop may be a clamp or some other type of stop fastened to the shaft that can be removed after the coupling installation is complete.

Step 7 Heat the coupling to a temperature between 200° and 250°F using a bearing heating method.

> **CAUTION**
> Be careful not to heat the coupling over 250° to 300°F because overheating will damage the coupling.

Step 8 Measure the inside diameter of the coupling to ensure that it has expanded enough to fit on the shaft. Using a temperature stick can give a more accurate gauge of the heat and expansion.

Step 9 Remove the coupling from the heating device, ensure the proper alignment and fit of the key and keyway, slip the coupling onto the shaft, and position it against the stop as quickly as possible before it begins to cool and shrink.

> **WARNING!**
> Wear protective gloves when handling the coupling to prevent burning your hands.

Step 10 Remove the stop from the shaft.

Step 11 Provide warning indicators to prevent other personnel from touching the hot coupling.

6.0.0 ◆ INSTALLING ROTATING EQUIPMENT

After the base plate has been installed, it is time to set the equipment on the pad with a rough alignment so that mounting holes can be drilled. This section describes the installation of rotating equipment, which involves setting the pump and setting the pump motor.

6.1.0 Setting the Pump

Once the base plate is set and the couplings are installed, the pump is ready to be placed in position on the base plate. This is only a rough alignment to drill and tap the bolt holes for the equipment. Final, precise alignment is accomplished later. The pump must be set on the equipment center line and placed to allow room for the motor.

Follow these steps to set the pump:

Step 1 Ensure that the base plate is clean and free of foreign matter that might interfere with the pump being set level.

Step 2 Place the pump on the base plate, aligning it with the equipment center lines (see *Figure 9*).

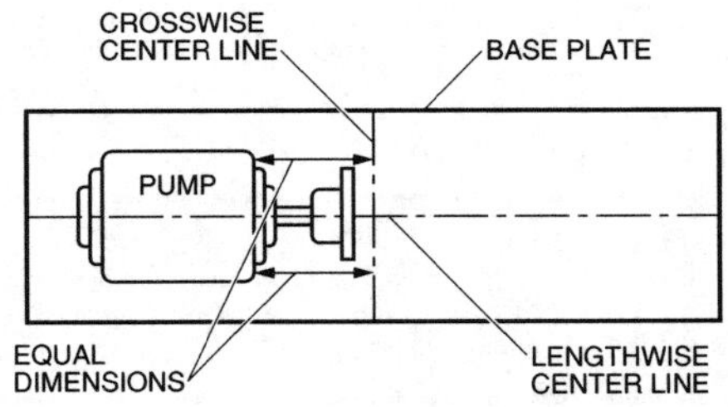

Figure 9 ◆ Aligning a pump with equipment center lines.

> **CAUTION**
> Thoroughly inspect all lifting equipment before lifting the pump. Replace any lifting equipment that is not serviceable.

> **NOTE**
> The pump must be centered on the lengthwise center line and perfectly parallel to the crosswise center line, leaving ample room for the pump motor.

Step 3 Place shims under the pump base to level the pump discharge and plumb the pump suction.

> **NOTE**
> Do not to move the pump out of alignment with the center line. If the pump moves, it will become bolt-bound and interfere with final alignment.

Safety

Discuss the consequences of overheating a coupling.

Stress the importance of wearing protective gloves when handling the coupling.

Teaching Tip

See the Teaching Tip for Section 5.0.0 at the end of this module.

Performance Testing

Have each trainee perform Performance Task 2 to your satisfaction. Fill out a Performance Profile Sheet for each trainee.

Homework

Have trainees review Sections 6.0.0–6.2.0 for the next session.

Materials

Ensure that you have everything required to teach this session.

Classroom

Refer to Figure 9. Explain how to safely place the pump on the base plate.

Safety

Ask a trainee to explain how to properly inspect lifting equipment.

Discuss the consequences of failing to inspect all lifting equipment.

Show Transparency 7 (Figure 10). Explain how to measure the horizontal parallel misalignment.

Ensure that trainees understand how to take readings using a dial indicator. Go over any questions trainees may have.

Step 4 Punch-mark the bolt hole locations using a transfer punch through the pump bolt holes.

Step 5 Remove the pump from the base.

Step 6 Drill and tap the bolt holes in the base plate.

Step 7 Place the pump back onto the base plate.

Step 8 Install the hold-down bolts.

6.2.0 Setting the Pump Motor

Once the pump is set and bolted, the pump motor is ready to be set. The motor must be set and accurately aligned with the pump center line before drilling and tapping the motor bolt holes into the base plate.

Follow these steps to set the pump motor:

Step 1 Place the motor on the base plate, aligning it with the lengthwise center line.

CAUTION
Thoroughly inspect all lifting equipment before lifting the motor. Replace any lifting equipment that is not serviceable.

Step 2 Move the motor toward the pump until the couplings are close enough together to set the gap between the couplings.

Step 3 Ensure that the motor is on the magnetic center if the motor contains sleeve bearings.

Step 4 Set the gap between the couplings using a thickness gauge.

Step 5 Align the couplings as close as possible by eyesight.

NOTE
The gap setting may vary, depending on the type of coupling used. Refer to the coupling manufacturer's specifications for the proper gap setting.

Step 6 Place a straightedge on the side of the couplings.

Step 7 Measure the horizontal parallel (side-to-side) misalignment using a thickness gauge (see *Figure 10*).

Step 8 Move the motor to correct the horizontal parallel misalignment.

NOTE
The horizontal parallel misalignment of the couplings must be minimal before the motor hold-down bolt holes are drilled and tapped, because once the bolt holes are drilled, there is very little side-to-side movement of the motor. After horizontal misalignment has been corrected as much as possible with a straightedge and thickness gauge, use a dial indicator to further align the coupling from side to side.

Step 9 Attach a dial indicator to the coupling.

Step 10 Set the dial indicator at the 3 o'clock position on the motor half of the coupling.

Step 11 Rotate the motor coupling until a zero reading appears on the dial.

Step 12 Rotate the pump half of the coupling until the dial is at the 9 o'clock position on the motor half of the coupling.

Step 13 Read the dial indicator.

Step 14 Adjust the motor and take readings at the 3 o'clock and 9 o'clock positions until zero readings on both sides of the coupling are achieved (see *Figure 11*).

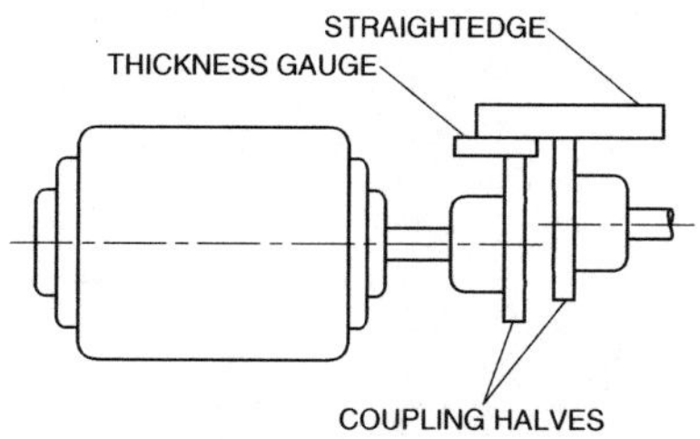

Figure 10 ◆ Measuring horizontal parallel misalignment.

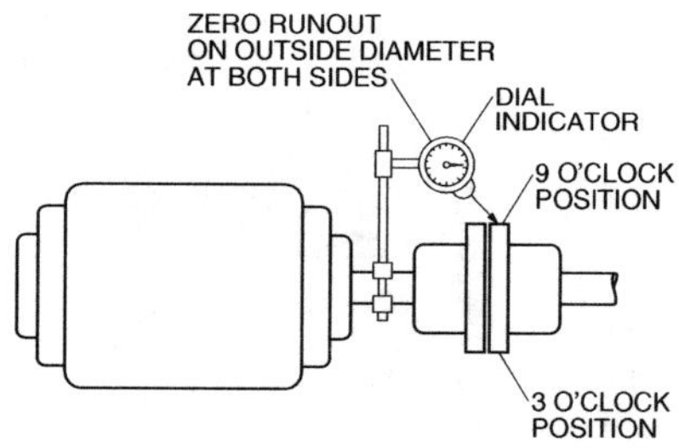

Figure 11 ◆ Taking readings using a dial indicator.

Instructor's Notes:

Step 15 Scribe the outlines of the holes carefully.

Step 16 Remove the motor.

Step 17 Find the motor hole center.

Step 18 Punch mark the hole centers.

Step 19 Remove the motor form the base plate.

CAUTION

Ensure that the base plate holes are clearly marked before moving the motor, or the entire alignment procedure will have to be repeated.

Step 20 Drill and tap the base plate holes.

Step 21 Place the motor back onto the base plate.

Step 22 Install the motor mounting bolts.

NOTE

The pump and motor do not need further adjustment because they will be properly aligned later.

7.0.0 ◆ STRESS RELIEVING AND ALIGNING ROTATING EQUIPMENT

Most pipeline systems are routed between storage tanks and pumps. It is important to properly align piping to pumps and other rotating equipment to avoid excessive stress on pipe, joints, and rotating equipment. When pipe and rotating equipment are not properly aligned, the stress that exists between these components causes leaks in the piping system, damage to the equipment, and excessive pressure and stress to the pipe hangers and supports. All of these conditions decrease the life of the piping system and equipment and also present serious safety hazards to those who work in the vicinity of the system.

Before stress relieving any pipe, you must check your company's policies and procedures. Some companies do not allow stress relieving of pipe.

7.1.0 Thermal Expansion

All materials change in size to some extent when the temperature changes. This is also true with piping materials, so when working with a pipe system you need to account for thermal expansion and contraction.

Under normal conditions, metallic piping will change several inches in a 100-foot length for a change of 300°F. If only one point in a pipeline were kept fixed during a change in length, movement would take place in perfect freedom, and no stress would be imposed on the pipe or the connection. However, piping systems have more than one fixed point. They are nearly always restrained at terminal points by anchors, guides, stops, rigid hangers, or sway braces. These restraining points resist expansion and put the line under stress, which causes it to deform when the temperature changes.

Stress imposed by expansion must be kept within the strength capability of the pipe and also of the points to which the pipe is anchored. Equipment, such as a pump or a turbine, that is fastened solid will be put under excessive stress if the pipe fastened to it expands toward it. This must be prevented by allowing expansion away from the equipment. Failure to do so can result in ruptured pipe or damage to the structure or the equipment to which the pipe is anchored.

All materials do not expand and contract at the same rate. This can cause problems when different types of materials are used in the same piping system. *Figure 12* shows the expansion of different types of pipes.

There are three accepted methods used to account for expansion and to avoid excessive stresses on the line: flexibility in the layout, installing expansion loops, and cold springing.

7.1.1 Flexibility in Layout

Flexibility can often be built into the layout of a piping system. If a pipeline is built with small-diameter pipe and if it will not have great temperature changes, the system can be designed to account for expansion. Flexibility in layout consists of making many changes in direction between anchor points (see *Figure 13*). These changes in direction allow for some thermal expansion of the line without damaging the joints and fittings.

7.1.2 Installing Expansion Loops

The second method used to account for expansion is the installation of expansion loops in the pipe run to provide extra flexibility between anchors (see *Figure 14*). These loops absorb the movement of the line caused by thermal expansion and contraction and sideways bowing. The pipe is anchored along its length and sometimes in the loop itself. Guides are placed at designated internals along the pipe run to ensure that the pipe moves toward the expansion loop as it expands.

Expansion loops are made with pipe bends or welded pipe and fittings. They should not be

Safety

Discuss problems that can result if the base plate holes are not clearly marked.

Homework

Have trainees prepare for testing during the next session.

Materials

Ensure that you have everything required for testing during this session.

Teaching Tip

See the Teaching Tip for Section 6.0.0 at the end of this module.

Performance Testing

Have each trainee perform Performance Task 3 to your satisfaction. Fill out a Performance Profile Sheet for each trainee.

Homework

Have trainees review Sections 7.0.0–7.1.3 for the next session.

Materials

Ensure that you have everything required to teach this session.

Classroom

Discuss problems that can result if the pipe and rotating equipment are not properly aligned.

Classroom

Explain how stress imposed by expansion affects the pipeline. Compare the three methods that are used to account for expansion and to avoid excessive stress.

Refer to Figure 12. Ensure that trainees understand how to interpret the data for different types of pipe. Ask whether trainees have any questions.

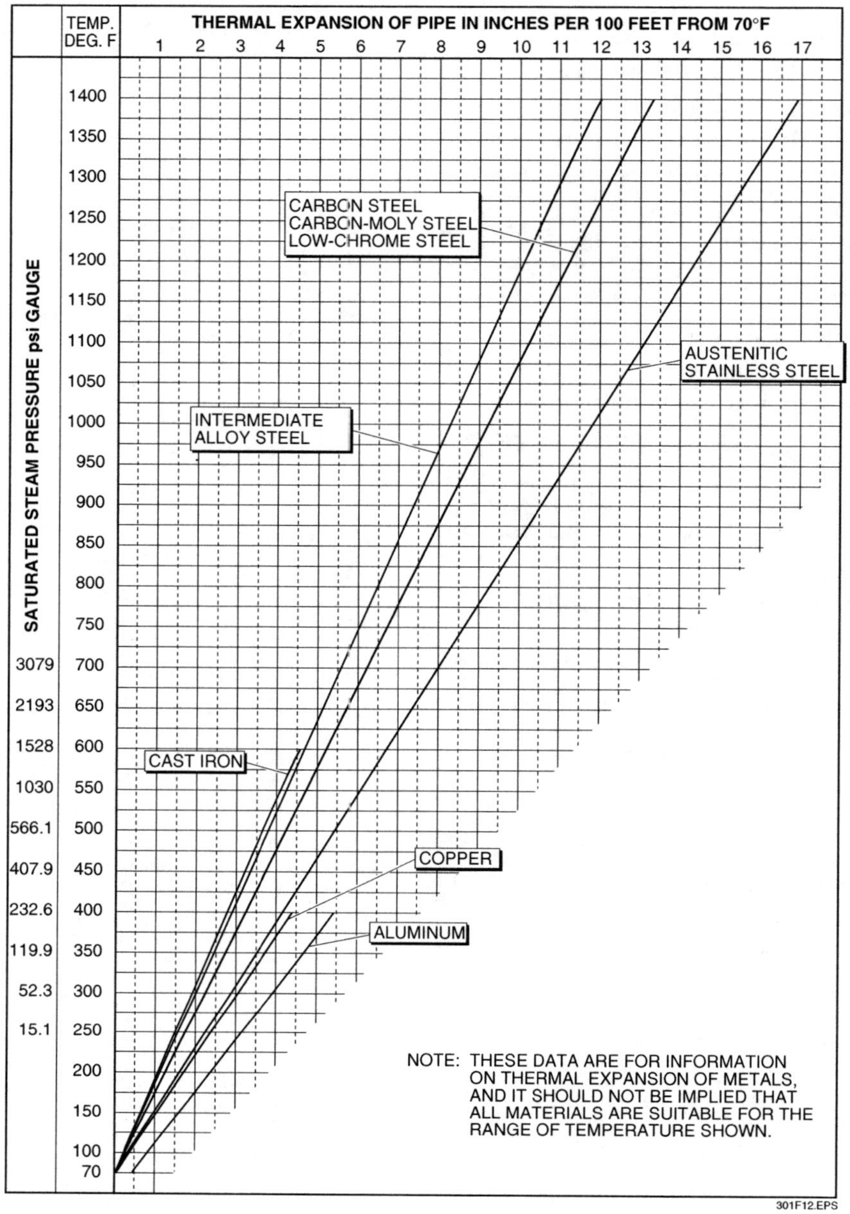

Figure 12 ◆ Expansion of different types of pipe.

Instructor's Notes:

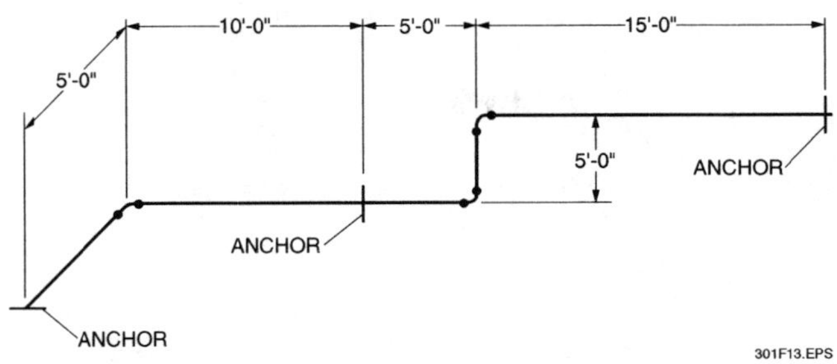

Figure 13 ◆ Example of flexibility in layout.

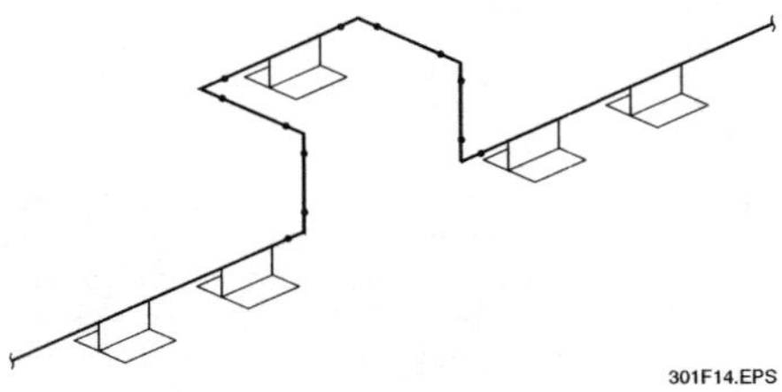

Figure 14 ◆ Expansion loop.

fabricated with flanges or threaded fittings because these types of joints cannot withstand the stress and will leak. Fabricated expansion U-loops of pipe welded to long-radius elbows are usually very simple in shape and can be fabricated in the shop according to the specifications on the layout drawing. The number and type of expansion loops used differs with each installation.

7.1.3 Cold Springing

The third way to deal with expansion is to use cold springing along with one of the other two methods. Cold springing is a method of shortening the pipe run when it is installed so that when the pipe heats up, it expands to its intended length (see *Figure 15*). Cold sprining also relieves the stress that thermal expansion exerts on the pipe.

The cold-spring gap refers to the difference between the length of the installed pipe and the length that the pipe should be. The location and the size of the cold-spring gap must be calculated accurately and the piping installed according to those calculations. Cold springing pipe at the wrong location or by the wrong amount can damage the pipe joints, anchors, valves, or attached equipment, so it is not advisable to guess at the cold-spring dimensions and locations.

There are various methods by which cold springing can be accomplished. One method is to bolt wood spacers very tightly between the flanges of a flanged line during fabrication. When the wood spacer is removed, the required cold-spring gap is created.

7.2.0 Aligning Pipe to Rotating Equipment

Your main objective is to connect the piping system to the rotating equipment without imposing any stress on the equipment. In order to do this, the pipe flange must be properly aligned with the equipment nozzle. Never force a pipe flange into alignment with the nozzle together.

If the pipe flange is not properly aligned to equipment flanges, excessive stresses are imposed on the equipment and the pipeline. You must be able to make the connection between the pipe flange and equipment nozzle without forcing the pieces together. If you have to use force to align the flange and equipment nozzle, the pipe will pull the pump and motor out of alignment and decrease the lives of both.

Pipe alignment to rotating equipment takes coordination of work activities between installing the pipe and installing the pump and motor. Before final pipe alignment can be made, the equipment must be set and **grouted** in place. When the equipment is set in place, a preliminary alignment between the pump and motor is accomplished before grouting is applied.

7.2.1 Grouting Baseplates

The two basic types of grout are nonshrink and epoxy. Nonshrink grout should be used under critical equipment or baseplates where settlement cannot be tolerated. Epoxy grout should be used in applications where the equipment foundation is subjected to chemical contamination or corrosion.

Audiovisual

Show Transparency 8 (Figure 13). Explain how changes in direction allow for thermal expansion.

Show Transparency 9 (Figure 15). Explain how cold springing relieves stress exerted on the pipe.

Classroom

Ask trainees to write down the three methods used to relieve stress. Have trainees define each method and explain how each method accounts for expansion.

Homework

Have trainees review Sections 7.2.0–7.2.2 and answer the Review Questions for the next session.

Materials

Ensure that you have everything required for the laboratory and testing during this session.

Classroom

Review the problems that can result if the pipe flange is not properly aligned.

Classroom

Explain how nonshrink and epoxy grout should be used. Review the three reasons for grouting under the base of machinery.

Ask a trainee to explain why concrete should never be used in place of grouting.

Laboratory

Set up stations with nonshrink and epoxy grout. Have trainees identify the appropriate application for each. Ask trainees to practice rodding the grout.

Classroom

Review the factors to consider when checking for misalignment. Go over any questions trainees have.

Teaching Tip

See the Teaching Tip for Section 7.0.0 at the end of this module.

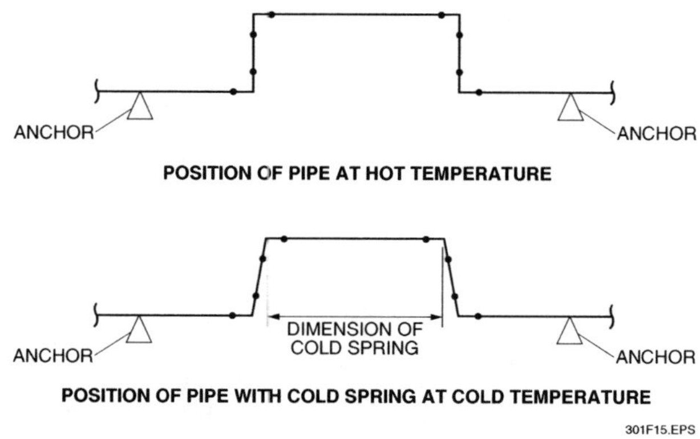

Figure 15 ◆ Examples of cold springing.

The three primary reasons for grouting under the bases of machinery are:

- To spread the impact of equipment operation over a larger area
- To provide higher compressive strength than reinforced concrete
- To give a more uniform support by filling in all the voids along the base of the equipment

Grout is also used for the installation of baseplates, anchor bolts, bearing plates, and other types of fasteners. Grout must be confined vertically until the drying process is complete. Equipment that is resting on isolators should not be grouted because the grouting defeats the purpose of the isolators.

Concrete should never be used in place of grouting. Grout expands as it sets up, thus filling voids under the bases and bedplates. Filling the voids provides more uniform distribution of the equipment weight and therefore makes the equipment alignment easier to maintain. In contrast, concrete shrinks as it dries, making alignment difficult to maintain.

Rodding and working the grout is the most common way to apply it. The forms used should be 6 inches wide on the side where the grout is to be poured with 2-inch sides elsewhere. The form should be high enough to force the grout to flow under the equipment. A piece of metal strapping can be used to work the grout under the equipment. Quality grouting depends on adequate flowability and quick placement of the grout. If possible, all vibration should be eliminated where the grout is to be poured because vibration causes the grout to set up improperly, thus weakening it.

7.2.2 Preliminary Alignment Between Equipment

The preliminary alignment between the equipment should have been completed prior to adding grout to the base but should be inspected again after the grout has been applied and cured.

Rotating shafts are usually aligned with the coupling halves in place. The measurements are taken from the coupling halves, and adjustments are made accordingly. There are two ways that the coupling halves must be aligned. First, the outer diameters, or rims, of the couplings must be lined up all the way around; then the faces of the couplings must be lined up. This is called rim-and-face alignment. When checking for misalignment, check for the following:

- Vertical offset
- Horizontal offset
- Vertical angularity
- Horizontal angularity

Instructor's Notes:

Summary

Trained and qualified technicians must plan, set up, and carry out the placement of rotating equipment. Technicians must have a strong understanding of the proper ways to move equipment, lift equipment, and adjust its placement after it is in place. They must also have a good understanding of how such equipment is connected, including wiring, plumbing, piping, and all other associated components.

Proper training in all aspects of installation, including alignment, schematics, component stress, and effects of temperature, is required. Mechanical technicians must be proficient in the setup and use of alignment gauges and measurement equipment associated with alignment. They must also be practiced at reading dials, gauges, and measurements to very small decimal points. It is only through training and experience that technicians are able to accomplish such work. Mechanical technicians must also have patience to overcome problems and obstacles when equipment does not fit easily.

Pipelines depend upon the dedication of mechanical technicians to ensure that equipment such as pumps, drivers, and other rotating equipment are set properly from the beginning so that their service life is extended with a minimum of unnecessary downtime.

Review Questions

1. The _____ is the key to effective operation of rotating equipment and must conform to the construction specifications that will provide the required support, levelness, and finish for the equipment that is going to be installed on it.
   a. equipment pad
   b. base plate
   c. shim
   d. wedge

2. The first step in installing rotating equipment is to _____.
   a. calculate the slope
   b. determine the equipment clearance
   c. set up and level the equipment base plate
   d. heat the bearings

3. Large bolts or all-thread rods that screw into the threaded holes near the corners of a base plate are called _____.
   a. anchor bolts
   b. jack bolts
   c. eyebolts
   d. locking screws

4. Wedges should always be used in _____.
   a. pairs
   b. threes
   c. fours
   d. packs

5. In the final, precise leveling of equipment, _____ are used.
   a. shim packs
   b. thick shims
   c. thin shims
   d. shims and wedges

6. Many technicians adjust shims by _____, which isn't a workable method for some installations.
   a. trial and error
   b. a level and feeler gauge
   c. a straightedge
   d. a thickness gauge

7. The first step in setting a pump and motor is to _____.
   a. wedge the nut to the shaft
   b. preheat the shaft
   c. set the gap between the couplings
   d. install couplings on the mating shafts

8. A setscrew or nut is used to lock the coupling to the shaft in a _____ installation.
   a. precision
   b. interference
   c. simple
   d. clearance

9. The coupling can be properly placed on the shaft by _____.
   a. driving it with a sledgehammer
   b. wedging it on
   c. lightly tapping it with a brass hammer
   d. screwing it onto the threads

10. To set a pump, it must be _____.
    a. centered on the lengthwise center line and parallel to the crosswise center line
    b. centered over both the lengthwise and crosswise center lines
    c. parallel to both the lengthwise and crosswise center lines
    d. parallel to the lengthwise center line and centered on the crosswise center line

Classroom

Emphasize the importance of understanding all aspects of installation when working with rotating equipment.

Go over the Review Questions for this module, and answer any questions trainees have.

Examination

Administer the Module Examination. Be sure to record the results of the Exam on Craft Training Report Form 200 and submit the results to the Training Program Sponsor.

Performance Testing

Have each trainee perform Performance Task 4 to your satisfaction. Fill out a Performance Profile Sheet for each trainee.

Ensure that all Performance Tests have been completed and Performance Profile Sheets for each trainee are filled out. Be sure to record the results of the testing on Craft Training Report Form 200 and submit the results to the Training Program Sponsor.

11. When aligning a motor, set the gap between the couplings using a _____.
    a. feeler gauge
    b. level
    c. dial indicator
    d. thickness gauge

12. To adjust the horizontal motor alignment, take dial indicator readings at the _____ o'clock positions until the dial reads zero at both.
    a. 3 and 9
    b. 6 and 12
    c. 8 and 10
    d. 2 and 10

13. Flexibility in layout consists of making many _____ between anchor points.
    a. fitting installations
    b. guides
    c. changes in direction
    d. straight lengths of pipe

14. Loops in the piping system that absorb the movement of the line caused by expansion and contraction are known as _____ loops.
    a. expansion
    b. contraction
    c. control
    d. thermal

15. Shortening the pipe run when it is installed so that it expands to its intended length when the pipe heats up is called _____.
    a. cold springing
    b. hot springing
    c. an expansion guide
    d. flexibility in layout

Instructor's Notes:

GLOSSARY

Trade Terms Introduced in This Module

Axial movement: Movement in the direction of a shaft axis.

Grout: A nonshrinking, hardening filler material used to fill the voids on and add strength to an equipment base.

Jack bolt: A bolt that is screwed through a plate and is used to raise or lower the plate.

Keyway: A slot cut in a shaft for a locking key.

Preheat: To heat a piece of metal before it is operated upon.

Rodding: An operation in which a rod is passed through a length of tubing, such as a rifle or pipework, to determine if the bore is clear.

Screed: A straight-edged wood or metal template fixed temporarily to a surface as a guide when plastering or concreting.

Shim: A small piece of metal used as a spacer to raise a base plate.

Shim pack: Several shims stacked together.

Stress: The load imposed on an object.

Wedge: A triangular piece of metal used to raise a base plate or other object.

MODULE 63301-02 — TEACHING TIPS

The following are suggested activities or instructional methods to help you teach the material in this AIG.

General

When you call on someone to answer a question, the rest of the class relaxes or even tunes out because they expect that the question and answer will take place only between you and the trainee you called on. Instead, use this technique to involve more trainees in answering questions and to keep them on their toes.

1. Ask trainees to define a term or explain a concept.
2. After one trainee has answered, ask a trainee seated nearby if the answer is right. Then ask whether a trainee in the back of the room agrees.
3. Ask trainees to explain why they think an answer is right or wrong.
4. Use the session to clear up incorrect ideas and encourage trainees to learn from their mistakes.

Section 3.4.0 *Precision Leveling Using Shims*

This exercise provides trainees with the opportunity to practice precision leveling. For this exercise, you will need base plates, anchor bolts, feeler gauges, shim packs, levels, straightedges, tools to tighten the anchor bolts, and calculators. Trainees will need their Trainee Guides and pencils and paper to take notes for discussion. Allow 30 minutes for this exercise.

1. Set up several stations with the necessary tools and materials to perform precision leveling. Divide the trainees into groups according to the number of stations.
2. Ask trainees to practice precision leveling, following the procedures in Section 3.4.0.
3. Observe and assess each group's work. Provide suggestions for accurate leveling.
4. Ask trainees to discuss the advantages and disadvantages of each of the leveling methods discussed.
5. Answer any questions trainees have about leveling.

Section 4.0.0 *Inspecting Equipment Prior to Installation*

This exercise will familiarize trainees with inspection techniques. For this exercise, you will need installation drawings; sample pumps and/or motors with physical damage; and sample base plates that are warped, damaged, or cracked. Trainees will need their Trainee Guides and pencils and paper to take notes for discussion. Allow 30 minutes for this exercise.

1. Display the sample installation drawings. Explain how to check the equipment against the drawings. Stress the importance of ensuring that the correct motor and pump are installed at the appropriate location.
2. Set up stations with the damaged base plates, pumps, and/or motors.
3. Ask trainees to visit each station, inspect the equipment, and identify the problems with each. Emphasize that trainees check for the following:
   - Physical damage
   - Proper lubrication
   - Proper rotation
   - Smooth turning
   - Radial runout
   - Smooth, clean surfaces
4. Review and discuss trainees' observations. Answer questions trainees have about inspecting equipment before installation.

Section 5.0.0 *Preparing Equipment for Installation*

Trainees have reviewed the basic methods of coupling installation. This exercise will allow trainees to practice the clearance and interference installation methods. For this exercise, you will need brass hammers, key and setscrew couplings, cleaning solvents, shop clothes, tapered shafts, locking nuts, micrometers, and equipment to preheat couplings and bearings. Trainees will need their Trainee Guides, pencils, and paper to take notes for discussion. Allow 45 minutes for this exercise.

1. Set up three stations to prepare equipment for installation. Provide the tools and materials necessary to perform the key and setscrew method, the tapered shaft and locking nut method, and the interference-fit method.
2. Ask trainees to visit each station and practice each of the clearance and interference installation methods. Observe and assess trainees' work.
3. Have trainees compare the methods and discuss any problems they encountered while performing each method.
4. Answer any questions trainees have about the clearance and interference methods.

Section 6.0.0 *Installing Rotating Equipment*

This exercise will give trainees the opportunity to observe the installation of rotating equipment. For this exercise, locate a nearby site where an installation is scheduled. Trainees will need their Trainee Guides and pencils and paper to take notes for discussion. Allow at least 60 minutes for this exercise.

1. Explain that trainees will observe the installation of rotating equipment.
2. Ask the operations personnel to explain how the base plate was set and the couplings installed in preparation for setting the pump and the pump motor.
3. Have trainees observe the procedure and consider the following:
   - Was the base plate clean and free of foreign matter?
   - Was the pump properly aligned?
   - How were shims placed?
   - Were hold-down bolts correctly installed?
   - How did the operations personnel ensure that the motor was on the magnetic center?
   - Was the horizontal parallel misalignment measured and corrected?
   - Were readings correctly taken using a dial indicator?
   - Were the motor mounting bolts properly installed?
4. Following the site visit, review and discuss trainees' answers.
5. Answer any questions trainees have about the installation of rotating equipment.

Section 7.0.0 *Stress Relieving and Aligning Rotating Equipment*

In this exercise, trainees will have the opportunity to review what they have learned about relieving stress on pipes and joints and aligning rotating equipment. For this exercise you will need sheets of acetate, an overhead projector, and transparency pens. Trainees will need their Trainee Guides and pencils and paper to take notes for discussion. Allow 20 minutes for this exercise.

1. On a sheet of acetate, write a list of problems that can occur on a line from excessive stress if the pipe and rotating equipment are not properly aligned.
2 Have trainees review each scenario and write a solution for each problem.
3. Determine whether there is more than one appropriate solution to each problem. Review and discuss trainees' answers.
4. Answer any questions trainees have about relieving stress and aligning rotating equipment.

MODULE 63301-02 — ANSWERS TO REVIEW QUESTIONS

Answer	Section
1. a	1.0.0
2. c	3.0.0
3. b	3.1.0
4. a	3.2.0
5. c	3.3.0
6. a	3.4.0
7. d	5.0.0
8. d	5.1.0
9. c	5.1.1
10. a	6.1.0
11. d	6.2.0
12. a	6.2.0
13. c	7.1.1
14. a	7.1.2
15. a	7.1.3

CONTREN™ LEARNING SERIES — USER UPDATES

The NCCER makes every effort to keep these textbooks up-to-date and free of technical errors. We appreciate your help in this process. If you have an idea for improving this textbook, or if you find an error, a typographical mistake, or an inaccuracy in NCCER's Contren™ textbooks, please write us, using this form or a photocopy. Be sure to include the exact module number, page number, a detailed description, and the correction, if applicable. Your input will be brought to the attention of the Technical Review Committee. Thank you for your assistance.

Instructors – If you found that additional materials were necessary in order to teach this module effectively, please let us know so that we may include them in the Equipment/Materials list in the Instructor's Guide.

Write: Curriculum Revision and Development Department
National Center for Construction Education and Research
P.O. Box 141104, Gainesville, FL 32614-1104

Fax: 352-334-0932

E-mail: curriculum@nccer.org

Craft ________________ Module Name ________________

Copyright Date ________ Module Number ________ Page Number(s) ________

Description ________________

____

____

____

(Optional) Correction ________________

____

____

(Optional) Your Name and Address ________________

____

____

Module 63302-02

Unit Alignment

MODULE OVERVIEW

This module explains how to recognize and correct equipment misalignment. Trainees will learn how to identify and correct soft foot and to perform conventional, rim-and-face dial indicator, reverse dial indicator, and laser alignment.

PREREQUISITES

Please refer to the Course Map in the Trainee Module. Prior to training with this module, it is recommended that the trainee shall have successfully completed the following:

Core Curriculum; Pipeline Core, Modules 66101-02 and 66102-02; Pipeline Mechanical Levels One and Two; Pipeline Mechanical Level Three, Module 63301-02

OBJECTIVES

Upon completion of this module, the trainee will be able to:

1. Recognize and describe the four types of equipment misalignment.
2. Identify the causes of soft foot.
3. Describe the major steps in performing conventional rim-and-face alignment.
4. Describe the major steps in performing reverse dial indicator alignment using the equation method of alignment.
5. Describe the major steps in performing reverse dial indicator alignment using the graphical chart method of alignment.
6. Describe the major steps in performing laser alignment.
7. Identify other laser alignment procedures that may be completed on the machinery trains depending on equipment needs.

PERFORMANCE TASKS

Under the supervision of the instructor, the trainee should be able to:

1. Recognize and describe the four types of equipment misalignment.
2. Describe the major steps in performing each of the following:
   - Conventional rim-and-face alignment
   - Reverse dial indicator alignment using the equation method of alignment
   - Reverse dial indicator alignment using the graphical chart method of alignment
   - Laser alignment

NCCER STANDARDIZED CRAFT TRAINING PROGRAM

The National Center for Construction Education and Research (NCCER) provides a standardized national program of accredited craft training. Key features of the program include instructor certification, competency-based training, and performance testing. The program provides trainees, instructors, and companies with a standard form of recognition through a National Craft Training Registry. The program is described in full in the *Guidelines for Accreditation*, published by the NCCER. For more information on standardized craft training, contact the NCCER by writing us at P.O. Box 141104, Gainesville, FL 32614-1104; calling 352-334-0911; or e-mailing info@nccer.org. More information may be found at our Web site, www.nccer.org.

HOW TO USE THIS ANNOTATED INSTRUCTOR'S GUIDE

Each page presents two sections of information. The larger section displays each page exactly as it appears in the Trainee Module. The narrow column ties suggested trainee and instructor actions to each page and provides icons to call your attention to material, safety, audiovisual, or testing requirements. The bottom of each page includes space for your notes.

If you see the Teaching Tip icon, that means there is a teaching tip associated with this section. Also refer to any suggested teaching tips at the end of the module.

SAFETY CONSIDERATIONS

Ensure that the trainees are equipped with appropriate personal protective equipment. Emphasize the importance of following all safety precautions and procedures when working with power tools.

PREPARATION

Before teaching this module, you should review the Module Outline, Objectives, Performance Tasks, and the Materials and Equipment List. Be sure to allow ample time to prepare your own training or lesson plan and gather all required equipment and materials.

MATERIALS AND EQUIPMENT LIST

Transparencies
Markers/chalk
Blank acetate sheets
Transparency pens
Pencils and scratch paper
Module Examinations*
Performance Profile Sheets*
Copies of Quick Quiz**
Overhead projector and screen
Whiteboard/chalkboard
Appropriate personal protective equipment
Copies of *49 CFR Part 192* (Gas) and/or *49 CFR Part 195* (Liquid)
Copies of your company policy and procedures manual
Calculators
Copies of *Figures 1, 2, 3, 4 , 23, 24,* and *25* with the titles covered
Copies of *Figure 33* with the callouts covered
Alignment simulators
Dial calipers
Straightedges
Scales
Rules
Feeler gauges
Chain indicator jigs
Dial indicators
Graphical alignment charts
Graph paper
Reverse indicator plotting guides
Sample alignment records

* Located in the Test Booklet.
**Located at the end of this module.

TEACHING TIME FOR THIS MODULE

An outline for use in developing your lesson plan is presented below. Note that each Roman numeral in the outline equates to one session of instruction. Each session has a suggested time period of 2½ hours. This includes 10 minutes at the beginning of each session for administrative tasks and one 10-minute break during the session. Approximately 40 hours are suggested to cover *Unit Alignment.* You will need to adjust the time required for hands-on activity and testing based on your class size and resources.

Topic	Planned Time
Session I. Types of Equipment Misalignment	
A. Vertical Offset	________
B. Horizontal Offset	________
C. Vertical Angularity	________
D. Horizontal Angularity	________
1. Performance Testing	________
Session II. Identifying and Correcting Soft Foot, Part One	
A. Parallel Air Gap	________
B. Bent Foot	________
Session III. Identifying and Correcting Soft Foot, Part Two	
A. Squishy Foot	________
B. Induced Soft Foot	________
C. Gaps without Soft Foot	________
Session IV. Performing Conventional Alignment, Part One	
A. Adjusting Horizontal Angularity	________
B. Adjusting Horizontal Offset	________
1. Laboratory	________
Session V. Performing Conventional Alignment, Part Two	
A. Adjusting Vertical Angularity	________
B. Adjusting Vertical Offset	________
1. Laboratory	________
Session VI. Performing Conventional Alignment, Part Three	
A. Adjusting Horizontal Angularity and Offset	________
B. Adjusting Vertical Angularity and Offset	________
1. Laboratory	________
Session VII. Performing Rim and Face Dial Indicator Alignment, Part One	
A. OD (Rim) Measurements	________
B. Face Measurements	________
C. Face and OD Measurements	________
Session VIII. Performing Rim and Face Dial Indicator Alignment, Part Two	
A. Horizontal Angularity Measurements	________
B. Horizontal Offset Measurements	________
Session IX. Performing Rim and Face Dial Indicator Alignment, Part Three	
A. Vertical Angularity and Offset Measurements	________
B. Checking Alignment Using Dial Indicators	________
Session X. Performing Reverse Dial Indicator Alignment, Part One	
A. Setting Up Reverse Dial Indicator Jigs	________
B. Performing Reverse Dial Indicator Alignment	________

Session XI. Performing Reverse Dial Indicator Alignment, Part Two

A. Charting Alignment ____________
   1. Laboratory ____________
B. Performing Vertical Alignment ____________
   1. Laboratory ____________
C. Recording Alignment ____________

Session XII. Performing Laser Alignment, Part One

A. Laser Theory ____________
B. Laser Safety ____________
C. Optalign® Laser Alignment ____________

Session XIII. Performing Laser Alignment, Part Two

A. Alignment Demonstration Rig ____________
B. Descriptive Characteristics ____________
C. Optalign® Laser System Capabilities ____________
D. Laser Limitations ____________
E. Determining Targets ____________

Session XIV. Performing Laser Alignment, Part Three

A. Laser/Detector Operation ____________
B. Alignment Procedures ____________

Session XV. Specific Laser Alignment Procedures, Part One

A. Aligning Machinery Trains ____________

Session XVI. Specific Laser Alignment Procedures, Part Two

A. Measuring Shaft Sag ____________
B. Aligning Vertical Machines ____________
C. Summary
   1. Summarize module ____________
   2. Answer questions ____________
D. Module Examination ____________
   1. Trainees must score 70% or higher to receive recognition from the NCCER.
   2. Record the testing results on Craft Training Report Form 200 and submit the results to the Training Program Sponsor.
E. Performance Testing ____________
   1. Trainees must perform each task to the satisfaction of the instructor to receive recognition from the NCCER.
   2. Record the testing results on Craft Training Report Form 200 and submit the results to the Training Program Sponsor.

Module 63302-02

Unit Alignment

Instructor's Notes:

Instructor's Notes:

Introduction to the Pipeline Operator Qualification Rule, API, and NCCER

It could be said that petroleum products keep America running. Petroleum products like oil and natural gas are the principal sources of energy in the United States and are used in many manufacturing and chemical processes. Petroleum products are primarily transported by pipeline. There are more than a million miles of liquid and natural gas pipeline in this country, ranging in diameter from two inches to four feet. These pipelines bring crude oil and natural gas from production fields to processing plants and refineries—and ultimately to the consumer.

Pipeline Operator Qualification

Petroleum products are considered hazardous, so pipeline personnel must take extra precautions and follow specific procedures when working with them. To help ensure the safe handling and transportation of petroleum products, the United States Department of Transportation (DOT) mandates safe practices as federal regulations.

One such regulation, the Operator Qualification (OQ) rule, requires pipeline operators to develop a written qualification plan to evaluate their personnel and their contract personnel (see *49 CFR Part 192 Subpart N—Gas*, and *49 CFR Part 195 Subpart G—Liquid*). The evaluation hinges on the operator's ability to perform certain covered tasks and to be able to recognize and respond to abnormal operating conditions that may be encountered while performing these covered tasks.

Covered Tasks

A pipeline performance task is deemed *covered* if the activity passes the following four-part test:

1. Is it performed on a pipeline facility?
2. Is it an operations or maintenance task?
3. Is it performed as a requirement of *49 CFR Part 195 (Liquid)* or *49 CFR Part 192 (Gas)*?
4. Does it affect the operation or integrity of the pipeline facility?

If the answer to all four questions is *yes*, then the activity is considered *covered* and must figure into the evaluation process of pipeline personnel.

American Petroleum Institute

The OQ rule and its related covered tasks provided the catalyst to develop a unified pipeline training and assessment program—so that pipeline operators and owners have ready access to the assessment and training tools they need. This effort was the direct result of the vision and leadership provided by The American Petroleum Institute (API) and API members.

Efforts to unify the industry are nothing new to API. Indeed, the impetus for forming the Institute in 1919 was the need to standardize engineering specifications for drilling and production equipment. Today, the most pressing issues revolve around public perceptions and government policies toward the industry. Speaking with one voice on these issues has become as essential as having interchangeable parts in the field. API's members determine the petroleum industry's positions on public policy as well as the standards that govern its day-to-day operations worldwide.

Consortium on Operator Qualification

API assembled a focus group whose primary purpose was to identify covered tasks and provide compliance recommendations to pipeline operators.

The Consortium on Operator Qualification (COOQ) became the guiding force for the identification and development of covered tasks for the liquid pipeline industry. Their work product, API publication 1161, *Guidance Document for the Qualification of Liquid Pipeline Personnel*, formed the basis for the resulting Pipeliner Training and Assessment Program.

The covered task list created in API publication 1161 has now evolved to include gas covered tasks and is maintained by NCCER. The most current covered task list can be downloaded at www.nccer.org.

National Center for Construction Education and Research

With the guidance and support of API, its members, and the Consortium on Operator Qualification, the National Center for Construction Education and Research (NCCER) was able to design and implement the Pipeliner Training and Assessment Program.

NCCER is a not-for-profit 501(c)(3) education foundation established in 1995 by the world's largest and most progressive construction and maintenance contractors and national contractor associations. NCCER was founded to address the

severe workforce shortage facing the construction and maintenance industries and to develop a standardized training process and curricula. Today, NCCER is supported by hundreds of leading construction and maintenance companies, manufacturers, and national associations and developed more than 35 construction and maintenance titles.

NCCER also maintains a National Registry, which contains the records of trainees and instructors who have received NCCER training. All completion information for every trainee is sent to the NCCER and filed within the Registry. The Registry can then confirm training and skills for workers as they move from company to company, state to state, or even within their own company.

Standardized Curriculum (NCCER's Contren™ Learning Series)

NCCER curricula are developed using the expertise of industry subject matter experts (SMEs). The use of SMEs as an integral part of every new development or revision ensures that the curriculum is current, industry-driven, and indicative of industry practices. The development and/or revision process is facilitated from start to finish by NCCER project managers, who work closely with the SMEs and a select team of technical writer(s) on the development and presentation of the curriculum.

Pipeline: Five Training Paths

The pipeline curriculum focuses on essential skills and competencies within five skill areas:

- Corrosion Control
- Electrical and Instrumentation
- Maintenance
- Mechanical
- Operations (Field and Control Center)

SMEs from the pipeline industry have determined the skills and competencies needed based on API's covered tasks and the consensus of the group/industry. Specific company information is generally not incorporated into the curriculum unless the SMEs determine that the information is the industry standard.

Training Modules and Covered Tasks

The smallest unit of instruction in the curriculum is a *module*. Each curriculum is composed of modules (the equivalent of stand-alone chapters) that are logically sequenced into complete *levels* of instruction. Every module is task-oriented and requires the successful completion of a written (knowledge-based) examination and a performance (hands-on) test. Modules may be individually selected and/or combined for efficient task training. Modules in the pipeline curricula, to the extent possible, will align with API-developed covered tasks. While not all modules will relate to covered tasks, all covered tasks will relate to at least one module.

Benefits of Training to the Pipeline Marketplace

Applying NCCER's standardized training process allows pipeline owners and contractors in the United States to establish the benchmark for acceptable skill and competency levels within the pipeline industry. This process will assist training organizations in developing and/or improving standardized pipeline training programs that effectively match their resources and operating conditions while responding to regulatory demands and to the needs of the industry. The process will also recognize training organizations that meet the minimum standards of competence, warranting regulatory, public, and professional confidence. This in turn will foster national unity within the pipeline industry for the mutual benefit of individual technicians, owners, contractors, and ultimately pipeline-users and society.

Additional Resources

To review the Pipeliner Training and Assessment Program Implementation and Resource Guide and the latest version of pipeline covered tasks, visit NCCER's Web site at www.nccer.org, or contact:

National Center for Construction Education and Research
PO Box 141104
Gainesville, FL 32614-1104
(888) 622-3720
Web site: www.nccer.org

For more information on API, please contact:

American Petroleum Institute
1220 L Street, Northwest
Washington, D.C. 20005
(202) 682-8000
Web site: www.api.org

To order *API Pub 1161*, contact Global Engineering Documents by calling 1-800-854-7149 or visiting their Web site at www.global.his.com.

Additional sources of information on the pipeline operator qualification regulation include:

DOT – Office of Pipeline Safety
www.ops.dot.gov

Transportation Safety Institute
www.tsi.dot.gov

Instructor's Notes:

COURSE MAP

This course map shows all of the modules in the third level of the Pipeline Mechanical curriculum. The suggested training order begins at the bottom and proceeds up. Skill levels increase as you advance on the course map. The local Training Program Sponsor may adjust the training order.

PIPELINE MECHANICAL LEVEL THREE

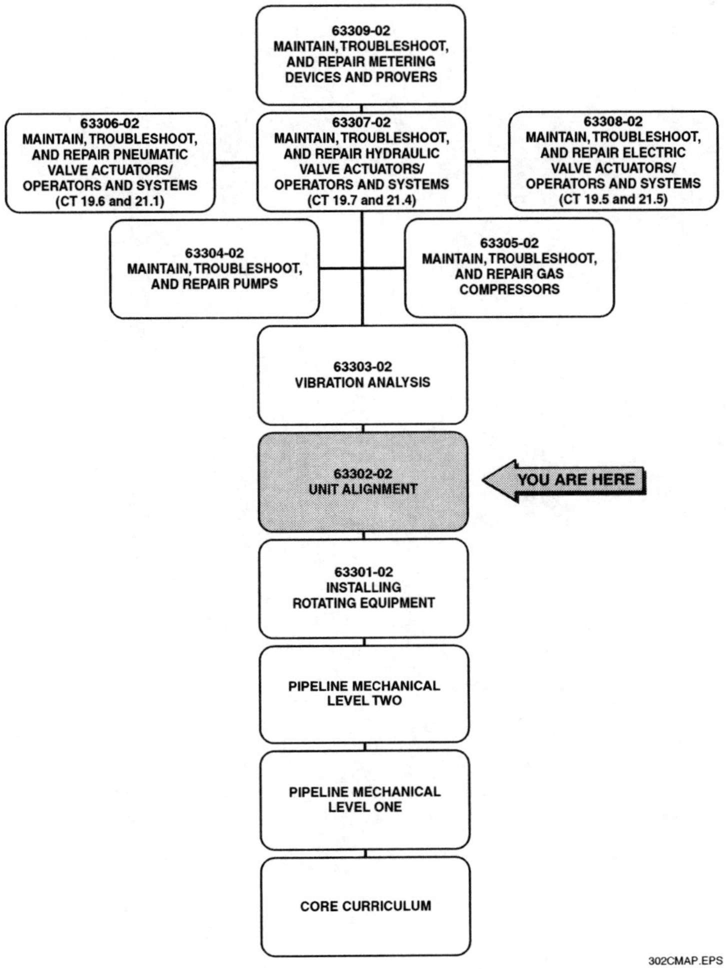

302CMAP.EPS

Homework

Assign reading of Module 63302-02.

Instructor's Notes:

MODULE 63302-02 CONTENTS

MODULE 63302-02 CONTENTS (Continued)

Instructor's Notes:

Figures

Instructor's Notes:

MODULE 63302-02

Unit Alignment

Objectives

When you have completed this module, you will be able to do the following:

1. Recognize and describe the four types of equipment misalignment.
2. Identify the causes of soft foot.
3. Describe the major steps in performing conventional rim-and-face alignment.
4. Describe the major steps in performing reverse dial indicator alignment using the equation method of alignment.
5. Describe the major steps in performing reverse dial indicator alignment using the graphical chart method of alignment.
6. Describe the major steps in performing laser alignment.
7. Identify other laser alignment procedures that may be completed on the machinery trains depending on equipment needs.

Prerequisites

Before you begin this module, it is recommended that you successfully complete the following: Core Curriculum; Pipeline Core, Modules 66101-02 and 66102-02; Pipeline Mechanical Level One; Pipeline Mechanical Level Two; Pipeline Mechanical Level Three, Module 63301-02.

Required Trainee Materials

1. Pencil and paper
2. Appropriate personal protective equipment

1.0.0 ◆ INTRODUCTION

Rotating equipment must be set up precisely straight; all motors and drive units must be lined up extremely accurately with the machines that they drive.

The job of setting up and straightening machinery is called alignment. It is very important that alignment be precisely done because the slightest misalignment of the shafts between the **prime mover** and the machine that it drives can cause extensive wear and damage to bearings and other parts of the machine and to the prime mover. Extensive misalignment can cause excessive equipment vibration. Equipment failures are a major maintenance expense, and misalignment is a major contributor to equipment failure.

When equipment is aligned, only the rotating shaft center lines are aligned. The positions of the feet, couplings, shaft surfaces, equipment housing, and bearings are not used to determine machine alignment. Shaft alignment is the positioning of two or more machines so that their rotational center lines are **collinear** at the point at which they are coupled.

There are several methods for aligning shafts of rotating equipment. The conventional method is the most basic and provides a good basis for the more complex methods.

Before any equipment is aligned, the shafts, couplings, and machine area must be cleaned and degreased. Accurate alignment is impossible through layers of grease and sludge that form on the machine. The area around the machine should be cleaned and a safety-approved solvent applied to the shafts, couplings, and machine feet after the equipment has been locked out and tagged out.

Materials

Ensure that you have everything required to teach the course. Check the Materials and Equipment List at the front of this module.

Classroom

Explain that terms shown in bold are defined in the Glossary at the back of this module.

Audiovisual

Show Transparency 1, Course Objectives.

Show Transparency 2, Performance Tasks.

Teaching Tip

See the general Teaching Tip at the end of this module.

Classroom

Discuss the importance of proper alignment. Explain problems that can result from extensive misalignment.

Ensure that trainees understand shaft alignment and what it means for the center lines to be collinear.

Classroom

Review the steps to be completed before performing alignment. Answer any questions trainees have.

Compare the four basic types of misalignment and explain that they can occur in combination.

Audiovisual

Show Transparency 3 (Figure 1). Explain when vertical offset occurs.

Classroom

Refer to Figure 2. Explain how to correct horizontal offset.

Refer to Figure 3. Review the cause of vertical angularity.

Audiovisual

Show Transparency 4 (Figure 4). Discuss problems caused by horizontal angularity.

Before performing alignment, the following steps should be completed:

Step 1 Inspect the base of the equipment for level, center line, and distortion.

Step 2 Set the coupling gap.

Step 3 Eliminate outside stress influences, such as pipe stress and conduit.

Step 4 Perform rough alignment of the equipment within the mechanical limits of the indicators (0.100 inch).

2.0.0 ◆ TYPES OF EQUIPMENT MISALIGNMENT

Rotating shafts are usually aligned with the couplings in place. The measurements are taken from the couplings, and adjustments are made accordingly. When aligning two coupling halves so that the shafts are aligned, there are two ways that they must be lined up. First the outer diameters, also called rims, of the couplings must be lined up all the way around; then the faces of the couplings must be lined up. For this reason, conventional alignment is also called rim-and-face alignment. Although there are only two ways that a coupling can be misaligned, there are also two ways of looking at each type of misalignment. This gives us four basic types of misalignment:

- Vertical offset
- Horizontal offset
- Vertical angularity
- Horizontal angularity

Although these are the four basic types of misalignment, any combination of the four may occur. You may have horizontal angularity and vertical offset, or you could have a situation that includes all four types.

2.1.0 Vertical Offset

Vertical offset (see *Figure 1*) occurs when one of the coupling halves is not in line with the other when viewed from the side. One coupling is higher than the other. In this type of misalignment, one of the units must be raised or lowered to align the couplings.

2.2.0 Horizontal Offset

Horizontal offset (see *Figure 2*) occurs when the outer diameters (ODs) of the couplings are misaligned from side to side as viewed from the top. One of the units must be moved to one side or the other to align the couplings.

2.3.0 Vertical Angularity

Vertical angularity (see *Figure 3*) means that the faces of the couplings are not square with one another when viewed from the side. One of the units must be tilted on its base to align the couplings.

2.4.0 Horizontal Angularity

Horizontal angularity (see *Figure 4*) means that the faces of the couplings are not square with one another when viewed from the top. One of the units must be rotated on its base to align the couplings.

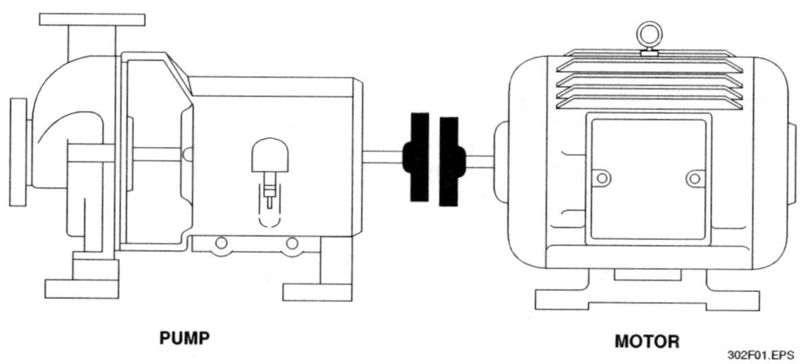

Figure 1 ◆ Vertical offset, side view.

Instructor's Notes:

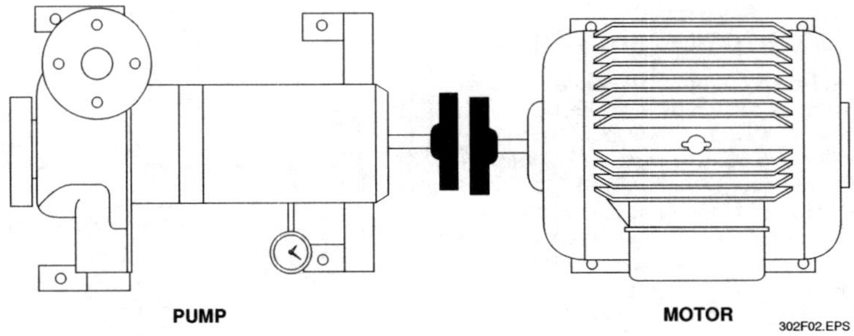

Figure 2 ◆ Horizontal offset, top view.

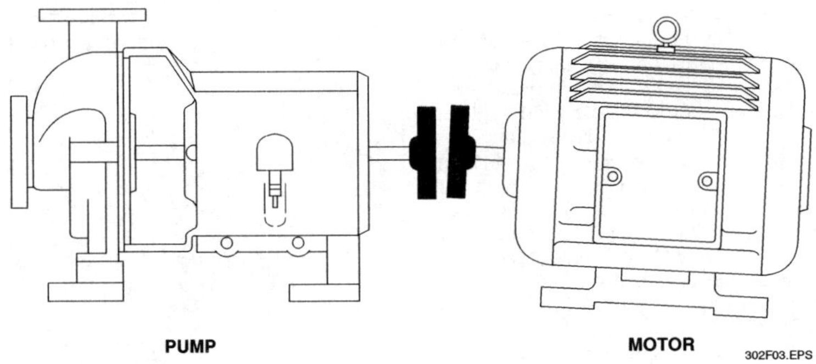

Figure 3 ◆ Vertical angularity, side view.

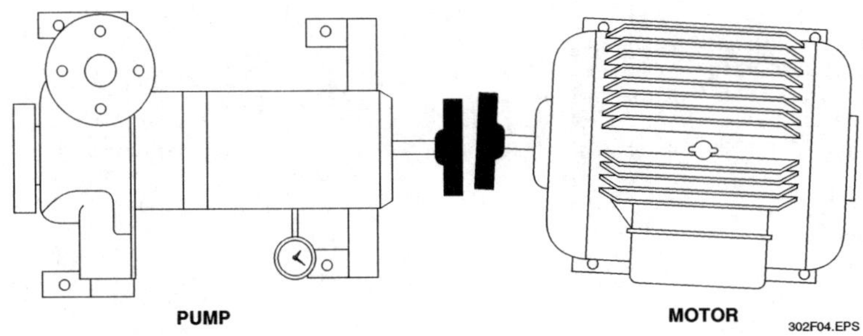

Figure 4 ◆ Horizontal angularity, top view.

UNIT ALIGNMENT — TRAINEE MODULE 63302-02 2.3

Classroom

Hand out copies of Figures 1 through 4 without their titles. Ask trainees to identify each type of misalignment and to explain when it occurs and how it is corrected. Go over any questions trainees have.

Performance Testing

Have each trainee perform Performance Task 1 to your satisfaction. Fill out a Performance Profile Sheet for each trainee.

Homework

Have trainees review Sections 3.0.0–3.2.0 for the next session.

Materials

Ensure that you have everything required to teach this session.

Classroom

Review conditions that can prevent a successful precision alignment.

Audiovisual

Show Transparency 5 (Figure 5). Discuss the common effects of soft foot.

Show Transparency 6 (Figure 6). Review the types of soft foot.

3.0.0 ◆ IDENTIFYING AND CORRECTING SOFT FOOT

Existing conditions can prevent a successful precision alignment. A good habit is to check for all the possible problems, such as soft foot, bent shafts, grossly misaligned equipment, and coupling stress. The machine base could also be weak, twisted, or on a bad foundation. Another possibility is that the **machine to be moved (MTBM)** could be bolt-bound, which means that the motor is out of the adjustment range so that the bolts are bent or difficult to loosen because of the side load. An MTMB is just as it sounds: a machine that is to be moved. In most cases it is the driver or motor that will be moved to line up with the **stationary machine (STAT)**. The time spent looking for these functional problems can prevent having to do the alignment twice. Some external factors that cause problems are thermal expansion, soft foot, and failure to calculate **indicator sag**.

Soft foot is a condition in which the tightening or loosening of the bolt(s) of a single foot distorts the machine frame. If the looseness indicates a soft foot, the condition should be analyzed to determine its cause before any corrective action is taken. The effects of soft foot vary because of the operating conditions and where the weakest conditions are. A machine with a soft foot that is bolted down firmly distorts and affects the operation of the **machinery train** in many ways. It also prevents proper alignment. The condition of the shims is critical to aligning the machine; therefore, the damaged shims should be discarded and new shims installed before alignment. *Figure 5* shows some common effects of soft foot.

One method of checking for soft foot uses a dial indicator mounted at one of the feet. To check for soft foot, zero the indicator, and loosen the hold-down bolt. If the indicator moves, there is a soft foot condition. Check the remaining feet, and write down all indicator movement at each foot. Insert the necessary shims to the soft foot, and recheck the condition. Indiscriminate shimming will often make things worse because not all soft feet are the same. Soft foot has various causes, some of which may not even be related to the machine itself. The following are some causes of soft foot:

- Parallel air gap
- Bent foot
- Squishy foot
- Induced soft foot
- Gaps without soft foot

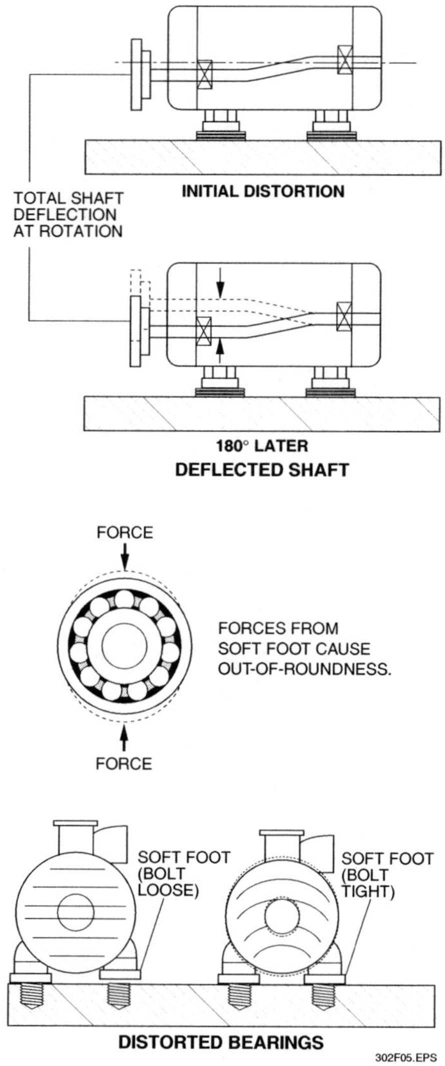

Figure 5 ◆ Effects of soft foot.

Soft foot can be difficult to find; it is therefore important to understand these conditions. *Figure 6* shows types of soft foot.

Instructor's Notes:

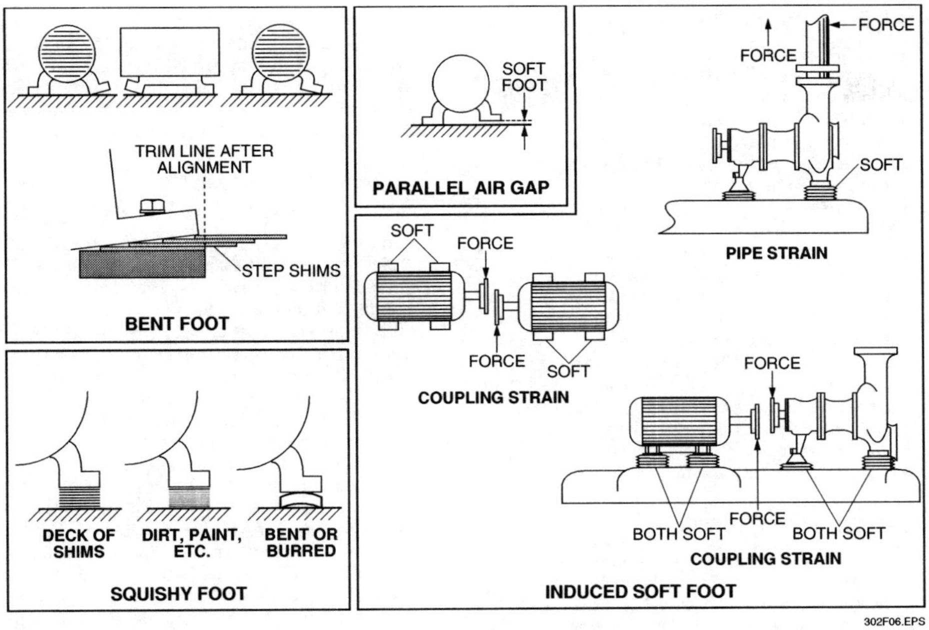

Figure 6 ◆ Types of soft foot.

3.1.0 Parallel Air Gap

Parallel air gap occurs when three mounting feet sit solid and flat but one foot does not touch. Possible causes are that either one leg is too short, one baseplate mounting pad is not level with the other three, or the shims under one foot are the incorrect size. Checking for soft foot shows which foot is short and how much it needs to be shimmed. Always shim one foot at a time; then proceed to check other feet, and shim as necessary. It is also advisable to confirm the reading using a feeler gauge before inserting and tightening the shims.

3.2.0 Bent Foot

Bent foot, or angular soft foot, occurs when the bottom of one foot is not parallel with the base. The feeler gauge readings show a slope from one corner to the other. The down corner may be touching, but when tightened, it acts as a lever and induces soft foot in the two opposing feet and sometimes in the fourth foot as well. This gives the machine the appearance of having three or four soft feet, but they disappear when the one bent foot is corrected.

The causes of bent foot include damaged machinery, bent or poorly machined baseplates, severe vertical **angularity** misalignment, feet that have been welded, or a settling foundation. The only corrections other than replacing the bent machine are to mill or grind the bent foot to square or to build a step shim (see *Figure 6*). If the foot is machined square, it will then be short and must be corrected as a parallel air gap. To build a step shim may be more expedient, but the following procedure should be completed:

Step 1 Fill any gap that exists under the entire foot so that one corner or edge of the foot is touching the base.

Step 2 Measure the largest remaining gap.

Step 3 Divide this gap by four or six, depending on the number of step shims, to obtain the step thickness.

Step 4 Select four to six shims of the step thickness, and insert them one at a time without lifting the machine until the fit is snug.

The purpose of using step shimming is to match the slope of the bent foot with a substantial wedge that will remain solid when the bolts are tightened. After rechecking the alignment, trim the outside of the step shims so that they are flush with the foot.

Classroom

Explain when parallel air gap occurs. Ask whether trainees have any questions.

Review the procedure to correct bent foot. Explain the purpose of using step shimming.

On the whiteboard/chalkboard, have several trainees draw examples of each type of soft foot. Ask trainees to identify each illustration.

Homework

Have trainees review Sections 3.3.0–3.5.0 for the next session.

Materials

Ensure that you have everything required to teach this session.

Classroom

Explain why it can be difficult to detect squishy foot. Review the causes of induced soft foot. Discuss how gaps without soft foot are detected.

On a sheet of paper, ask trainees to list the symptoms of each type of soft foot. Have trainees provide a solution for each problem. Answer any questions trainees have.

Teaching Tip

See the Teaching Tip for Sections 3.0.0–3.5.0 at the end of this module.

Homework

Have trainees review Sections 4.0.0–4.2.0 for the next session.

3.3.0 Squishy Foot

Squishy foot clearly shows as a soft foot measurement on the laser or dial indicator, but when measured using a feeler gauge, little or no gap is indicated. This is because there is dirt, grease, paint, or rust between the shims, because the shims are bent or burred, or because there are too many shims under the foot.

All shims have some imperfection and allow some compression when the bolts are tightened. This microscopic spring effect accumulates at the rate of 0.0003 inch per shim and increases as the shims are reused. As a standard rule, no more than four shims should be used under one foot.

Cleaning the base and foot area and altering the condition and number of shims can eliminate squishy foot. Scrape, sand, brush, and wipe off all surfaces under and around the feet. Always use new, clean shims and no more than four in a pack. Install the shim pack, tighten the mounting bolts, and recheck the alignment.

3.4.0 Induced Soft Foot

Induced soft foot shows up as two soft feet on the same side or end of the machine. The feeler gauge indicates a gap that is parallel or nearly parallel. Secondary symptoms are that the condition does not improve, the condition becomes worse, or another foot becomes much worse after shimming the gap amount. These are caused by external forces on the machine. Coupling strain and pipe strain are the two most common causes of induced soft foot. If the coupling is difficult to bolt up, expect an induced soft foot until alignment is improved. Other external forces that can cause induced soft foot are:

- Overhung machines or attachments
- Belt, gear, or chain loads
- Hoses or stressed conduit
- Structural bracing attached to the machine
- Jack bolts tight against the base

The solution for induced soft foot is to remove the external force. Document measurements before and after the corrective action to confirm that the problem is diagnosed completely.

3.5.0 Gaps without Soft Foot

Gaps without soft foot occur when there is a visual gap under the foot before tightening and none after tightening, but the measurement does not show a soft foot or misalignment. The laser or dial indicator may indicate a relatively small soft foot, but the feeler gauge indicates a much larger gap. This may be caused by the base moving, the foot flexing without bending the machine, or the base being cracked, causing the foot to be loose. Shimming to fill the gap is cosmetic; the logical solution is to replace the base or machine housing.

4.0.0 ◆ PERFORMING CONVENTIONAL ALIGNMENT

Conventional alignment is performed using straight edges and feeler gauges. All equipment is aligned in the same way, whether it is a motor and a pump, a motor and a gearbox, or a gearbox and a machine. The shaft center lines are aligned. To accomplish this, precision-made couplings are mounted on the shafts, and the alignment measuring procedures are performed on the couplings. In this section, alignment is presented on an alignment simulator that consists of two shafts with couplings mounted in bearings. The simulator represents equipment to be aligned. This equipment includes a driver and a driven component. The driver, such as a motor, is the MTBM. The driven equipment, such as a pump, is the stationary machine (STAT), or the machine receiving power from the driver. The bearings of the simulator are mounted on baseplates that are attached to a frame. *Figure 7* shows the alignment simulator.

Aligning couplings using the conventional alignment method is a matter of repeatedly measuring misalignment and making adjustments to bring the couplings into precise alignment. Aligning a STAT and MTBM consists of performing the following procedures:

- Adjusting horizontal angularity
- Adjusting horizontal offset
- Adjusting vertical angularity
- Adjusting vertical offset
- Adjusting horizontal offset and angular alignment
- Adjusting vertical offset and angular alignment

4.1.0 Adjusting Horizontal Angularity

For alignment purposes, the coupling faces are designated four positions (see *Figure 8*). These are the 12, 3, 6, and 9 o'clock positions, which are the positions at which all measurements are taken. The 12 o'clock position is the top of the coupling; the 6 o'clock position is the bottom of the coupling; and the 3 and 9 o'clock positions are the sides of the coupling as viewed from the motor side of the coupling. These positions are used in all methods of alignment.

Materials

Ensure that you have everything required for the laboratory during this session.

Classroom

Refer to Figure 7. Discuss the necessary adjustments made when performing conventional alignment.

Instructor's Notes:

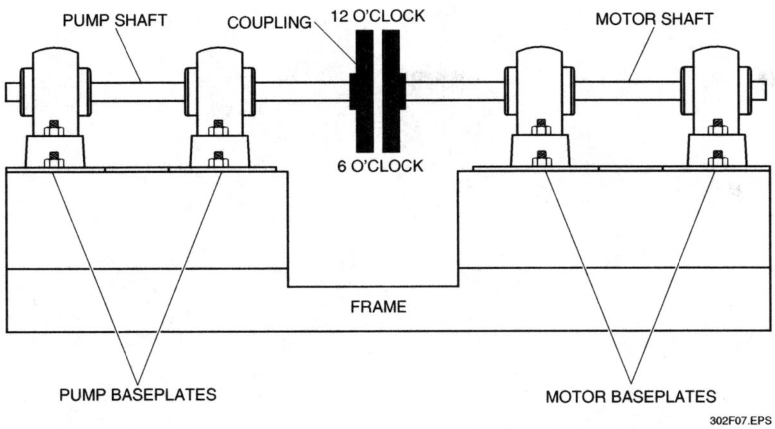

Figure 7 ◆ Alignment simulator.

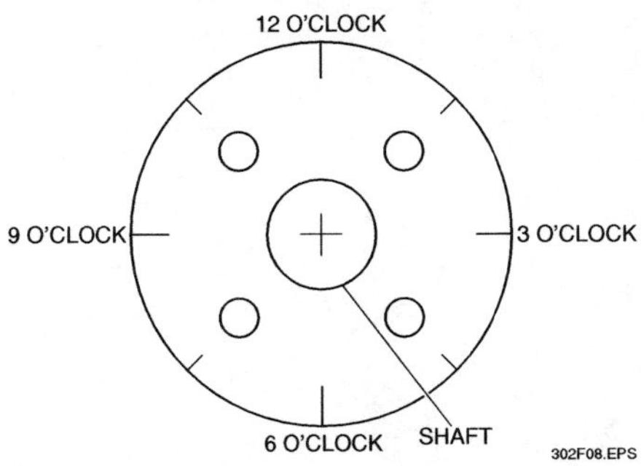

Figure 8 ◆ Positions of coupling face.

When adjusting horizontal angularity, the measurements are taken at the 3 and 9 o'clock positions. For this procedure, assume that the alignment problem is caused by soft foot. Soft foot occurs when one or more feet of the STAT or MTBM are not set flat on the base. This can cause problems when you are trying to determine the thickness of shims needed to align the couplings. When the machine is moved or shimmed, the amount of movement is proportional to the distance moved times the length between the foot and the coupling. Follow these steps to adjust the horizontal angularity:

Step 1 Make marks to be used as match marks across the edge of both couplings using chalk or a grease pencil (see *Figure 9*).

NOTE

These match marks will be used in all of the conventional alignment procedures that follow. These marks help ensure that the couplings are in the exact same position for all measurements. These marks should always be kept in line with each other and at the 12 o'clock position.

Step 2 Check for soft foot on all feet.

Step 3 Check the OD and the face alignment visually from the top view to determine if the STAT and MTBM are close to alignment.

Step 4 Loosen the MTBM lug bolts and move the MTBM base using jack bolts until the couplings look as though they are aligned.

NOTE

By visually checking the couplings and adjusting the MTBM, you can roughly align the coupling enough to make precise measurements. The baseplate may need to move in two directions to align the coupling face and OD.

Step 5 Turn the coupling until the match marks are lined up and at the 12 o'clock position.

Step 6 Measure the gap between the coupling faces at the 3 o'clock position using a feeler gauge.

Step 7 Measure the gap between the faces at the 9 o'clock position (see *Figure 10*).

Audiovisual

Show Transparency 7 (Figure 8). Review the positions of the coupling face used for alignment purposes.

Classroom

Explain how to adjust horizontal angularity. Ensure that trainees understand how to make match marks.

Audiovisual

Show Transparency 8 (Figure 10). Explain how to measure the gap between the faces using a feeler gauge.

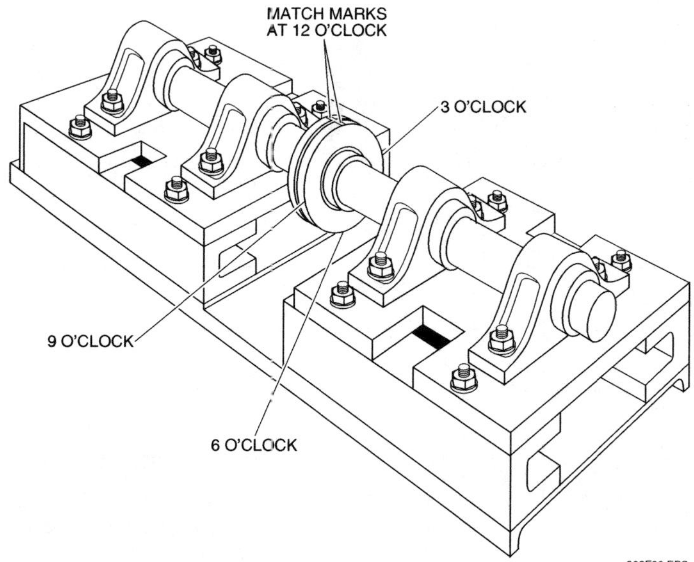

Figure 9 ◆ Match marks.

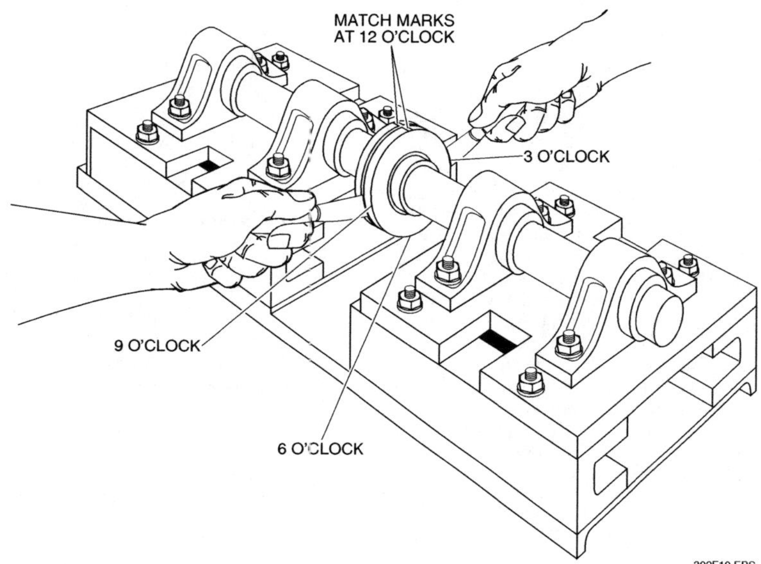

Figure 10 ◆ Measuring the face, top view.

Instructor's Notes:

Step 8 Determine which of the two measurements is the largest.

Step 9 Move the baseplate using the jack bolts so that the gaps at the 3 and 9 o'clock positions are even.

When moving the plate into position, the front of the plate will need to go in one direction and the rear in the other direction, depending on which gap is the largest. *Figure 11* shows adjusting the face alignment if the gap at the 3 o'clock position is wider than that at the 9 o'clock position. If the gap is larger at the 9 o'clock position, the plate should be moved in the opposite direction.

4.2.0 Adjusting Horizontal Offset

After the horizontal angularity has been adjusted, you are ready to adjust the horizontal offset. This will bring the outside edges of the two couplings in line from the top view. Care should be taken not to change the face alignment as the OD alignment is being adjusted. Follow these steps to adjust the horizontal offset:

Step 1 Ensure that the lug bolts are snug.

> **NOTE**
> The lugs should be tight enough to prevent excessive movement when adjustments are made but should also allow adjustments.

Step 2 Place a precision straightedge across the coupling halves at the 9 o'clock position so that it is flat against the side of one coupling half (see *Figure 12*).

Step 3 Slip a feeler gauge into the gap between the straightedge and the side of the other coupling half to measure the amount of misalignment (see *Figure 13*).

> **NOTE**
> The gap may be very small. Be sure to use the thinnest feeler gauge that will slip into the gap. If there is no gap at all, the coupling halves are aligned and need no adjustment in that view.

Step 4 Move the MTBM base using the mounting bolts to align the coupling halves and eliminate the gap (see *Figure 14*).

> **NOTE**
> The front and the rear of the MTBM must be moved the same amount and in the same direction to prevent dislocating the face alignment. Make small adjustments, and check the misalignment after each adjustment.

Step 5 Continue to check and adjust the alignment until the coupling halves are aligned (see *Figure 15*). The coupling halves are aligned when there is no gap between the straightedge and the side of the coupling half.

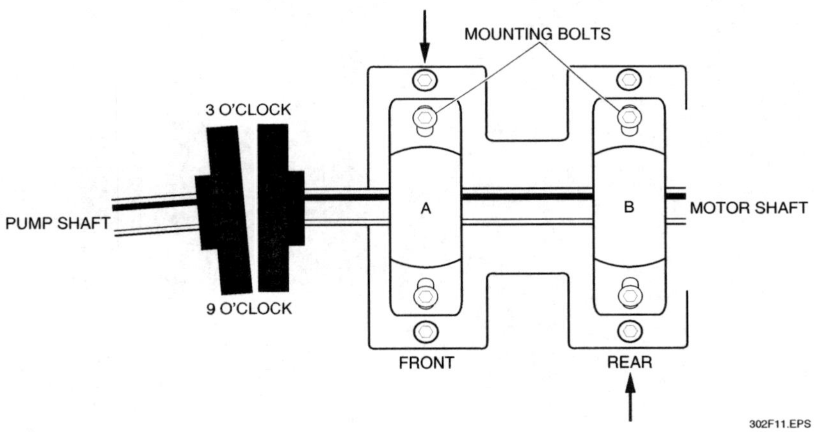

Figure 11 ◆ Adjusting face alignment, top view.

Classroom

Refer to Figure 11. Explain how to move the plate into position when adjusting face alignment.

Review the procedures to adjust horizontal offset. Go over any questions trainees have.

Refer to Figure 12. Explain how to place a precision straightedge across the coupling halves.

Audiovisual

Show Transparency 9 (Figure 13). Explain how to check misalignment using a straightedge and feeler gauge.

Laboratory

Have two trainees demonstrate the steps to adjust horizontal angularity and horizontal offset. Ask trainees to observe the procedures and discuss what was done correctly and/or incorrectly.

Homework

Have trainees review Sections 4.3.0–4.4.0 for the next session.

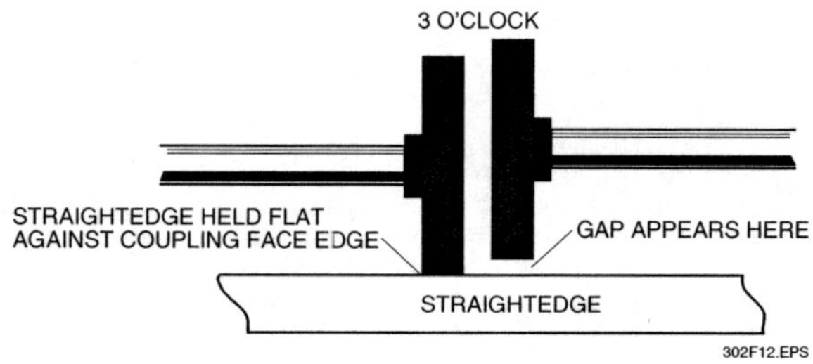

Figure 12 ◈ Placing the straightedge.

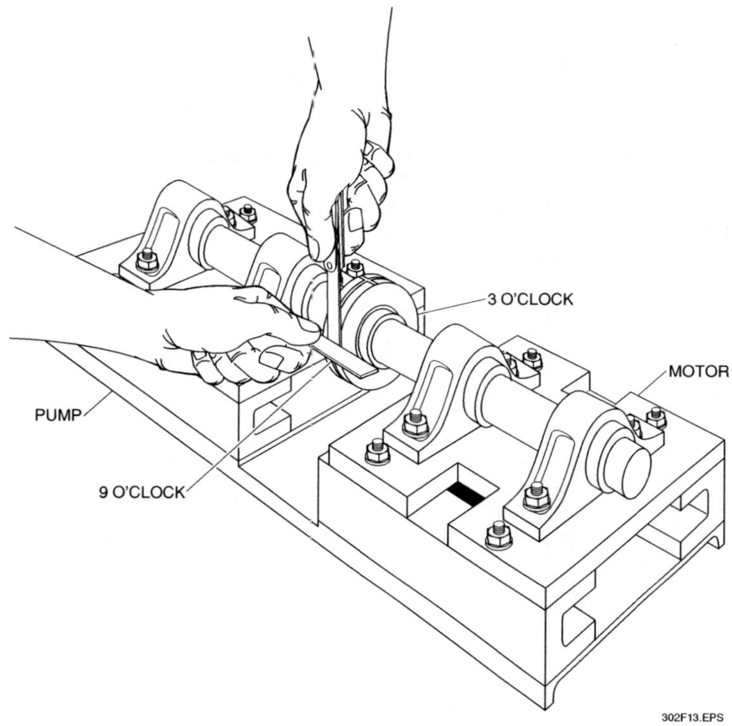

Figure 13 ◈ Checking misalignment using a straightedge and feeler gauge.

Instructor's Notes:

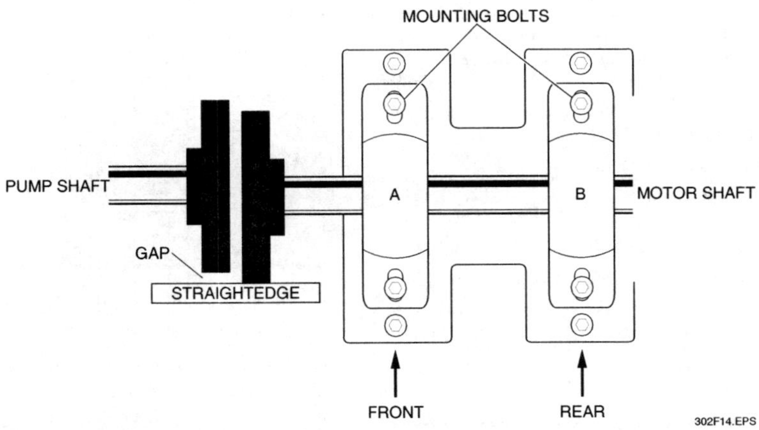

Figure 14 ◆ Adjusted base.

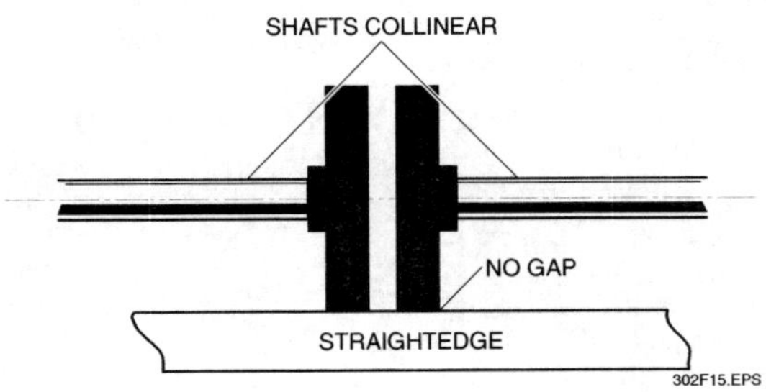

Figure 15 ◆ Aligned coupling.

4.3.0 Adjusting Vertical Angularity

Once the coupling has been aligned horizontally, you are ready to perform the vertical alignment. Vertical alignment is a little more complicated than horizontal alignment because the MTBM must be raised, lowered, or tilted to bring the coupling halves into alignment. This procedure requires that you measure the misalignment, calculate the shim thickness needed, and install the shim(s) to correct the misalignment. Adjusting vertical angularity also requires that the STAT be tilted on its base to correct misalignment. Remember that the MTBM, or driver, is always adjusted to the STAT, or driven. Follow these steps to adjust vertical angularity:

Step 1 Line up the match marks and position them at 12 o'clock.

Step 2 Measure the gap between the coupling halves to the nearest 0.001 inch at the 12 o'clock position using dial calipers, and record the measurement.

Step 3 Measure the gap to the nearest 0.001 inch at the 6 o'clock position, and record the measurement.

Step 4 Subtract the smaller measurement from the larger measurement to determine the difference, and record the difference. Also note at which position the larger measurement was taken.

Step 5 Measure the diameter of the coupling to the nearest ¼ inch, and record the measurement.

Materials

Ensure that you have everything required for the demonstrations and laboratory during this session.

Classroom

Explain why vertical alignment can be more complicated than horizontal alignment. Review the procedure to adjust vertical angularity.

Demonstration

Demonstrate how to measure the gap between coupling halves using dial calipers.

Audiovisual

Show Transparency 10 (Figure 16). Explain how to measure the distance between the center of the lugs using a scale or rule.

Demonstration

On the whiteboard/ chalkboard, demonstrate how to calculate shim thickness.

Classroom

Review the procedures to adjust vertical offset. Ask whether trainees have any questions.

Step 6 Measure the distance between the centers of the lugs to the nearest ¼ inch using a scale or rule, and record the measurement (see *Figure 16*).

Step 7 Calculate the shim thickness needed using the following formula:

$$\frac{\text{Distance between foot center}}{\text{Coupling diameter}} \times \text{difference in gaps} = \text{shim thickness}$$

Example:

Distance between feet = 12.5 inches

Coupling diameter = 3 inches

Difference in gaps = 0.014 inch

$$\frac{12.5 \text{ in.}}{3 \text{ in.}} \times 0.014 \text{ in.} = 0.158 \text{ in.}$$

> **NOTE**
> The formula conveys the anticipated result but must be checked by measurement after the alignment is complete.

Step 8 Determine which feet to put the shims under. Check the measurement made in Steps 2 and 3. If the larger measurement was taken at the 12 o'clock position, the shims go under the feet farthest away from the coupling (rear feet). If the larger measurement was at the 6 o'clock position, the shims go under the feet nearest the coupling (front feet).

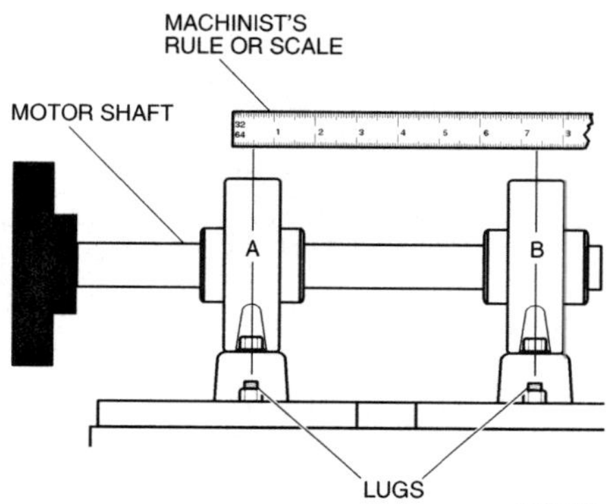

Figure 16 ◆ Measuring between lugs.

Step 9 Install shims of the required thickness under the proper feet.

> **NOTE**
> Install shims in pairs. An equal thickness of shims must be installed under each of the two feet in order to raise the MTBM evenly. When installing the shims, loosen only two bolts at a time to prevent disturbing the top view adjustments that have already been made.

Step 10 Measure the coupling gap at the 12 and 6 o'clock positions using a feeler gauge.

> **NOTE**
> If the gap measurements at the 12 and 6 o'clock positions are the same, the vertical angularity is properly aligned. If they are not the same, repeat Steps 1 through 10 until the measurements are the same.

4.4.0 Adjusting Vertical Offset

Once the vertical angularity is correct, you are ready to adjust the vertical offset. If this procedure is properly performed without disturbing any of the previous adjustments, the alignment will be complete. Follow these steps to adjust the vertical offset:

Step 1 Line up the match marks, and position them at 12 o'clock.

Step 2 Place a straightedge on the coupling at the 12 o'clock position.

Step 3 Measure the distance between the straightedge and the lower coupling half to determine how wide the gap is.

> **NOTE**
> If there is no gap, the coupling is properly aligned and no further adjustment is needed. If there is a gap, it should be measured to the nearest 0.0001 inch.

Step 4 Determine the thickness of the shims using another set of feeler gauges.

Step 5 Loosen the two front lug bolts on the lower unit.

Instructor's Notes:

> **NOTE**
> Always loosen either both front lugs or both rear lugs. Never loosen all four lugs at once because the unit may move and negate all of your previous adjustments.

Step 6 Raise the front of the unit carefully, and slip shims of the proper thickness under both front lugs.

Step 7 Tighten the two front lugs.

Step 8 Loosen the two rear lugs.

Step 9 Raise the rear of the unit carefully, and slip shims of the proper thickness under the rear lugs.

Step 10 Tighten the two rear lugs.

Step 11 Recheck the gap using a straightedge and feeler gauge.

> **NOTE**
> If there is no gap, the coupling halves are properly aligned and need no further adjustment. If the gap is out of tolerance with the equipment specifications, repeat Steps 1 through 11 until the gap is correct.

Step 12 Recheck all four types of alignment to ensure that they have not changed during the alignment procedures.

> **NOTE**
> If any of the alignment is found to be incorrect, start at that point and repeat all procedures until the entire alignment job is complete and correct.

4.5.0 Adjusting Horizontal Angularity and Offset

You have learned to adjust the horizontal angularity and offset alignment as two separate procedures. You may have realized that it is possible to adjust both the angularity and offset at the same time. To combine the procedures, measure both horizontal alignments (angular and offset) at the same time, then adjust and continually check the angularity as you are adjusting the offset. Whenever you make an offset adjustment, check the angularity using a straightedge. This way when the offset is correct, the angularity will also be correct, and you will have achieved both alignments with fewer adjustments.

4.6.0 Adjusting Vertical Angularity and Offset

It is also possible to combine the vertical angularity and offset alignments into one procedure. It is a little more complicated and requires more mathematical calculations than performing the procedures separately, but it will save time and work. You have already used some of the math to calculate the shim thickness needed to correct vertical angularity.

In the previous procedure, placement of the shims raised or lowered the coupling half depending on where you placed the shims. Shims placed under the lugs nearest the coupling half raised the coupling half; shims placed under the lugs farthest from the coupling half lowered it. *Figure 17* shows the effect of shim placement.

The procedure for adjusting the vertical angularity depends on whether the coupling must be raised or lowered to correct the side view face alignment. If there is a wider face gap at the bottom, the coupling is raised. If there is a wider face gap at the top, the coupling is lowered.

4.6.1 Raising the Coupling

When you adjust the vertical angularity, you are affecting the offset alignment as well. There is a way to figure out exactly what effect the vertical angularity adjustment will have on the offset alignment. To do this, measure the vertical angularity and the offset misalignment at the same time. Then use those two measurements to calculate the total shim thickness needed to align the coupling. Assuming that the horizontal alignment has already been adjusted, you are ready to adjust the vertical angularity and offset.

A sample alignment problem is used to help you understand this procedure. *Figure 18* shows this procedure. Follow these steps to adjust the vertical angularity and offset alignment:

Step 1 Measure the coupling OD.

> **NOTE**
> In the sample, it is 4 inches.

Step 2 Measure the vertical angularity and offset misalignments, and record the measurements.

Laboratory

Set up stations with dial calipers, straightedges, scales, and rules. Ask trainees to practice measuring the distance between couplings and lugs.

Homework

Have trainees review Sections 4.5.0–4.6.2 for the next session.

Materials

Ensure that you have everything required to teach this session.

Classroom

Explain how to combine the procedures for adjusting horizontal angularity and offset.

Explain that combining the procedures to adjust vertical angularity and offset depends on whether the coupling is raised or lowered.

Refer to Figure 17. Review the effects of shim placement on the coupling half.

Audiovisual

Show Transparency 11 (Figure 18). Discuss the procedure to adjust the vertical angularity and offset alignment when raising the coupling.

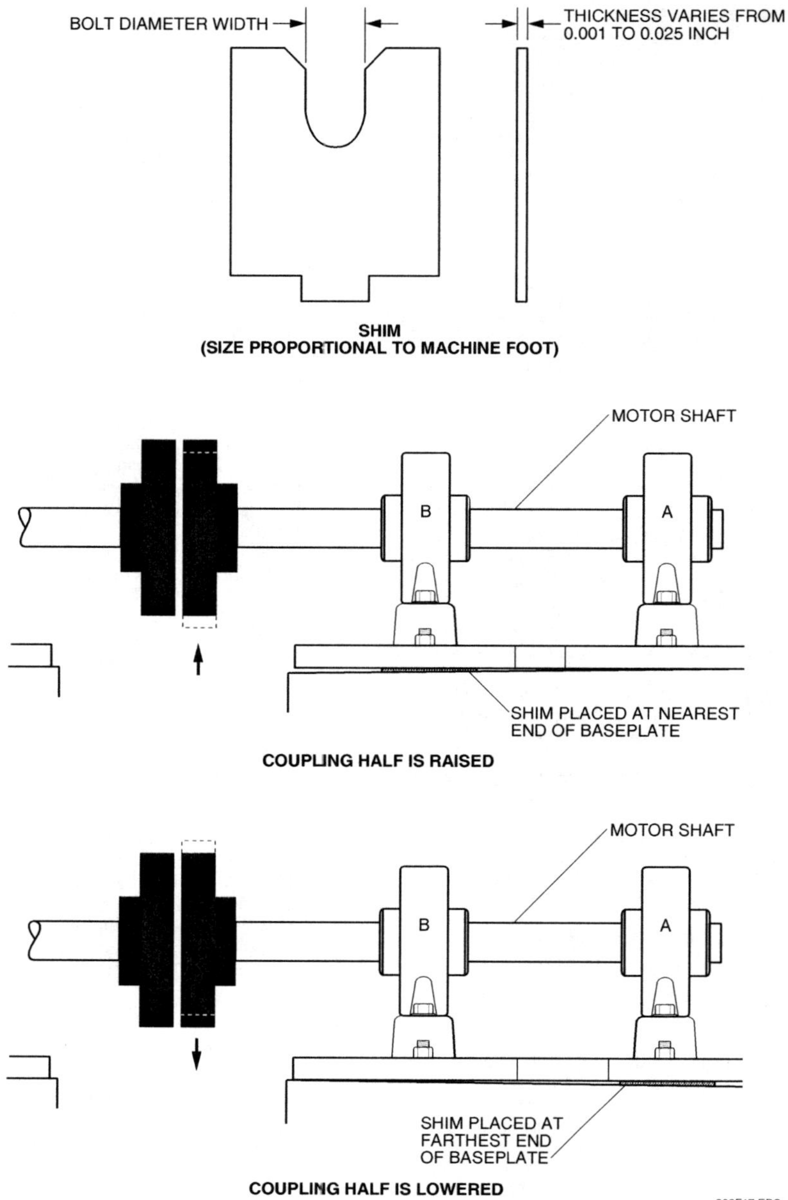

Figure 17 ◆ Effect of shim placement.

Instructor's Notes:

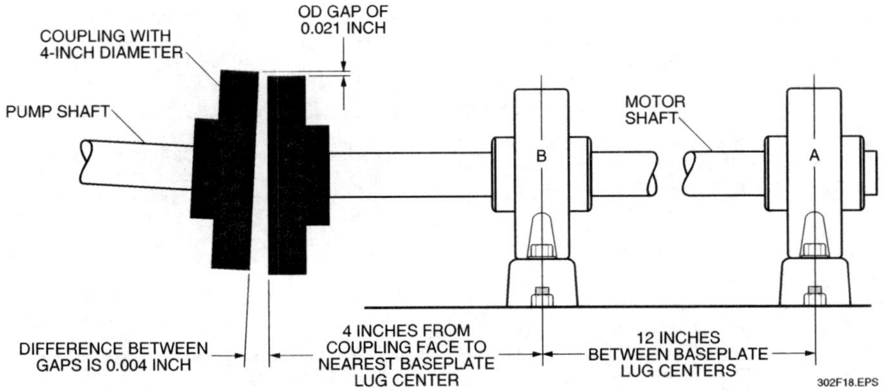

Figure 18 ◆ Raising the coupling, sample alignment problem.

> **NOTE**
> In the sample, they are 0.004 inch and 0.021 inch, respectively.

Step 3 Calculate the shim thickness needed to correct the vertical angularity, using the following formula and record the answer:

$$\frac{\text{Distance between foot centers}}{\text{Coupling diameter}} \times \text{difference in gaps} = \text{shim thickness}$$

$$\frac{12 \text{ in.}}{4 \text{ in.}} \times 0.004 \text{ in.} = 0.012 \text{ in.}$$

Step 4 Determine which end of the baseplate the shims must be put under to correct the vertical angularity.

> **NOTE**
> If the larger coupling gap is at the 12 o'clock position, the shims go under the lug bolts farthest away from the coupling (rear lugs). If the larger gap is at the 6 o'clock position, the shims go under the lugs nearest the coupling (front lugs). In the sample problem the shims are placed under the lug marked B. This is because the largest gap is at the bottom of the coupling.

Step 5 Determine whether the MTBM coupling half is going to be raised or lowered by the placement of the shims.

> **NOTE**
> If the shims are to be placed under the front feet, or those nearest the coupling, the coupling will be raised. If they are to be placed under the rear feet, or those farthest from the coupling, the coupling will be lowered.

Step 6 Determine how much the coupling half will be raised when the shims are placed. When the shims are placed under the front feet, the coupling half will be raised. To find out how much it will be raised, use the following formula:

(Distance from coupling face to center of farthest lug ÷ distance between center feet)
× shim thickness under nearest foot
= distance coupling will be raised

$(16 \text{ in.} \div 12 \text{ in.}) \times 0.012 \text{ in.} = 0.016 \text{ in.}$

Step 7 Subtract the amount that the coupling half will raise from the original vertical offset misalignment found in Step 2. This is the shim's thickness required under both ends of the plate to correct the vertical offset misalignment.

> **NOTE**
> For the sample problem, this is 0.005 inch.

Classroom

Review the formula to calculate shim thickness. Answer any questions trainees have.

Demonstration

On the whiteboard/chalkboard, demonstrate how to calculate the amount that the coupling half will be raised.

Classroom

Review the procedure to correct vertical offset misalignment by lowering the coupling.

Audiovisual

Show Transparency 12 (Figure 19). Explain how to measure vertical angularity and offset misalignment.

Demonstration

Demonstrate how to calculate the distance that the coupling half will be lowered.

Step 8 Install shims of the required thickness to correct the face misalignment, as found in Step 3, plus the required thickness to correct the vertical offset misalignment, as found in Step 7, under both front feet.

Step 9 Under the rear feet only, install the required thickness of shims to correct the vertical offset misalignment.

4.6.2 Lowering the Coupling

When the coupling half must be lowered to correct the vertical offset misalignment, the procedure is different. In this case, the shims are placed under the rear lugs to raise the rear of the MTBM and lower the coupling half. As with the previous procedure, a sample alignment problem (see *Figure 19*) is used to help you understand this procedure:

Step 1 Measure the coupling OD.

> **NOTE**
> In the sample, it is 4 inches.

Step 2 Measure the vertical angularity and offset misalignment, and record the measurements.

> **NOTE**
> In the sample, they are 0.004 inch and 0.021 inch, respectively.

Step 3 Calculate the shim thickness needed to correct the vertical angularity, using the formula, and record the results:

$$\frac{12 \text{ in.}}{4 \text{ in.}} \times 0.004 \text{ in.} = 0.012 \text{ in.}$$

Step 4 Determine which foot the shims must be put under to correct the vertical angularity.

> **NOTE**
> Since the larger coupling gap was at the 12 o'clock position, the shims go under the lug bolts farthest away from the coupling. In the sample problem, the shims would be placed under the lug marked A. This is because the largest gap is at the top of the coupling.

Step 5 Determine whether the MTBM coupling half is going to be raised or lowered by the placement of the shims.

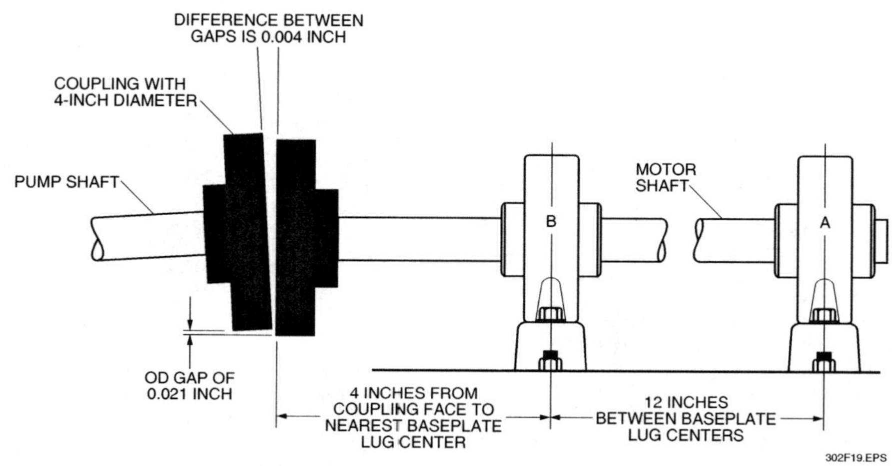

Figure 19 ◆ Lowering the coupling, sample alignment problem.

Instructor's Notes:

> **NOTE**
> In this case, the coupling half will be lowered because the largest gap is at the top of the coupling.

Step 6 Calculate the distance that the coupling half will be lowered when the shims are installed to correct the face misalignment, using the following formula:

Distance from coupling face half to nearest lug center ÷ distance between lug centers × shim thickness = side view misalignment

(4 in. ÷ 12 in.) × 0.012 in. = 0.004 in.

Step 7 Add the amount that the coupling will be lowered to the original offset misalignment to determine the thickness of shims needed to correct the offset misalignment.

> **NOTE**
> For the sample, 0.004 inch + 0.021 inch = 0.025-inch shim to correct the offset misalignment.

Step 8 Install shims of the required thickness under the rear lugs to correct the vertical angularity and offset misalignment.

Step 9 Under the front lugs only, install the required thickness of shims needed to correct the offset misalignment.

Step 10 Tighten all lug bolts.

Step 11 Recheck both the horizontal and vertical alignments.

5.0.0 ◆ PERFORMING RIM AND FACE DIAL INDICATOR ALIGNMENT

The most accurate and widely used method of conventional alignment is the reverse dial indicator alignment method. The dial indicator is a precision instrument that produces much more accurate measurements than a straightedge and feeler gauge.

> **CAUTION**
> The dial indicator is very fragile and must be handled with great care. Be careful not to bump or drop it.

The adjustments that are made to correct misalignment using the dial indicator method are the same as those made with the rim and face method. The difference is that the measurements are taken with the more accurate dial indicator. The accuracy is ensured by following a standard procedure for alignment. The procedures for aligning couplings using dial indicators include:

- Setting up dial indicators
- Taking horizontal measurements
- Taking vertical measurements
- Taking face and OD measurements

5.1.0 Setting Up Dial Indicators

In order to take accurate measurements of the coupling, the dial indicators must be correctly set up. They must be rigidly attached and properly adjusted for the procedure being performed. The dial indicator must have free movement from stop to stop to give accurate readings. The bracket or support that holds the dial indicator must be measured for sag, and this measurement must be used to correct the reading. Always take at least two different readings to ensure that the setup is valid and accurate. The area of the measurement must be clean and the brackets strong and adjusted snugly to give the dial indicator a solid mounting.

A chain indicator jig consists of a block, a chain, an adjustment bolt, and a wing nut (see *Figure 20*). A dial indicator jig is used to support the dial indicator in the proper position. The block is drilled and tapped to facilitate mounting of the dial indicators and is notched to fit the shaft. The chain is attached to the adjusting bolt, which fits in a hole through the block. There is a hook or pin on the opposite side of the block on which to hook the chain.

A reverse dial indicator jig consists of a clamping device with bolts, a mounting pin, and an indicator support device. Some consider this type of jig to be more accurate over long distances because it more rigid and has less indicator sag than other jigs.

In some applications, the motor and pump must be mounted some distance apart. In this case, a jackshaft must be used. A jackshaft is an extension shaft that is used to connect the STAT and the MTBM when the space between them is longer than the stock shafts. When a jackshaft is used, the alignment jig must span this distance without any deflection and be rigid enough to support the dial indicator without sagging (see *Figure 21*).

Determining indicator sag is critical to any indicator alignment setup. Gravity causes the extension bracket to droop by a very slight amount, so all brackets have some degree of sag.

Laboratory

On the whiteboard/chalkboard, provide sample alignment problems. Ask trainees to practice the procedures to adjust vertical angularity and offset by raising and lowering the coupling.

Teaching Tip

See the Teaching Tip for Section 4.0.0 at the end of this module.

Homework

Have trainees review Sections 5.0.0–5.1.3 for the next session.

Materials

Ensure that you have everything required to teach this session.

Safety

Emphasize that the dial indicator is a fragile instrument. Explain how to properly use the dial indicator.

Classroom

Review the procedures for aligning couplings using a dial indicator. Explain how to properly set up a dial indicator.

Audiovisual

Show Transparency 13 (Figure 20). Review the components in a chain indicator jig.

Classroom

Refer to Figure 21. Review the effects of the jackshaft on indicator sag.

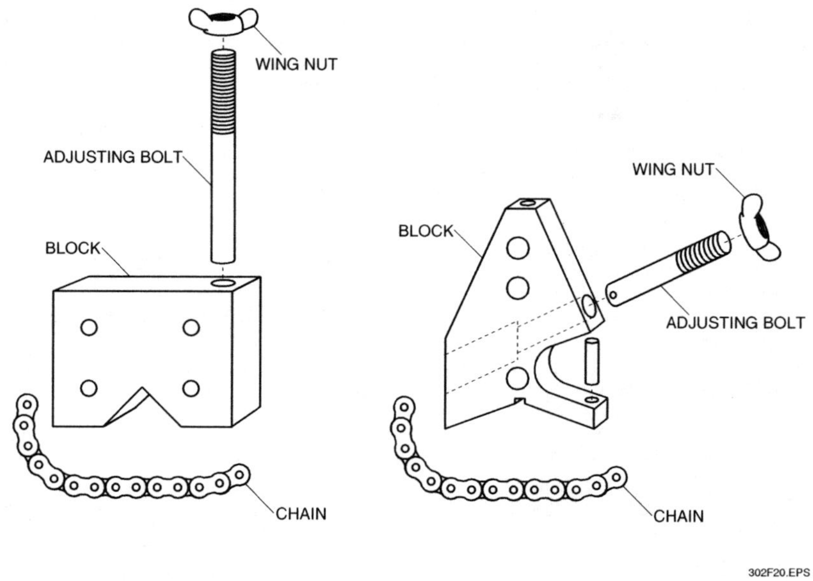

Figure 20 ◆ Chain indicator jigs.

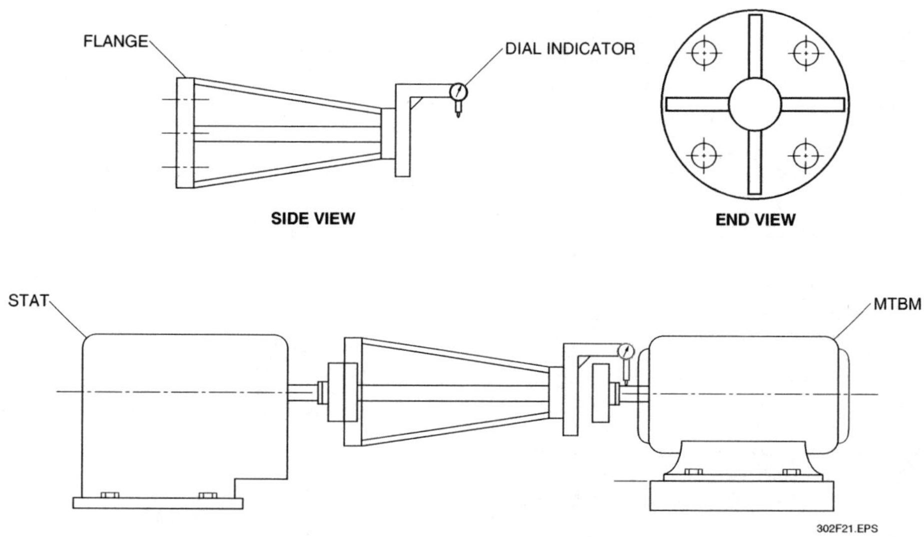

Figure 21 ◆ Jackshaft alignment jig.

Instructor's Notes:

This amount could be as small as 0.001 inch or up to 0.050 inch, depending on the span and rigidity of the fixture. In either case, the error induced by this sagging is enough to make true centerline alignment impossible.

The bracket sags towards the shaft in the 12 o'clock position. When rotated 180 degrees to the 6 o'clock position, that same amount of sag pulls away from the shaft, thereby doubling the amount of total movement and error that must be added to the alignment calculations.

To measure sag, mount the entire measurement fixture, including the brackets, bars, and indicators, on the machine shaft or off the machine on a piece of straight pipe about 2 inches in diameter. Adjust the fixture, and position the brackets until they are the same distance apart as they will be when they are mounted on the machinery. With the indicators at the 12 o'clock position, zero the dials. Rotate the shaft until the indicators are at the 6 o'clock position, and record this value. Subtract this number from the alignment total calculated from the four readings. *Figure 22* shows indicator sag.

Just as with the rim and face method, there are three major types of measurements to be taken:

- OD (rim)
- Face
- Face and OD

5.1.1 OD (Rim) Measurements

OD misalignment is also called parallel misalignment. To check OD misalignment, the dial indicator is set up on the outside rim of the coupling. The indicator bracket is usually fastened to the MTBM coupling half; the indicator is set to take readings from the STAT coupling half. Follow these steps to set up a dial indicator to take OD measurements:

Step 1 Attach a jig to the MTBM coupling half.

> **NOTE**
> The jig should be rigidly attached so that it does not move or slip on the coupling half.

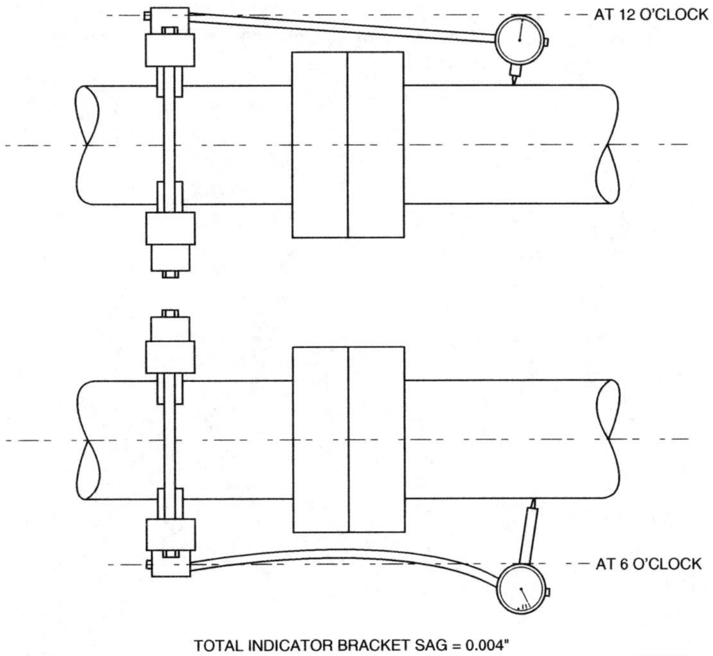

Figure 22 ◆ Indicator sag.

Audiovisual

Show Transparency 14 (Figure 22). Explain how to determine indicator sag.

Classroom

Review the three major types of measurements to be taken when performing rim and face dial indicator alignment.

Classroom

Refer to Figure 23. Discuss the procedures to set up a dial indicator to take OD measurements. Go over any questions trainees have.

Hand out copies of Figures 23 through 25 without their titles. Ask trainees to identify the dial indicator setup for OD, face, and face and OD measurements.

Homework

Have trainees review Sections 5.2.0–5.2.2 for the next session.

Materials

Ensure that you have everything required to teach this session.

Classroom

Explain how to take horizontal angularity measurements using a dial indicator.

Step 2 Attach appropriate hardware to the jig to extend it past the STAT coupling half.

Step 3 Attach one dial indicator to the jig hardware so that the indicator rides on the OD of the STAT coupling half (see *Figure 23*).

5.1.2 Face Measurements

Setting up the dial indicator for face measurements is the same as for rim measurements except that the indicator rides on the face of the STAT coupling half instead of on the rim (see *Figure 24*).

5.1.3 Face and OD Measurements

Dial indicators can be set up to take both face and OD measurements at the same time. The setup is the same as above except that two dial indicators are set on the coupling at the same time, one for face measurements and one for OD measurements (see *Figure 25*).

5.2.0 Taking Horizontal Measurements

The same horizontal measurements can be taken using dial indicators as were taken with the straightedge and feeler gauge, and the adjustments for alignment are made in the same way. The difference is that the dial indicator gives much more accurate measurements. The horizontal measurements taken include the following:

- Horizontal angularity
- Horizontal offset

5.2.1 Horizontal Angularity Measurements

Follow these steps to take the horizontal angularity measurements using a dial indicator:

Step 1 Set up a dial indicator for horizontal angularity measurements.

Step 2 Press in the plunger until the indicator needle completes one full turn to set the dial indicator's freedom of movement, or travel.

> **NOTE**
> Pressing in the indicator plunger one full turn of the dial indicator ensures that the indicator has at least one full turn of travel in both the in and out directions.

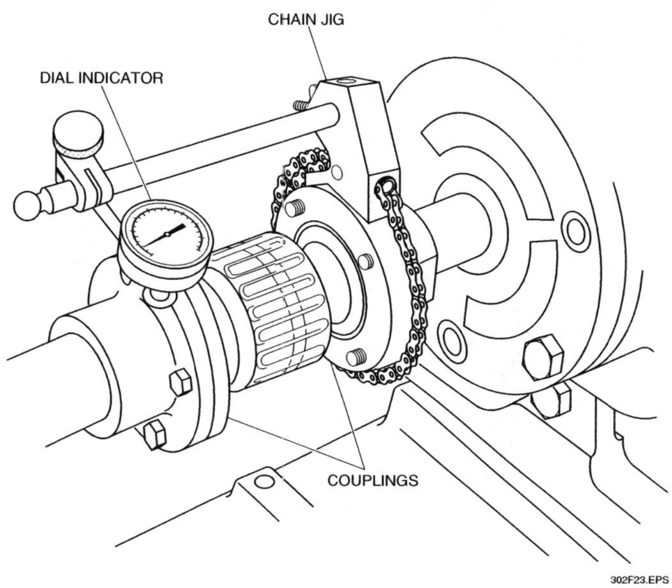

Figure 23 ◆ Dial indicator setup for OD (rim) measurements.

Instructor's Notes:

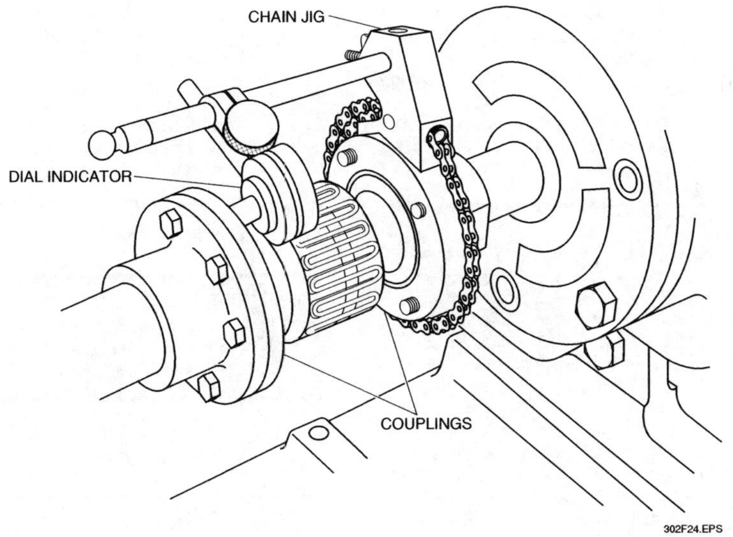

Figure 24 ◈ Dial indicator setup for face measurements.

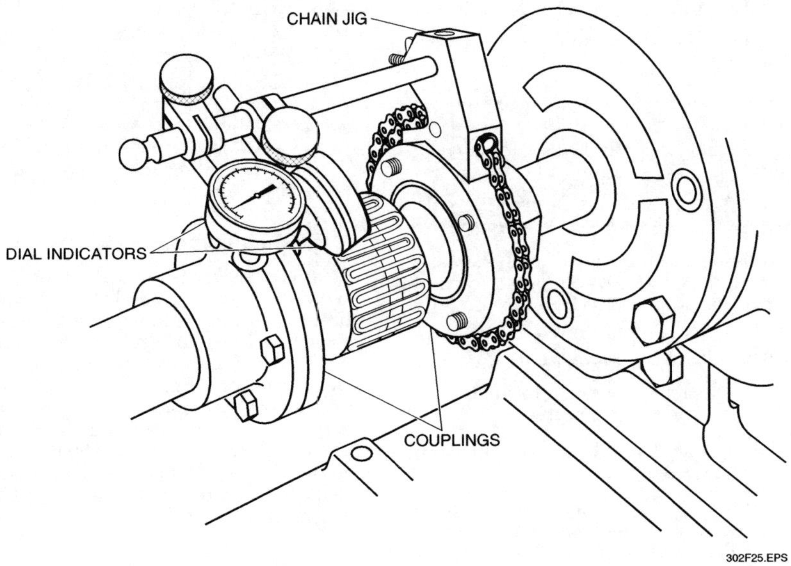

Figure 25 ◈ Dial indicator setup for OD and face measurements.

Safety

Discuss the consequences of failing to move the dial indicator away from the coupling.

Classroom

Compare the procedure for taking horizontal offset measurements using a dial indicator with those for taking horizontal angularity measurements.

Safety

Ask a trainee to explain why the dial indicator should be moved away from the coupling before making adjustments.

Homework

Have trainees review Sections 5.3.0–5.3.2 for the next session.

Materials

Ensure that you have everything required to teach this session.

Classroom

Discuss the differences between taking vertical and horizontal measurements using dial indicators. Ask whether trainees have any questions.

Step 3 Turn both coupling halves bolted together until the dial indicator is at the 3 o'clock position.

Step 4 Zero the dial indicator.

Step 5 Turn both coupling halves still bolted together until the indicator is at the 9 o'clock position.

> **NOTE**
> There are two things to watch for as you rotate the indicator from the 3 to the 9 o'clock position. One is whether the indicator moves in the positive or negative direction. This indicates where the widest gap is. If the dial moves in the negative direction, the widest gap is at the 3 o'clock position. If it moves in the positive direction, the widest gap is at the 9 o'clock position. This tells which way the MTBM must be moved to correct the misalignment. The second thing is whether the indicator needle moves more than one complete turn. If this occurs, the dimension is greater than the total indicator range. If this is the case, you should be able to see the misalignment by looking at it, and you need to rough-align the coupling until it is within the range of the dial indicator.

Step 6 Take a reading at the 9 o'clock position.

Step 7 Adjust the MTBM to correct the misalignment.

> **CAUTION**
> Move the dial indicator away from the coupling before making adjustments to prevent the indicator from being damaged.

Step 8 Continue to measure and adjust the alignment until a zero reading is achieved at both the 3 and 9 o'clock positions.

5.2.2 Horizontal Offset Measurements

Follow these steps to take the horizontal offset measurements using a dial indicator:

Step 1 Set up the dial indicator for horizontal offset measurements.

Step 2 Press in the plunger until the dial indicator needle completes one full turn to set the dial indicator travel.

Step 3 Turn both coupling halves bolted together until the dial indicator is at the 3 o'clock position.

Step 4 Zero the dial indicator.

Step 5 Turn both coupling halves bolted together until the dial indicator is at the 9 o'clock position.

Step 6 Take a reading at the 9 o'clock position.

Step 7 Divide the reading by two.

> **NOTE**
> When measuring the horizontal offset, the reading on the dial indicator is actually two times the amount of misalignment.

Step 8 Adjust the MTBM to correct the horizontal offset misalignment.

> **CAUTION**
> Move the dial indicator away from the coupling before making adjustments.

> **NOTE**
> The procedures for making adjustments are the same as those used in the conventional method.

Step 9 Continue to measure and adjust the offset misalignment until you achieve a zero reading at both the 3 and 9 o'clock positions.

5.3.0 Taking Vertical Measurements

The procedure for using dial indicators to measure vertical angularity and offset are the same as those for measuring horizontal angularity and offset except that you set the indicator at zero at the 12 o'clock position and rotate the coupling halves together to the 6 o'clock position to take your readings. Remember that the angularity measurements are taken at face value, and the offset measurements are divided by two.

Use the same formulas for making vertical calculations as you used in the conventional method of alignment.

Instructor's Notes:

5.3.1 *Taking Angularity and Offset Measurements*

You can take the angularity and offset measurements at the same time by setting up the dial indicators on both the face of the coupling half and the OD of the coupling half.

Follow these steps to check the vertical angularity and offset measurements:

Step 1 Set up two dial indicators on the face and the OD of the STAT coupling half.

Step 2 Turn both coupling halves bolted together until both indicators are at the 12 o'clock position.

Step 3 Zero both dial indicators.

Step 4 Turn both coupling halves bolted together until both indicators are at the 6 o'clock position.

Step 5 Take both the face reading and the OD reading.

Step 6 Calculate the shim thickness needed to correct the angularity and offset misalignments using the shim thickness formula.

Step 7 Install the shims in the appropriate places.

Step 8 Continue to take readings and make adjustments until you achieve a zero indicator reading at both the 12 and 6 o'clock positions.

5.3.2 *Checking Alignment Using Dial Indicators*

Once all adjustments have been made, recheck the alignment in all positions. This is much easier using the dial indicators than using the rim-to-face method. Follow these steps to check alignment using the dial indicators:

Step 1 Set up two dial indicators on the face and the OD of the STAT coupling half.

Step 2 Turn both coupling halves bolted together until both indicators are at the 12 o'clock position.

Step 3 Zero both dial indicators.

Step 4 Turn both coupling halves bolted together until both indicators are at the 3 o'clock position.

Step 5 Take both angularity and offset readings.

Step 6 Turn both coupling halves bolted together until both indicators are at the 6 o'clock position.

Step 7 Take both angularity and offset readings.

Step 8 Turn both coupling halves bolted together until both indicators are at the 9 o'clock position.

Step 9 Take both angularity and offset readings.

Step 10 Turn both coupling halves bolted together until both indicators are at the 12 o'clock position.

Step 11 Take both angularity and offset readings.

NOTE

If the alignment is correct, you should have zero indicator readings on the angularity and offset measurements at all four positions. If any of the positions do not produce a zero reading, further alignment is needed.

6.0.0 ◆ PERFORMING REVERSE DIAL INDICATOR ALIGNMENT

The alignment of shafts using the reverse dial indicator method is simple if you follow each step and avoid shortcuts or guessing. This method can be adapted to many situations. Close attention to **thermal growth**, elimination of soft foot, and factoring out the indicator sag helps ensure a successful alignment.

6.1.0 Setting Up Reverse Dial Indicator Jigs

Setting up the reverse dial indicator jig is somewhat complicated and can be done in two ways. Two jigs and two dial indicators are used. Because two jigs are used together, this is referred to as a complex reverse dial indicator jig. One jig is clamped around the driving coupling half, or MTBM, and the other is clamped around the driven coupling half, or STAT. The two jigs can be positioned so that the mounting pin of one jig is in line with the flat on top of the other jig or on the same side (see *Figure 26*).

Another method is to mount the dial indicators 180 degrees apart from each other. *Figure 27* shows mounting the complex reverse dial indicator jigs on opposite sides.

One indicator and its hardware are mounted on the MTBM coupling jig so that the indicator contacts the flat on top of the STAT coupling jig. The other indicator and its hardware are mounted on the STAT coupling jig so that the indicator contacts the flat on top of the MTBM coupling jig.

Classroom

Ask trainees to write on a sheet of paper the correct settings of the dial indicator when checking alignment and taking horizontal and vertical angularity and offset measurements.

Teaching Tip

See the Teaching Tip for Section 5.0.0 at the end of this module.

Homework

Have trainees review Sections 6.0.0–6.2.0 for the next session.

Materials

Ensure that you have everything required to teach this session.

Classroom

Discuss the two methods for setting up the reverse dial indicator jig.

Audiovisual

Show Transparency 15 (Figure 26). Explain how to mount two reverse dial indicator jigs on the same side.

Classroom

Refer to Figure 27. Compare opposite-side mounting with same-side mounting.

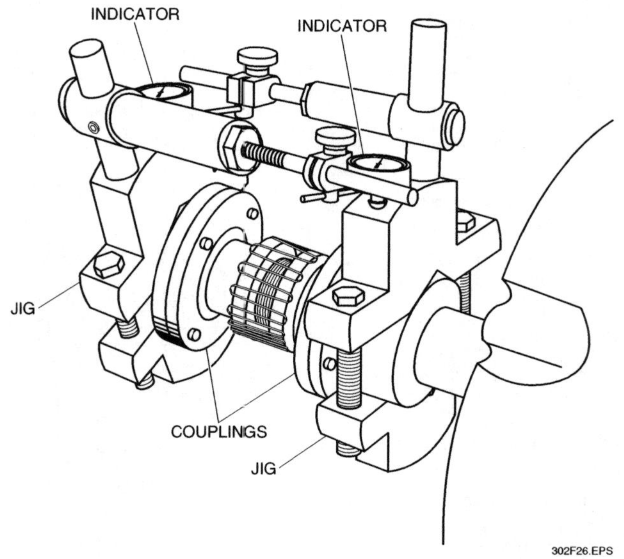

Figure 26 ◆ Same-side mounting.

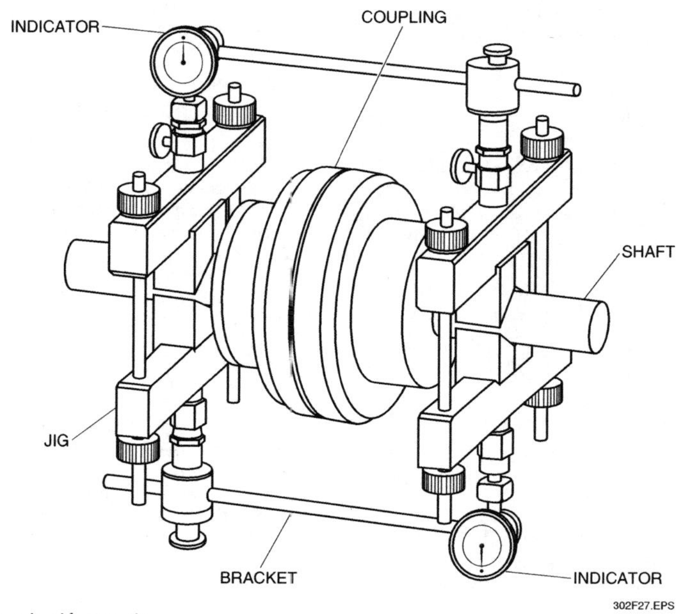

Figure 27 ◆ Opposite-side mounting.

Instructor's Notes:

6.2.0 Performing Reverse Dial Indicator Alignment

Before starting reverse dial indicator alignment, lock and tag out the machinery to ensure that it will not start up during this procedure. After the reverse dial indicator jigs are mounted in the preferred position, the technique of reverse dial indicator alignment must be chosen.

Chart alignment involves taking the indicator readings and plotting them on an accurate graph to arrive at a corrective movement or shim thickness. The other method involves taking the indicator readings and inserting them into a mathematical formula. Both of these methods also require measuring the distances from feet to couplings. *Figure 28* shows distance measurements.

Both methods require practice and attention to detail. It is very easy to misread the direction of travel on the dial indicator and have a positive value recorded as a negative. Make it a habit to read every position at least twice. Always measure and note the indicator sag for each setup.

6.2.1 Charting Alignment

Charting alignment on a graph requires a chart divided into thousandths of an inch vertically for the indicator readings and into whole inches horizontally for the distance measurements. The plotting of points on this chart correlates to the shim thickness or distance to be moved. The chart can be customized as needed.

Figure 29 shows a graphical alignment chart. Follow these steps to perform chart alignment:

Step 1 Measure and chart the dimensions from foot to foot and feet to couplings shown as distances A, B, C, D, E, and F in *Figure 28*.

Step 2 Measure the indicator sag, and record the value on the chart.

Step 3 Note on the chart whether the reading is made on a cold or hot motor.

Step 4 Zero the indicator on the side of the coupling opposite from you.

Step 5 Rotate both shafts together in the correct direction of rotation until the indicator is on your side of the coupling.

Step 6 Record the final reading for MTMB to STAT (M-S) as the top-to-bottom indicator reading (TIR) on the graph.

Step 7 Multiply M-S TIR by one-half; record the result on the left side of the graph paper.

> **NOTE**
> Plot the number above the STAT if it is positive and below if it is negative.

Step 8 Zero the indicator that is now on the side opposite you.

Step 9 Rotate the shafts in the correct direction of rotation until the indicator is on your side.

Step 10 Record the number as TIR S-M on the graph paper.

Step 11 Multiply TIR S-M by one-half, and plot this number at a distance to the right that is equal to the distance between the indicators from the previously plotted number.

Step 12 Connect the two dots with a straight line. Use the numbers at the bottom of the graph to show the location of the feet on the MTBM.

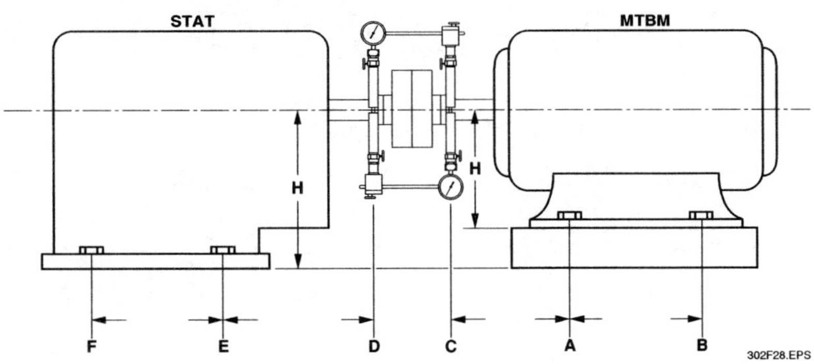

Figure 28 ◆ Distance measurements.

Audiovisual

Show Transparency 16 (Figure 28). Review the techniques for taking indicator readings and calculating distance measurements. Answer any questions trainees have.

Homework

Have trainees review Sections 6.2.1–6.2.3 for the next session.

Materials

Ensure that you have everything required for laboratories during this session.

Classroom

Review the procedures to perform chart alignment.

Laboratory

Distribute copies of a graphical alignment chart. Ask trainees to practice plotting points on the chart using variables you provide.

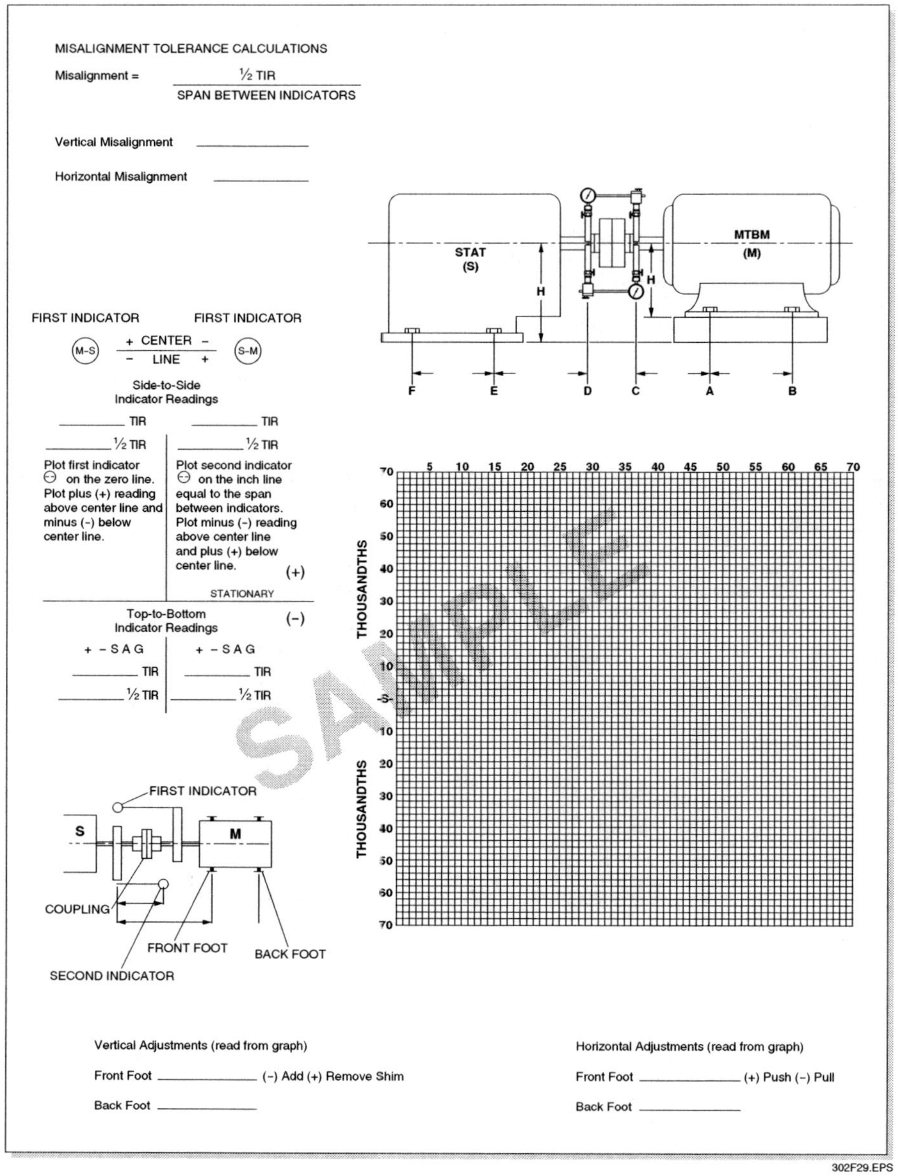

Figure 29 ◆ Graphical alignment chart.

Instructor's Notes:

6.2.2 *Performing Vertical Alignment*

Follow these steps to perform vertical alignment:

Step 1 Indicate the direction at the tip of the arrow and the top of the graph paper as *UP*.

Step 2 Set the dial indicator on top to read the positive value of indicator sag.

Step 3 Rotate the shafts 180 degrees, and note the direction of rotation and the final reading.

Step 4 Record the reading as TIR M-S on the graph paper.

Step 5 Multiply TIR M-S by one-half, and plot this value on the left side of the graph.

Step 6 Set the dial indicator that is now on top to read the positive value of sag.

Step 7 Rotate the shafts 180 degrees. Note the direction of rotation and the final reading.

Step 8 Record the reading as TIR S-M on the graph.

Step 9 Multiply TIR S-M by one-half, and plot this value at a distance equal to the span between the dial indicators to the right of the previously plotted number.

Step 10 Connect the dots with a straight line. Use the numbers on the side of the graph to show the location of the feet on the MTBM.

Step 11 Make the proper alignment corrections.

Step 12 Continue to calculate and plot shims until the machines are aligned.

Step 13 Check the horizontal alignment, and correct it if required.

The alignment can also be represented graphically on a reverse dial indicator plotting guide (see *Figure 30*).

Explain how to perform vertical alignment. Go over any questions trainees have.

Refer to Figure 30. Ask trainees to practice recording dial indicator readings using the plotting guide.

REVERSE INDICATOR PLOTTING GUIDE			
REVERSE READING		PLOT LOCATION	ABSOLUTE VALUES
MTBM - STAT	STAT - MTBM	STAT D C MTBM	
+	–	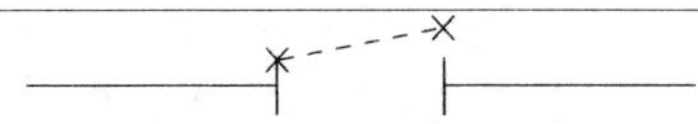	R - N < N - R
+	–	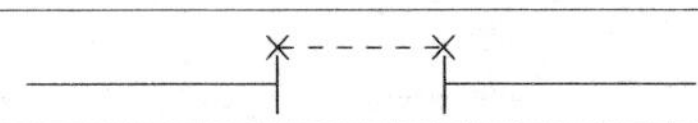	R - N = N - R
+	–	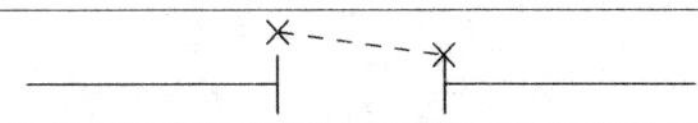	R - N > N - R
+	+	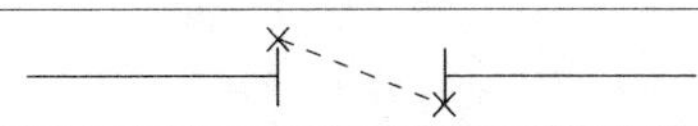	
+	0	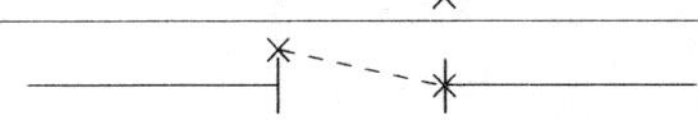	
–	+	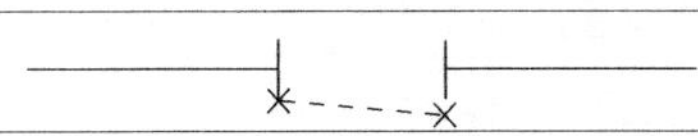	R - N < N - R
–	+	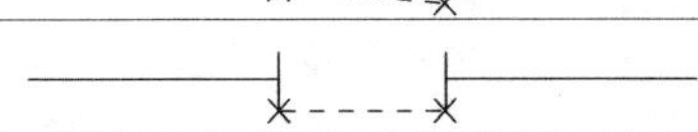	R - N = N - R
–	+	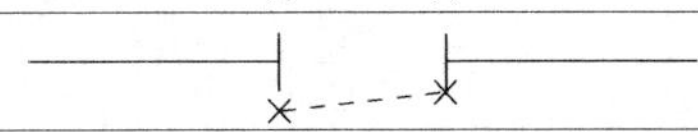	R - N > N - R
–	–	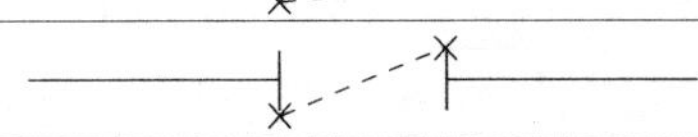	
–	0	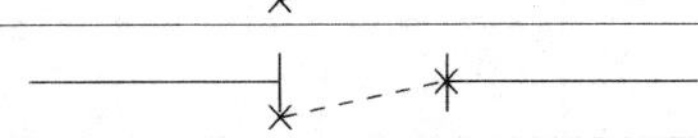	

N=STAT R=MTBM

302F30.EPS

Figure 30 ◆ Reverse dial indicator plotting guide.

Classroom

Refer to Figure 31. Discuss the data that must be documented when recording alignment.

Teaching Tip

See the Teaching Tip for Section 6.2.0 at the end of this module.

Homework

Have trainees review Sections 7.0.0–7.3.0 for the next session.

6.2.3 Recording Alignment

The absolute necessity of recording and maintaining records of all alignment is to ensure that a logical system can be productive and not cause more problems than it solves. Every alignment should be documented on a standard form that is distributed to all departments and stored in an area where it is accessible. *Figure 31* shows a sample alignment record.

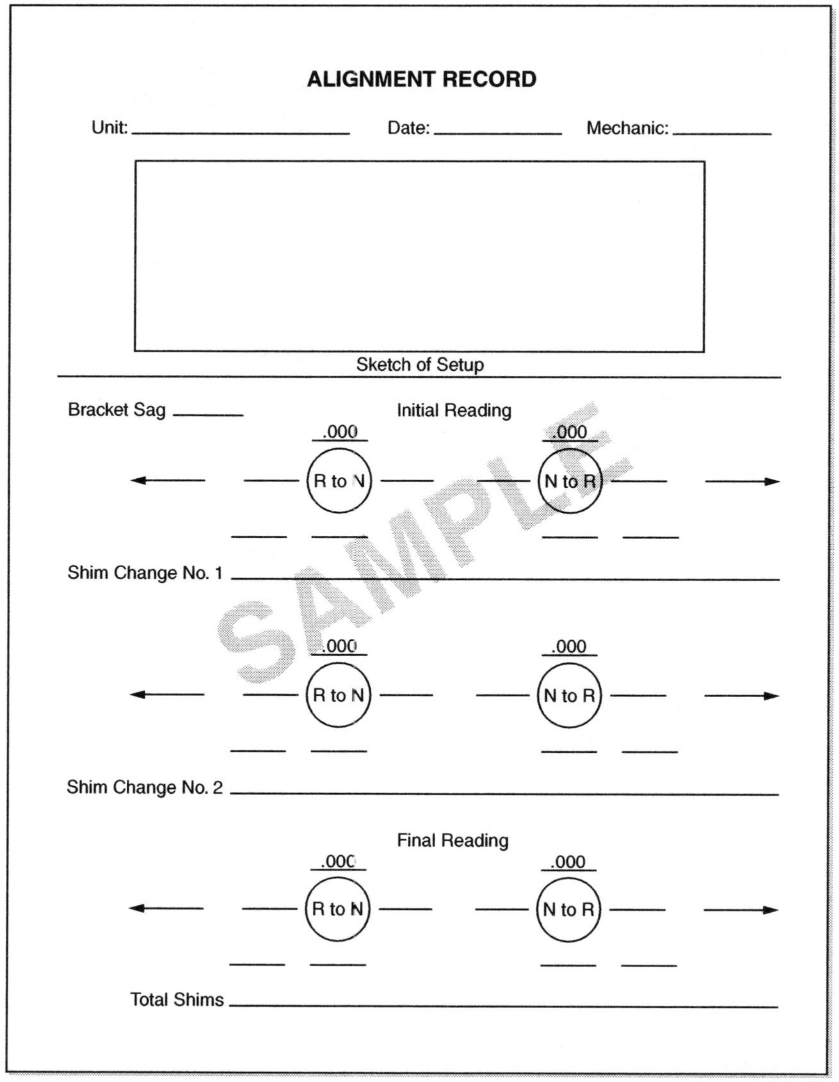

ALIGNMENT RECORD

Unit: ______ Date: ______ Mechanic: ______

Sketch of Setup

Bracket Sag ______

Initial Reading

.000 R to N

.000 N to R

Shim Change No. 1 ______

.000 R to N

.000 N to R

Shim Change No. 2 ______

Final Reading

.000 R to N

.000 N to R

Total Shims ______

SAMPLE

302F31.EPS

Figure 31 ◆ Sample alignment record.

Instructor's Notes:

7.0.0 ◈ PERFORMING LASER ALIGNMENT

When machinery is installed, only the rotating shaft center lines of machines are aligned. The positions of the feet, couplings, shaft surfaces, machine housing, and bearings are not used to determine machine alignment. It is important to understand that alignment refers to the positions of two center lines of rotation and that the machined center line may or may not be the same as the rotational center line. Therefore, we can define shaft alignment as positioning two or more machines so that their rotational center lines are collinear at the coupling point under operating conditions.

The coupling point refers to the fact that vibration caused by misalignment originates at the point of power transmission, which is the coupling. It does not mean that the couplings are being aligned. The shaft center lines are being aligned, and the coupling center is the measuring point. Dynamic factors after start-up, such as wear, thermal growth, dynamic loads, and support structure shifts, make it important to align machinery under operating conditions.

Although rough alignment can be made using straightedges, the only way to determine the rotational center line is to rotate the equipment. Any system whose alignment measurements are taken without turning both shafts is aligning surfaces, not center lines or rotation.

One of the fastest and most accurate methods for aligning machinery is using a laser alignment tool. Before performing laser alignment, you must perform the following five steps:

Step 1 Inspect the base of the equipment for level, center line, and distortion.

Step 2 Check for soft foot.

Step 3 Set the coupling gap.

Step 4 Eliminate outside stress influences, such as pipe stress and conduit.

Step 5 Perform rough alignment of the equipment within the mechanical limits of the indicators (0.1.00 inch).

7.1.0 Laser Theory

A laser beam is an amplified and **collimated beam** of coherent light energy that travels perfectly straight in a vacuum and performs reasonably well in the atmosphere. The cause of imperfect performance in the atmosphere is humidity or dust in the air because each tiny drop of water acts as a prism that diffracts part of the laser beam. The emitter source in each laser and the amount of power applied to it determine the application for which it is used.

There are three classes of lasers, and each laser is classified according to it intensity. Most class I lasers are under 5 milliwatts and are used for optical alignment, reading bar codes, or measuring distance. Class II lasers have more power and can also be used for alignment and measuring, printing on objects, precision soldering, or cutting delicate tissue in microsurgery applications. Class III lasers are used in metal cutting, welding, and other processes in which precise control of intense power is necessary. Lasers are either continuous beam or pulsed beam and can be visible or invisible, depending on the emitter source.

The optical alignment beams discussed in this section are very low power and are relatively safe, but they are limited in range and in their ability to penetrate smoke or mist. *Figure 32* shows the optical alignment laser.

7.2.0 Laser Safety

The Occupational Safety and Health Administration (OSHA) provides standards for laser use, as do some states. The laser manufacturer publishes these standards in the owner's manual or training guide. Some states require laser warning signs to be posted in any area in which a laser is being used.

> **WARNING!**
> Because of their intense, collimated energy beam, lasers can damage the eye irreparably. Always wear polycarbonate safety glasses when working around lasers and never look directly at a laser beam. Additionally, never point the beam at mirrors or other reflective surfaces because this could send the beam directly into an unprotected eye.

Another safety consideration is the power source because of the inherent inefficiency of a laser. To produce a small output of laser energy, a large input of electrical energy is required; therefore, any laser with a substantial amount of power has a large source of electrical energy.

7.3.0 Optalign® Laser Alignment

Several laser alignment systems are available, and each has its advantages. The laser levels and transits used in site layout and surveying are similar in operation. This section uses information based on the Optalign® laser alignment system (see *Figure 33*),

Materials

Ensure that you have everything required to teach this session.

Classroom

Explain the coupling point and how it relates to vibration.

Discuss the procedures to complete before performing laser alignment.

Audiovisual

Show Transparency 17 (Figure 32). Review the three classes of lasers. Ask whether trainees have any questions.

Safety

Discuss problems that can result from exposure to a laser's energy beam. Review safety considerations when working with lasers.

Audiovisual

Show Transparency 18 (Figure 33). Review the components of the Optalign® laser alignment system.

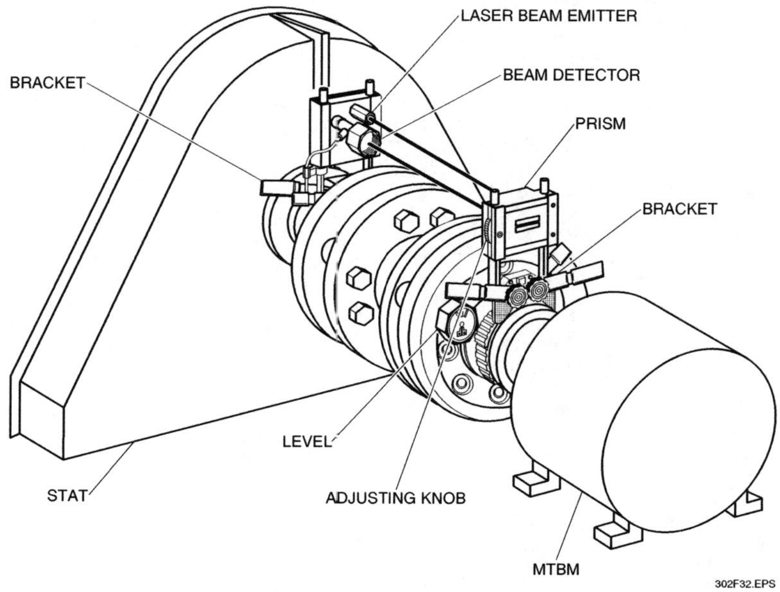

Figure 32 ◈ Optical alignment laser.

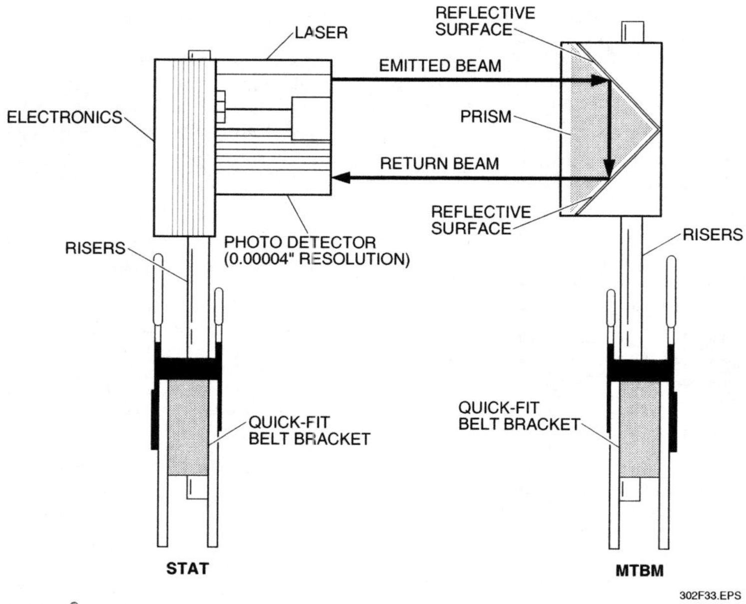

Figure 33 ◈ Optalign® laser alignment system.

Instructor's Notes:

which is a class II, 5-milliwatt, semiconductor laser. The Optalign® alignment laser and copyrighted material from the *Optalign® Training Manual* by Gale Evans and Pedro Casanova is used in this section with permission from Ludeca, Inc. The major components are the laser/detector, prism, computer, belt brackets, and level. The prism reflects the beam back to the detector or target, and the angularity and offset are calculated in the computer, which gives a numerical and graphic readout.

7.3.1 Alignment Demonstration Rig

If possible, it is good to have a table-top demonstration rig for alignment. This can be a simple unit that has all of the adjustments a pump and motor may have. Most maintenance shops have spare pillow block bearings, couplings, and plates that can be assembled to provide a hands-on teaching device. This device should be similar to the rig shown in *Figure 34*, or it may include specific needs for your experience.

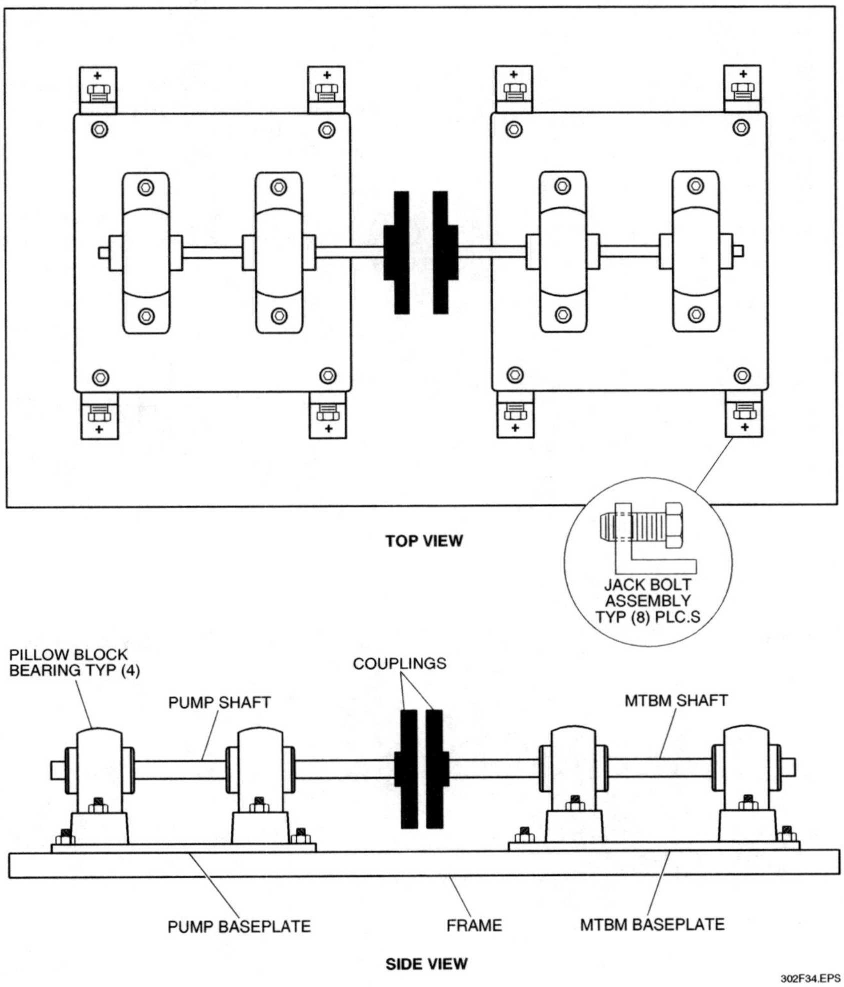

Figure 34 ◆ Alignment demonstration jig.

Homework

Have trainees review Sections 7.3.1–7.3.5 for the next session.

Materials

Ensure that you have everything required to teach this session.

Audiovisual

Show Transparency 19 (Figure 34). Explain how to use an alignment demonstration jig.

![Classroom]

Review the terms used to describe misalignment.

![Audiovisual]

Show Transparency 20 (Figure 35). Explain how offset and angularity affect combined misalignments.

Show Transparency 21 (Figure 36). Discuss the basic dimensional definitions.

7.3.2 Descriptive Characteristics

Correct and complete expression of misalignment is necessary before any meaningful tolerance or target discussion can take place. Actual misalignment is almost always a combination of angularity misalignment and offset misalignment. It is therefore necessary to describe any misalignment in a plane as a value of offset and a value of angularity.

The position of a rotating shaft center line in space can be completely described by showing both vertical and horizontal views. These views are sometimes called the side view and top view, and they are also called the elevation and plan view. Each view displays offset and angularity. The combination of offset and angularity is the most likely situation. *Figure 35* shows combined misalignments.

Since two views describe the location of a shaft and each view can have offset and angularity, four terms are used to describe misalignment. Misalignment can be described by the following terms:

- Vertical offset (VO)
- Horizontal offset (HO)
- Vertical angularity (VA)
- Horizontal angularity (HA)

The basic dimensional definitions are horizontal to vertical and front to back. The stationary machine, or STAT, refers to the machine that does not move and stays firmly bolted. The machine to be moved, or MTBM, is where all the adjustments take place. Each term listed above describes the position of the MTBM in relation to the STAT using the center line of the STAT as zero. *Figure 36* shows the dimensional definition.

The convention for determining positive and negative is defined in three planes, rotational, vertical, and horizontal, and combinations of the three. The rotational directions use the clock to coordinate position. Clockwise direction is defined by looking along the shafts from the MTBM toward the STAT. The clock is used to describe which quadrant the offset is in and the

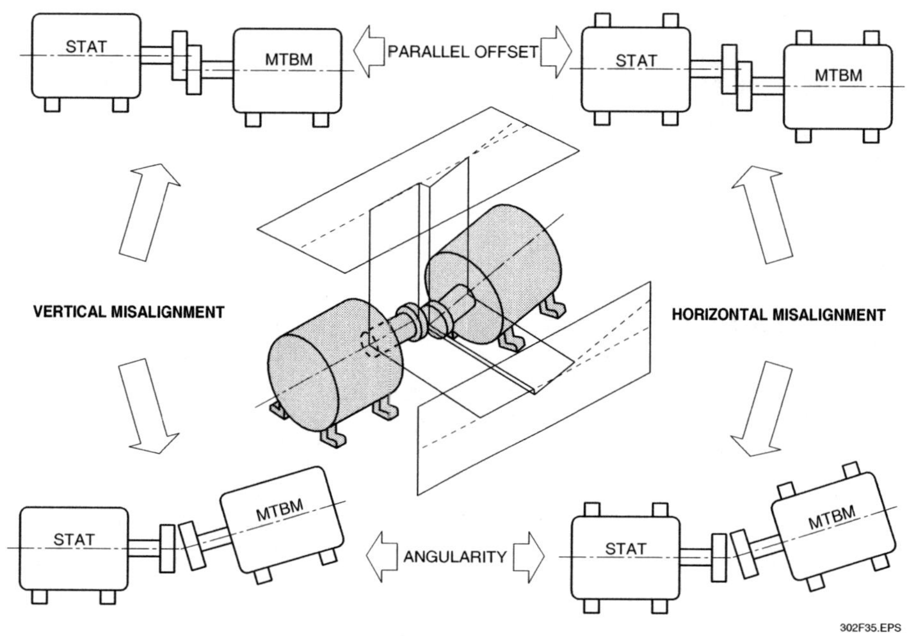

Figure 35 ◆ Combined misalignments.

Instructor's Notes:

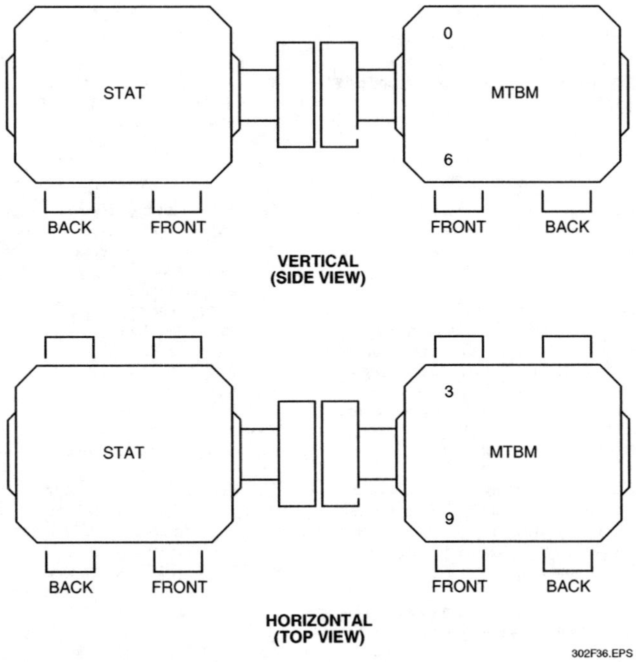

Figure 36 ◆ Dimensional definition.

number value based on Cartesian coordinates. From 12 o'clock to 3 o'clock are positive values and from 3 to 6 o'clock, 6 to 9 o'clock, and 9 to 12 o'clock are negative values.

Positive offsets occur when the MTBM is high or vertical toward 3 o'clock or horizontal compared to the STAT. Negative offsets are when the MTBM is low or vertical or toward 9 o'clock or horizontal compared to the STAT.

Angularity is defined by where the coupling gap is open. Open at 12 o'clock is positive VA; open at 3 o'clock is positive HA. Likewise, negative VA means that the coupling gap is open at 6 o'clock; negative HA means that the coupling gap is open at 9 o'clock. *Figure 37* shows the laser detector target.

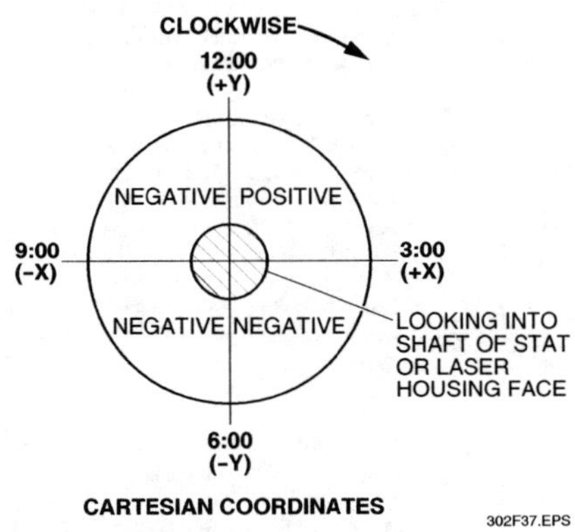

Figure 37 ◆ Laser detector target.

7.3.3 Optalign® Laser System Capabilities

Normally, vibration has no effect on the laser because the beam travels so rapidly. Each measuring pulse travels at the speed of light across the span and back, which is 40 inches in 0.0000000034 seconds. During the transit time, even under severe vibration, the laser and prism will not move far. The beam deflection or lack thereof due to uniform vibration is not a problem. If the STAT and MTBM are mounted on separate foundations or for some other reason are vibrating out of phase with each other, then it is possible that the reading

Classroom

Explain why vibration has no effect on lasers. Answer any questions trainees have.

Classroom

Refer to Figure 38. Review the Optalign® computer keys.

On the whiteboard/ chalkboard, draw a variety of Optalign® symbols. Ask trainees to identify and write down the meaning of each.

catches the machines in different positions with respect to each other and picks up an error. The maximum potential for error is approximately equal to the maximum amplitude from one machine to another.

The input to the computer consists of measurements made by the laser and distance keyed in by the operator. The computer performs the calculations that result in a shim thickness or move distance on the display. *Figure 38* shows the Optalign® computer keys.

ON/OFF: Used to turn computer on and off. Must be held at least 1 second to activate. Erases all dimensions and results if pressed when on.

SLASH: If pressed immediately after initial ON, toggles Optalign between metric and inch systems. Otherwise used to enter fractions of an inch; only halves, quarters, and eighths are permitted. If no denominator is pressed before enter is pressed, the denominator defaults to ten. (i.e., the last digit entered becomes tenths of an inch.)

DIM: Used to access the *dimensions mode*, where the five required dimensions are entered: laser to prism, laser to front foot, front to back foot, coupling center to prism, and working diameter.

M: The computer enters the *measure mode*, clears all previous results (but not dimensions), displays coarse beam position coordinates, and waits for at least three measurements. Once the M key is pressed, measurements must be taken. No other functions, except F1, F2, and F3, work until three or more measurements have been entered.

BEG/END: Used to start and end the process of readjusting the prism for extending the detector's measuring range. Only used while in the *measure mode* (M) or in the *move mode.* Do not rotate shafts between pressing BEG/END the first and second times.

COUPLING: Press the coupling key to produce coupling condition results. The coupling key is also used in conjunction with ENT and RCL to advance or recall the condition results.

FOOT CORRECTIONS: In *measure mode*, this key is used to end measurements, run alignment calculations, and display foot corrections. (Also note its usage in the vertical machines (F5) function).

ENT: In *dimension mode,* pressing the ENT key enters the displayed value and proceeds to next screen. Used to enter all data, measurements, and readings, and to advance through all results. Used to zero the display in *soft foot mode.* Used to advance to next step in the *move mode.*

RCL: In *dimension entry mode*, pressing the RCL key recalls previous dimensions. In *foot correction results mode,* it recalls the previous result. In *coupling result mode*, it recalls the previous coupling results.

MOVE: Used to start the MOVE function. (MOVE, ENT, rotate shafts to 1:30, 0 0, ENT). To exit the move function, press M.

SOFT FOOT: Used to enter the soft foot function and to measure the effects of machine frame distortion.

TARGET: Used to enter target specifications mode and for deliberate entry of misalignment conditions at the coupling.

MINUS: Used to enter negative numbers.

CLR: Clears only displayed targets or dimensions. CLR also sets the vertical and horizontal coordinates to zero when the computer is in fine coordinate measure mode (F1). CLR can be used to delete a clock position entry in the *measure mode* (M) when used in place of ENT.

F1: Actual coordinate values function for X- and Y-axes.

F2: Semiautomatic clockwise measurements.

F3: Semiautomatic counterclockwise measurements.

F4: Thermal growth at the foot function.

F5: Vertical flange-mounted machines.

302F38.EPS

Figure 38 ◆ Optalign® computer keys.

Instructor's Notes:

7.3.4 Laser Limitations

Some conditions can limit or stop a laser beam either in a plant or in the field. Heavy concentrations of dust, vapor, or other airborne contaminants can block or hinder the beam. The display shows an OFF message if the particles are extremely concentrated. The solution to this problem is to shoot the beam inside of a length of PVC pipe or cardboard tube. Strong sunlight, infrared light, and welding flash directly striking the detector cause an error (ERR) message to appear. Using a shade or hood or shooting the beam through a pipe will solve the problem. Theoretically, heat waves, steam, or temperature variations, if severe, can distort or vary the beam but in reality, if you can see through the disturbance for the distance the beam has to travel, any beam variation is unlikely.

7.3.5 Determining Targets

Alignment targets are defined as deliberate, initial misalignment specifications that compensate for differences in machine shaft positions between alignment conditions and operating conditions. Alignment targets are properly expressed as VO, HO, VA, and HA at a coupling center.

In order to specify targets, one of two data sets must be known. One is the machine growth characteristics supplied by the equipment manufacturer and is based on a dimensional difference between average, optimal, and running temperatures and ambient temperature at rest. This is calculated based on given metal characteristics and is theoretical. The other data set is based on actual measurements made when the machine is cold and hot. These are more realistic and simple to perform with the Optalign® system.

All known growth characteristics and their values must be converted into offsets and angles at the coupling. These are entered into the Optalign® computer, which will then compensate for thermal growth when calculating the alignment. The computer display has a keyboard for entering the dimensions and a display that shows the movement or shim thickness at each foot. *Figure 39* shows the Optalign® computer. Follow these steps to enter targets into the Optalign® computer:

Step 1 Determine the targets.

Step 2 Press the coupling key.

Step 3 Enter the coupling-to-prism distance.

Step 4 Enter the coupling diameter.

NOTE

This diameter is used only for computing the VA and HA, so any convenient working diameter can be entered. Most facilities and aligners standardize with 10 inches.

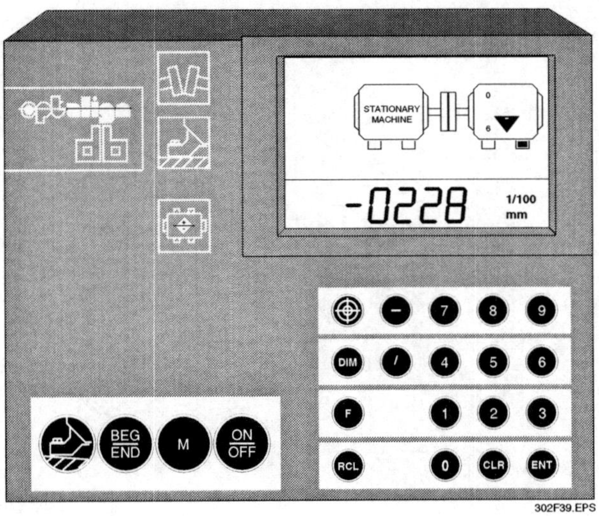

Figure 39 ◆ Optalign® computer.

Classroom

Review considerations that affect the distance laser beams can travel.

Explain how to determine and enter targets into an Optalign® computer.

Classroom

Explain why you must take hot alignment readings as soon as possible after the shafts stop rotating.

Homework

Have trainees review Sections 7.4.0–7.5.3 for the next session.

Materials

Ensure that you have everything required to teach this session.

Classroom

Explain how lasers can measure offsets between two points.

Audiovisual

Show Transparency 22 (Figure 40). Explain how a detector senses the motion of the laser beam.

Step 5 Enter the vertical offset target (VOt), horizontal offset target (HOt), vertical angularity target (VAt), and horizontal angularity target (Hat) one at a time, observing the proper sign, positive or negative.

Hot alignment readings must be taken as soon as possible after the shafts stop rotating, no more than 3 to 5 minutes. In order to achieve this short time, preparations must be made in advance and the readings must be taken by an experienced aligner. Although the term *hot readings* is commonly used, a more technically correct term is *operating readings* because the machine may heat or cool depending on the process.

Targets for cold alignment are cold conditions minus hot conditions. The actual cold temperature does not matter. However, alignment must be performed with the equipment stable at the cold temperature. If the machines are cooling or heating while the alignment is being done, the readings will probably be wrong.

7.4.0 Laser/Detector Operation

When the prism moves up and down, the return beam moves up and down twice the distance the prism moves. This is how the laser can measure offsets between two points. The offset measured is the offset of the prism relative to the laser. Because the prism is rotated in the horizontal plane, the detector can sense the up-and-down and side-to-side motion of the beam. *Figure 40* shows laser/detector movement.

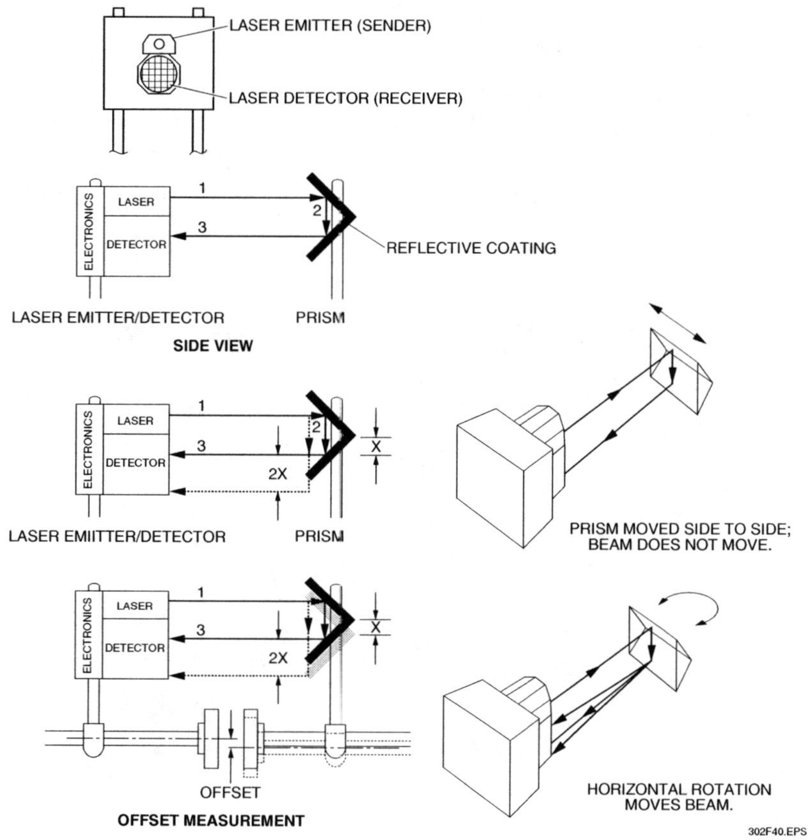

Figure 40 ◆ Laser/detector movement.

Instructor's Notes:

In *Figure 40*, the angle the return beam moves is twice the angle the prism moves. If the beam is zeroed at the 12 o'clock position and then read at the 6 o'clock position, the horizontal beam movement is one leg of a 90-degree triangle. The distance from the laser to the prism is the other leg. The angle defined by these two legs is two times the actual angular misalignment between the shafts.

For example, determine the clockwise and counterclockwise directions using the clock on the laser housing face. At each position, obtain X and Y coordinates by pressing the H/V key after entering the measure mode (M). Note the XO and YO are the X and Y readings at the 12 o'clock position (0 degrees), X3 and Y3 at the 3 o'clock position, X6 and Y6 at the 6 o'clock position, and X9 and Y9 at the 9 o'clock position. For all alignment readings in which a single axis of rotation exists, the computer uses the following verification rule:

XO + X6 = X3 + X9

YO + Y6 + Y3 + Y9

In the following equations, the constant 4.75 is an approximation of the lens factor associated with the laser-prism detection system. For offsets, positive results mean that the MTBM is high or on the 3 o'clock side of the STAT. For angularities, positive results mean that the coupling is more open at the top, or 3 o'clock, side. In these equations, the angularity result is an angle expressed in mils per inch. The angle is equal to the rise over the run and is expressed in mils per inch. For alignment, angles are often expressed in mils per 10 inches rather than mils per inch. To convert to a coupling gap difference, multiply the angle by the working diameter. *Figure 41* shows offset and angle.

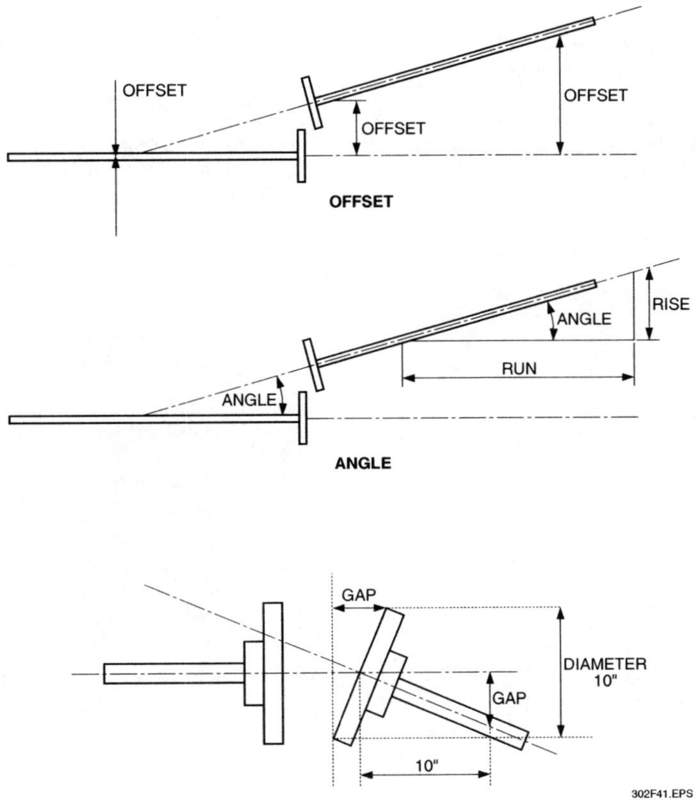

Figure 41 ◆ Offset and angle.

Demonstration

On the whiteboard/chalkboard, demonstrate how to determine clockwise and counterclockwise directions. Ensure that trainees understand the verification rule.

Audiovisual

Show Transparency 23 (Figure 41). Explain how angles are expressed for purposes of alignment. Go over any questions trainees have.

Classroom

Review the abbreviations used in the equations to determine which foot to shim and how much to move or shim. Explain how the equations are used.

On the whiteboard/ chalkboard, write the abbreviations used in the equations. Ask trainees to identify each and to solve sample equations using variables you provide.

Audiovisual

Show Transparency 24 (Figure 44). Explain how the dimensions needed to define the offset angularity relate to the laser target.

Classroom

Review the steps to set up the laser alignment equipment.

Safety

Explain why the equipment should not be started while an alignment is being performed.

The Optalign® computer calculates the VO, HO, VA, and HA using the equations shown in *Figure 42*. The Optalign® computer displays both HA and VA as gap differences using the user-supplied coupling diameter. Angularity targets must be entered as gap differences for the same coupling diameter. If no user coupling diameter is supplied, the Optalign® computer uses 206.2 inches, the working diameter at which a gap difference of 0.001 inch equals 1 arc second (3,600 arc seconds = 1 degree). *Figure 43* shows equations used to determine which foot to shim and how much to move or shim.

The key to the abbreviations used in the equations is as follows:

- Positive shim results indicate to add shims.
- Negative shim results indicate to remove shims.
- Positive move results means move toward 3 o'clock.
- Negative move results means move toward 9 o'clock.
- VOt = Vertical offset target
- VAt = Vertical angularity target
- HOt = Horizontal offset target
- HAt = Horizontal angularity target
- CD = Coupling diameter entered into Optalign®
- LF = Laser-to-front-foot distance
- LP = Laser-to-prism distance
- FB = Front-foot-to-back-foot distance
- CP = Coupling center-to-prism distance

$$VO = \frac{(Y0 - Y6)}{2} + \frac{CP(X9 - X3)}{2(LP + 4.75)} \text{ (mils)}$$

$$HO = \frac{(Y3 - Y9)}{2} + \frac{CP(X0 - X6)}{2(LP + 4.75)} \text{ (mils)}$$

$$VA = \frac{(X9 - X3)}{2(LP + 4.75)} \text{ (mils/inch)}$$

$$HA = \frac{(X0 - X6)}{2(LP + 4.75)} \text{ (mils/inch)}$$

302F42.EPS

Figure 42 ◆ VO, HO, VA, and HA equations.

The dimensions needed to define the offset angularity are in the equations in *Figure 42*. The dimensions relate to the laser target as to what the detector reads and computes as the total movement or shim thickness. *Figure 44* shows the dimensions of targets.

7.5.0 Alignment Procedures

The alignment procedures consist of performing a rough alignment, setting up the laser equipment, and performing an initial alignment.

7.5.1 Rough Alignment

Before laser alignment is attempted, the STAT and MTBM must be aligned as closely as possible by hand. As a rough rule, the coupling offset should be less than 3/32 inch, and the angles between the coupling faces should be near parallel or within 10 mils per inch of the coupling diameter. This ensures that the shafts are within range of the laser and that all the major causes of shaft misalignment have been corrected before setting up the laser.

7.5.2 Setting Up Laser Equipment

Follow these steps to set up the laser alignment equipment:

Step 1 Ensure that the machines are shut down, locked out, and tagged out per company policy.

> **WARNING!**
> If the equipment is started up while an alignment is being performed, serious injury can occur.

$$\text{Front shim} = \frac{(Y0 - Y6)}{2} + \frac{(X9 - X3)(LF - LP)}{2(LP + 4.75)} + VOt - \frac{VAt(LF - LP + CP)}{CD}$$

$$\text{Back shim} = \frac{(Y9 - Y0)}{2} + \frac{(X9 - X3)(LF - LP + FB)}{2(LP + 4.75)} + VOt - \frac{VAt(LF + FB - LP + CP)}{CD}$$

$$\text{Front move} = \frac{(Y9 - Y3)}{2} + \frac{(X0 - X6)(LF - LP)}{2(LP + 4.75)} + HOt - \frac{HAt(LF - LP + CP)}{CD}$$

$$\text{Back move} = \frac{(Y9 - Y3)}{2} + \frac{(X0 - X6)(LF - LP)}{2(LP + 4.75)} + HOt - \frac{HAt(LF + FB - LP + CP)}{CD}$$

302F43.EPS

Figure 43 ◆ Equations used to determine which foot to shim and how much to move or shim.

Instructor's Notes:

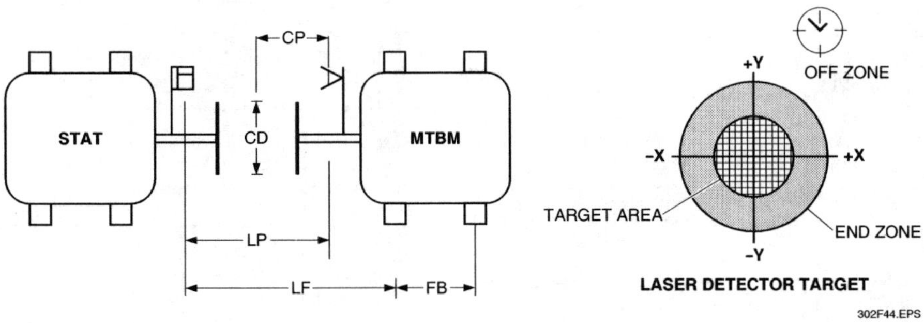

Figure 44 ◆ Dimensions of targets.

Step 2 Mount the laser and prism on the shafts just behind each coupling half.

Step 3 Clamp the brackets firmly in place directly across from each other.

Step 4 Ensure that the beam hits the detector.

> **NOTE**
> If the beam does not hit the detector, use the beam finder to fine-tune the prism location. The beam finder is a hand-held target that locates the laser beam.

7.5.3 Initial Laser Alignment

After the laser and prism are in place and square to each other, an initial alignment procedure is performed. Follow these steps to check the initial alignment:

Step 1 Enter the machinery and bracket dimensions in the Optalign® computer.

Step 2 Press the M key to zero the system.

Step 3 Take readings at four positions.

> **NOTE**
> These readings are normally taken at the 12, 3, 6, and 9 o'clock positions.

Step 4 Press the coupling key and record the four values.

Step 5 Press the foot key and record the four values.

Step 6 Determine if the dimensions at all corners are within specifications. If the dimensions are not within specifications, move or shim the equipment the required amount.

Step 7 Take another reading to confirm that the alignment is correct.

8.0.0 ◆ SPECIFIC LASER ALIGNMENT PROCEDURES

After the initial laser alignment is completed, specific alignment procedures can then be performed depending on the needs of a particular job. Specific alignment procedures are discussed in the following sections:

- Aligning machinery trains
- Measuring shaft sag
- Aligning vertical machines

8.1.0 Aligning Machinery Trains

Aligning machinery trains is designed for more than two machines. The objective is to move the machines as little as possible. Draw a scale graphical plot of how much movement is available with each machine. This method avoids the problem of having to lower a machine with insufficient or zero shims and minimizes the chance of the machine becoming bolt-bound.

The vertical corrections are performed first and are rechecked before the horizontal adjustments are determined. This procedure is for zero-zero alignment; however, it can also be used with intentional misalignment targets. Zero-zero alignment occurs when all the sight lines line up to no variation and the display shows zeroes at every placement.

Classroom

Explain how to check the initial alignment. Ask whether trainees have any questions.

Teaching Tip

See the Teaching Tip for Section 7.0.0 at the end of this module.

Homework

Have trainees review Sections 8.0.0–8.1.0 for the next session.

Materials

Ensure that you have everything required to teach this session.

Classroom

Review the problems you can prevent by drawing a scale graphical plot.

Classroom

Review the procedures to align a machinery train. Answer any questions trainees may have.

On the whiteboard/chalkboard, ask a trainee to write the steps to align a machinery train. Have trainees correct any inaccuracies and reorder the steps as needed.

Homework

Have trainees review Sections 8.2.0–8.3.0 and answer the Review Questions for the next session.

If one machine is difficult to move, consider this machine as the STAT. Follow these steps to align a machinery train:

Step 1 Draw the configuration of your machinery train on graph paper true to scale, and indicate the front and back feet of each machine.

Step 2 Use the machine on the far left as your base center line, and plot this as a horizontal line on your graph paper.

> **NOTE**
> The machine on the far left is machine A; the next one to the right is machine B.

Step 3 Mount the laser on machine A and the prism on machine B.

Step 4 Determine the position of machine B relative to machine A.

> **NOTE**
> This is setup AB.

Step 5 Plot on the graph paper the vertical corrections for the front and back feet of machine B relative to machine A.

Step 6 Draw a connecting line between these two points, and extend the line in both directions.

Step 7 Mount the laser on machine B and the prism on machine C.

Step 8 Determine the relative position of machine C to machine B.

> **NOTE**
> This is setup BC.

Step 9 Plot the vertical corrections for the front and back feet of machine C relative to machine B.

Step 10 Draw a connecting line between these two points, and extend the line in both directions.

Step 11 Repeat Steps 7 through 10 for each of the remaining machines.

> **NOTE**
> These are setups CD, DE, and so on.

Step 12 Draw a scale drawing of all the machines, showing exactly how much movement is possible in all directions.

> **NOTE**
> This optimizes the bolt clearances and defines the adjustment.

Step 13 Use the graph to determine the correct values for adjusting the machines to the ideal condition, and adjust the machines accordingly.

Step 14 Repeat Steps 1 through 11 while checking the misalignment at each coupling. Compare the vertical offsets and angularities with the permissible tolerances.

> **NOTE**
> If the alignment is within tolerance, proceed to Step 15. If the alignment is not within tolerance, repeat Steps 12 and 13.

Step 15 Redraw the configuration of your setup of machinery on new graph paper true to scale, again indicating the front and back feet of each machine.

Step 16 Repeat Steps 2 through 11, but this time, only plot the horizontal relationship between the machines.

Step 17 Draw a scale drawing of all the machines that shows exactly how much movement is possible in all directions.

Step 18 Use the graph or rise/run calculations to determine the correct values for adjusting the machines to the ideal condition, and adjust the machine accordingly.

Step 19 Repeat Steps 2 through 11 while checking the misalignment at each coupling. Compare the horizontal offset and angularities with the permissible tolerances.

> **NOTE**
> If the tolerances are not correct, repeat Steps 17 through 19.

Instructor's Notes:

8.2.0 Measuring Shaft Sag

Measuring shaft sag is used to determine the normal deflection of jackshafts or long spacers caused by gravitational forces. The sag is greatest in the center of the jackshaft and occurs only in the vertical plane. The Optalign® laser will not measure permanent deflection or a bent shaft.

Deflection of jackshafts occurs in static condition and may increase or decrease in dynamic condition depending on the operating speed of the machine. This procedure measures the sag in a shaft in static condition as a result of its own dead weight. Follow these steps to measure shaft sag:

Step 1 Loosen the coupling bolts on the rigid couplings sufficiently to loosen the couplings.

> **NOTE**
> It is not necessary to loosen the bolts of flexible couplings.

Step 2 Mount the laser on one end of the jackshaft and the prism on the other end (see *Figure 45*).

Step 3 Press the ON/OFF key to turn on the system.

Step 4 Press the / key to switch to inches.

Step 5 Press the M key.

Step 6 Adjust the laser and prism to 0.00.0 in the 12 o'clock position; then press the M, O, and ENT keys.

Step 7 Turn the laser/prism to the 3 o'clock position; the press the M, 3, and ENT keys.

Step 8 Turn the laser/prism to the 6 o'clock position; then press the M, 9, and ENT keys.

Step 9 Return the laser/prism to the top, or 12 o'clock, position, and confirm the zero setting.

Step 10 Press the RUN key.

Step 11 Enter the distance of the laser to the prism (LP).

Step 12 Enter the distance of the laser to the front foot (LF) as zero.

Step 13 Enter the distance of the front foot to the back foot (FB) as zero.

Step 14 Record the displayed front foot correction.

> **NOTE**
> This value is four times the value of the sag or halfway between the laser and prism.

8.3.0 Aligning Vertical Machines

Because the lateral alignment can change very easily when shimming, the user must first make shim corrections to correct angular misalignment and then measure the alignment once more before proceeding with a horizontal move to correct the offset misalignment.

Vertical machines are aligned using the special function F5. For this procedure, a 0-degree (12 o'clock) reference point is chosen. The angular positions are based on this reference. *Figure 46* shows aligning vertical machines. The machine dimensions needed to align vertical machines are:

- Distance from the laser to the prism
- Distance from the laser to the coupling center
- Greatest diameter (Dmax) between two fastening bolts in the shim plane

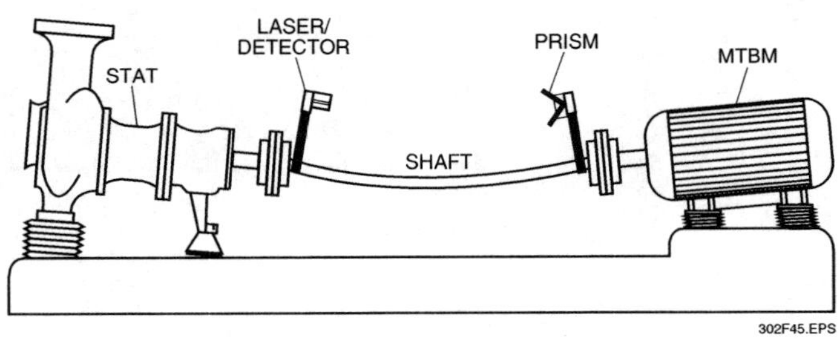

Figure 45 ◆ Measuring shaft sag.

Ensure that you have everything required for testing during this session.

Explain how to use shaft sag measurements to determine the normal deflection of jackshafts or long spacers.

Show Transparency 25 (Figure 45). Discuss the procedures to measure shaft sag.

Audiovisual

Show Transparency 26 (Figure 46). Review the machine dimensions needed, and explain how to align vertical machines. Ask whether trainees have any questions.

Teaching Tip

See the Teaching Tip for Section 8.0.0 at the end of this module.

Follow these steps to align vertical machines:

Step 1 Turn on the system.

Step 2 Measure the machine dimensions listed above, and record them.

Step 3 Enter the laser-to-prism distance.

Step 4 Enter the laser-to-front-foot distance as the distance from the laser to the coupling center.

Step 5 Enter the maximum bolt circle diameter (Dmax) for the front-to-back-foot distance.

Step 6 Take measurements in at least three of the four positions.

Step 7 Press F5 to activate the vertical shimming function. After pressing F5, the angle 000.0 is displayed.

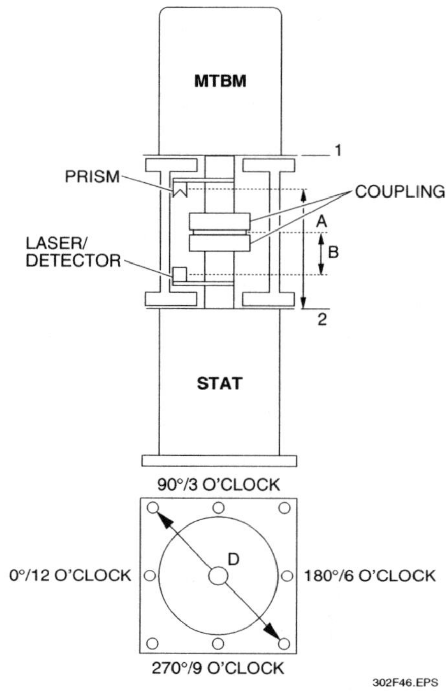

Figure 46 ◆ Aligning vertical machines.

> **NOTE**
> The computer internally calculates the relative position of the MTBM to the STAT on the entire circumference where the greatest distance must be lowered. This amount is later added to the respective corrective value so that negative corrective values are eliminated. If lowering is desired because the entire machine would otherwise stand too high, zero must be entered as Dmax.

Step 8 Enter the respective diameter and angular position of each fastening bolt using the RCL and ENT keys.

> **NOTE**
> Repeatedly pressing the ENT key causes the corrective value for each bolt to be displayed automatically. If the plus sign (+) appears, insert shims; if the minus sign (–) appears, remove shims.

Step 9 Shim the machinery where the computer specifies.

Step 10 Take alignment measurements in at least three of the four positions, and repeat the vertical shimming procedure until angularity is within tolerance.

> **NOTE**
> Once the results indicate the angularity is within tolerance, proceed with horizontal moves.

Step 11 Press the RUN key to view the horizontal corrections of the MTBM.

> **NOTE**
> As in all alignments with the Optalign® laser, the coupling function can be used to enter coupling target specifications and to determine compliance with tolerances.

Instructor's Notes:

Summary

Achieving proper unit alignment is one of the most important procedures related to the installation and repair of the components of machinery. Proper alignment of shafts and other components of a drive train results in lower energy consumption, reduced wear, less downtime, and a longer service life for the equipment. Achieving proper alignment involves training, skill, and patience because some procedures can be frustrating and time consuming.

Mechanical technicians must become proficient in the use of measuring devices, such as dial indicators and feeler gauges. Advanced technology, such as laser equipment and computers, is also available to achieve the required result. Most of the various methods and tools for alignment that must be mastered by mechanical technicians require extensive training and understanding. Mechanical technicians must first have a strong understanding of the machinery involved in the testing. They must then have appropriate training in the setup and use of the test equipment and indicators.

Accuracy is of the utmost importance in alignment. Accuracy in measurement, gauging, and testing are perfected through training, practice, and experience. Learning the skills to be able to read and interpret measuring devices requires discipline and determination.

All of the skills of measurement, setup and use of testing equipment and gauges, reading instruments accurately, and interpreting the data are included in the work of unit alignment. To become a qualified mechanical technician who is able to accomplish alignment tasks, all these skills must be mastered through experience. When you have completed this module, you will have developed skills in alignment that will enable you to progress toward becoming a trained and qualified mechanical technician.

Review Questions

1. When equipment is aligned, the _____ are aligned.
   a. gears
   b. equipment bases
   c. equipment housings
   d. rotating shaft center lines

2. The results of _____ shafts include equipment vibration and failure.
   a. round
   b. unscored
   c. rotating
   d. misaligned

3. Shaft alignment is defined as positioning two or more machines so that their rotational center lines are _____ at the coupling point under operating conditions.
   a. coexistent
   b. collinear
   c. parallel
   d. constant

4. The two basic types of shaft misalignment are _____.
   a. length and width
   b. offset and angularity
   c. top and bottom
   d. front and back

5. The outer diameter of a coupling is also called the _____.
   a. rim
   b. face
   c. radius
   d. circumference

6. Shaft alignment involves aligning the _____ of couplings.
   a. rims and feet
   b. feet and faces
   c. rims and faces
   d. rims and gaps

7. One of the units, the driver or the driven, must be tilted on its base to correct _____ misalignment.
   a. vertical angularity
   b. horizontal angularity
   c. vertical offset
   d. horizontal offset

Classroom

Emphasize the importance of accuracy when performing unit alignment.

Go over the Review Questions for this module, and answer any questions trainees have.

8. One of the units, the driver or the driven, must be rotated on its base to correct _____ misalignment.

   a. vertical angularity
   b. horizontal angularity
   c. vertical offset
   d. horizontal offset

9. Before laser alignment is attempted, the coupling offset should be less than _____.

   a. $\frac{3}{32}$ inch
   b. 0.010 inch
   c. 8 mm
   d. 15 degrees

10. Soft foot, bent shafts, grossly misaligned equipment, and coupling stress can prevent a successful _____.

    a. production run
    b. PM report
    c. alignment
    d. reading

11. A condition in which tightening or loosening bolt(s) of a single foot distorts the machine frame is _____.

    a. squishy foot
    b. bent foot
    c. shaking foot
    d. soft foot

12. Bent foot can be corrected using _____.

    a. step shims
    b. a hammer
    c. extra grout
    d. welding

13. A condition that shows up as soft foot on the laser but in which feeler gauges indicate no gap is called _____.

    a. flat foot
    b. shot foot
    c. moist foot
    d. squishy foot

14. As a standard rule, no more than _____ shims should be used under one foot.

    a. two
    b. four
    c. six
    d. eight

15. External forces on a machine create a condition called _____ soft foot.

    a. indicated
    b. internal
    c. indescribable
    d. induced

16. When the coupling is divided into four positions for aligning purposes, the top of the coupling is in the _____ o'clock position.

    a. 12
    b. 9
    c. 6
    d. 3

17. Match marks on the coupling are always kept at the _____ o'clock position.

    a. 12
    b. 9
    c. 6
    d. 3

18. When one foot of a pump or motor does not sit flat on the baseplate, the condition is called _____.

    a. squishy foot
    b. bent foot
    c. shaking foot
    d. soft foot

19. When adjusting vertical angularity and the coupling gap is larger at the 12 o'clock position, the shims go under the lugs that are _____.

    a. nearest the coupling
    b. farthest from the coupling
    c. on the right side of the motor
    d. on the left side of the motor

20. When adjusting vertical angularity and the coupling gap is larger at the 6 o'clock position, the shims go under the lugs that are _____.

    a. nearest the coupling
    b. farthest from the coupling
    c. on the right side of the motor
    d. on the left side of the motor

21. The most accurate and widely used method of conventional alignment is the _____ method.

    a. straightedge and feeler gauge
    b. reverse dial indicator
    c. square and rule
    d. plumb bob and rule

Instructor's Notes:

22. A good reverse dial indicator jig is designed to minimize _____ sag.
    a. pipe hanger
    b. coupling
    c. indicator
    d. air hose

23. When setting up a dial indicator, the indicator is usually set to take readings on the _____.
    a. STAT coupling half
    b. MTMB coupling half
    c. MTMB base
    d. STAT lugs

24. Two indicators mounted either opposite each other or 180 degrees apart are called a _____ reverse dial indicator jig.
    a. simple
    b. quadratic
    c. static
    d. complex

25. The two ways to process reverse alignment measurements are graphical charts and _____.
    a. mathematical equations
    b. flow charts
    c. statistical process control
    d. spreadsheets

26. Charting alignment on a graph requires a chart that is divided into thousandths of an inch vertically for the indicator readings and into whole inches horizontally for the _____ measurements.
    a. time
    b. flow
    c. distance
    d. angle

27. The absolute necessity of recording and maintaining records of all alignment is to ensure that a(n) _____ can be productive and not cause more problems than it solves.
    a. pump
    b. equipment replacement
    c. coupling
    d. logical system

28. For machinery installation, only the _____ of different machines are aligned.
    a. feet
    b. shaft surfaces
    c. bearings
    d. rotating shaft center lines

29. An amplified and collimated beam of coherent light energy that travels perfectly straight is a _____.
    a. flash
    b. welding arc
    c. laser beam
    d. quartz halogen

30. A beam of radiation or matter whose rays or particles are nearly parallel is _____.
    a. soft
    b. hard
    c. collimated
    d. prismic

31. The Optalign® laser beam is pulsed instead of continuous and is a class _____ beam.
    a. I
    b. II
    c. III
    d. IV

32. Actual misalignment is virtually always a combination of _____ misalignment.
    a. inside and outside
    b. vibrational and rotational
    c. seasonal and frictional
    d. angularity and offset

33. Vertical offset, vertical angularity, horizontal offset, and horizontal angularity define shaft _____ accurately.
    a. misalignment
    b. sag
    c. torque
    d. rpm

34. The expansion of materials caused by a rise in temperature and that affects alignment is called thermal _____.
    a. sizing
    b. growth
    c. seizure
    d. current

35. When the laser prism is moved up or down, the return beam moves _____ the distance the prism moved.
    a. half
    b. three times
    c. twice
    d. one tenth

Examination

Administer the Module Examination. Be sure to record the results of the Exam on Craft Training Report Form 200 and submit the results to the Training Program Sponsor.

Performance Testing

Have each trainee perform Performance Task 2 to your satisfaction. Fill out a Performance Profile Sheet for each trainee.

Ensure that all Performance Tests have been completed and Performance Profile Sheets for each trainee are filled out. Be sure to record the results of the testing on Craft Training Report Form 200 and submit the results to the Training Program Sponsor.

GLOSSARY

Trade Terms Introduced in This Module

Angularity: The angle between two center lines.

Collimated beam: A beam of radiation or matter whose rays or particles are nearly parallel.

Collinear: Two or more shaft center lines with no offset or angularity between them.

Indicator sag: An error condition in which the weight of the indicator causes the post or bracket that it is mounted on to sag a small amount; when taking readings at the 12 and 6 o'clock positions, this error doubles and can affect the alignment unless measured and compensated for. It is also called bracket sag.

Machinery train: A group of separate machines connected to each other by a processor or physically coupled to transfer power to each other.

Machine to be moved (MTBM): A machine that is to be moved; in most cases, this is the driver or motor that will be moved to line up with the STAT.

Prime mover: The piece of equipment, such as an electric motor or gasoline engine, that provides the driving force for another piece of equipment, such as a pump.

Stationary machine (STAT): The driven machine or machine receiving the power that is used as the reference line to which the MTBM is aligned.

Thermal growth: The expansion of materials with a rise in temperature.

Instructor's Notes:

ACKNOWLEDGMENTS

Acknowledgments

NCCER would like to thank the following company for permitting the use of their training manual.

Optalign® Training Manual. Gale Evans and Pedro Casanova. Miami, FL: Ludeca, Inc. 305-591-8935.

MODULE 63302-02 — TEACHING TIPS

The following are suggested activities or instructional methods to help you teach the material in this AIG.

General

When you call on someone to answer a question, the rest of the class relaxes or even tunes out because they expect that the question and answer will take place only between you and the trainee you called on. Instead, use this technique to involve more trainees in answering questions and to keep them on their toes.

1. Ask trainees to define a term or explain a concept.
2. After one trainee has answered, ask a trainee seated nearby if the answer is right. Then ask whether a trainee in the back of the room agrees.
3. Ask trainees to explain why they think an answer is right or wrong.
4. Use the session to clear up incorrect ideas and encourage trainees to learn from their mistakes.

Sections 3.0.0–3.5.0

Identifying and Correcting Soft Foot

This Quick Quiz will familiarize trainees with the types of soft foot. You will need photocopies of the quiz provided on the following page. Trainees will need pencils. If you allow trainees to use the Trainee Guide, decrease the amount of time you give them to complete the quiz.

1. Make as many photocopies as you will need of the quiz.
2. Give trainees 5 to 10 minutes to complete the quiz.
3. Go over the answers to the quiz.
4. Ask trainees whether they have questions about the types of soft foot.

Answers to Quick Quiz

1. c. Parallel air gap
2. e. Gaps without soft foot
3. b. Squishy foot
4. a. Bent foot
5. d. Induced soft foot

Quick Quiz *Identifying and Correcting Soft Foot*

For each description listed, identify the type of soft foot that the text best describes. Write the corresponding letter in the blank provided.

_____ **1.** This occurs when three mounting feet sit solid and flat but one foot does not touch the base.

_____ **2.** This occurs when there is a visual gap under the foot before tightening and none after tightening, but the measurement does not show a soft foot or misalignment.

_____ **3.** This occurs when a soft foot measurement shows up on the laser or dial indicator, but when measured using a feeler gauge little or no gap is indicated.

_____ **4.** This occurs when the bottom of one foot is not parallel with the base.

_____ **5.** This occurs when two soft feet show up on the same side or end of the machine, but the feeler gauge indicates a gap that is parallel or nearly parallel.

a. Bent foot
b. Squishy foot
c. Parallel air gap
d. Induced soft foot
e. Gaps without soft foot
f. Sag foot

Section 4.0.0 *Performing Conventional Alignment*

Trainees have reviewed the procedures to adjust horizontal and vertical angularity and offset. In this exercise, trainees will practice these adjustments. For this exercise, you will need several alignment simulators, feeler gauges, dial calipers, straightedges, scales, and rules. Trainees will need their Trainee Guides, calculators, and pencils and paper to take notes for discussion. Allow 45 minutes for this exercise.

1. Set up several stations with alignment simulators and tools to perform conventional alignment. Divide the trainees into small groups according to the number of stations.
2. Have each group practice conventional alignment following the procedures for adjusting:
   - Horizontal angularity
   - Horizontal offset
   - Vertical angularity
   - Vertical offset
   - Horizontal angularity and offset
   - Vertical angularity and offset
3. On the whiteboard/chalkboard, ask each group to write their solution to one of the procedures. (If possible, have each group present a different procedure.) Review and discuss trainees' answers.
4. Go over any questions trainees have about performing conventional alignment.

Section 5.0.0 *Performing Rim and Face Dial Alignment*

This exercise provides trainees with the opportunity to perform rim and face dial alignment. For this exercise, you will need chain indicator jigs and dial indicators. Trainees will need their Trainee Guides and pencils and paper to take notes for discussion. Allow 30 minutes for this exercise.

1. Set up several stations with chain indicators and dial indicators.
2. Ask trainees to practice setting up the dial indicators. Ensure that trainees understand how to measure sag.
3. On the whiteboard/chalkboard, ask several trainees to write the procedures to take horizontal and vertical measurements.
4. Have trainees determine whether the procedures are written correctly and make any necessary corrections.
5. Answer any questions trainees have about performing rim and face dial alignment.

Section 6.2.0 *Performing Reverse Dial Indicator Alignment*

This exercise will familiarize trainees with the procedures for performing reverse dial indicator alignment. For this exercise, you will need graph paper, graphical alignment charts, reverse dial indicator plotting guides, and sample alignment records. Trainees will need their Trainee Guides and pencils and paper to take notes for discussion. Allow 30 minutes for this exercise.

1. On the whiteboard/chalkboard, provide sample distance measurements for a hypothetical STAT and MTBM.
2. Have trainees perform the chart alignment using the data you provide.
3. Using the reverse indicator plotting guide, have trainees perform and record vertical alignment.
4. Answer any questions trainees have about performing reverse dial indicator alignment.

Section 7.0.0 *Performing Laser Alignment*

This exercise will give trainees the opportunity to describe laser alignment procedures. For this exercise you will need an overhead projector, sheets of acetate, and transparency markers. Trainees will need their Trainee Guides and pencils and paper to take notes for discussion. Allow 20 minutes for this exercise.

1. Display a copy of *Figure 33* with the callouts covered.
2. Have trainees identify the components in the Optalign® laser alignment system.
3. Have trainees explain the role of each component and list the steps to perform laser alignment.
4. Review and discuss trainees' answers. Go over any questions trainees have about performing laser alignment.

Section 8.0.0 *Specific Laser Alignment Procedures*

This exercise will allow trainees to determine when certain specific alignment procedures may be required. For this exercise, trainees will need their Trainee Guides and pencils and paper to take notes for discussion. Allow 20 minutes for this exercise.

1. Describe several scenarios for which specific laser alignment procedures may be required.
2. Based on your descriptions, ask trainees to explain which procedure should be performed and why it is needed for the particular scenario.
3. Discuss trainees' answers and review the steps for performing each procedure.
4. Answer any questions trainees have about specific laser alignment procedures.

MODULE 63302-02 — ANSWERS TO REVIEW QUESTIONS

Answer	Section
1. d	1.0.0
2. d	1.0.0
3. b	1.0.0
4. b	2.0.0
5. a	2.0.0
6. c	2.0.0
7. a	2.3.0
8. b	2.4.0
9. a	7.5.1
10. c	3.0.0
11. d	3.0.0
12. a	3.2.0
13. d	3.3.0
14. b	3.3.0
15. d	3.4.0
16. a	4.1.0
17. a	4.1.0
18. d	4.1.0
19. b	4.3.0
20. a	4.3.0
21. b	5.0.0
22. c	5.1.0
23. a	5.1.1
24. d	6.1.0
25. a	6.2.0
26. c	6.2.1
27. d	6.2.3
28. d	7.0.0
29. c	7.1.0
30. c	7.1.0
31. b	7.3.0
32. d	7.3.2
33. a	7.3.2
34. b	6.0.0
35. c	7.4.0

CONTREN™ LEARNING SERIES — USER UPDATES

The NCCER makes every effort to keep these textbooks up-to-date and free of technical errors. We appreciate your help in this process. If you have an idea for improving this textbook, or if you find an error, a typographical mistake, or an inaccuracy in NCCER's Contren™ textbooks, please write us, using this form or a photocopy. Be sure to include the exact module number, page number, a detailed description, and the correction, if applicable. Your input will be brought to the attention of the Technical Review Committee. Thank you for your assistance.

Instructors – If you found that additional materials were necessary in order to teach this module effectively, please let us know so that we may include them in the Equipment/Materials list in the Instructor's Guide.

Write: Curriculum Revision and Development Department
National Center for Construction Education and Research
P.O. Box 141104, Gainesville, FL 32614-1104

Fax: 352-334-0932

E-mail: curriculum@nccer.org

Craft ______ Module Name ______

Copyright Date ______ Module Number ______ Page Number(s) ______

Description ______

______

______

______

(Optional) Correction ______

______

______

(Optional) Your Name and Address ______

______

______

Module 63303-02

Vibration Analysis

Vibration Analysis
Annotated Instructor's Guide

Module 63303-02

MODULE OVERVIEW

This module discusses the causes of vibration and explains vibration analysis. Trainees will learn to identify the different kinds of basic vibration test equipment and to explain vibration monitoring and field balancing.

PREREQUISITES

Please refer to the Course Map in the Trainee Module. Prior to training with this module, it is recommended that the trainee shall have successfully completed the following:

Core Curriculum; Pipeline Core, Modules 66101-02 and 66102-02; Pipeline Mechanical Levels One and Two; Pipeline Mechanical Level Three, Modules 63301-02 and 63302-02

OBJECTIVES

Upon completion of this module, the trainee will be able to:

1. Explain the causes of vibration.
2. Explain vibration analysis.
3. Identify and explain the different kinds of basic vibration test equipment.
4. Explain vibration monitoring.
5. Explain field balancing of machines.

PERFORMANCE TASK

Under the supervision of the instructor, the trainee should be able to:

1. Identify and explain the different kinds of basic vibration test equipment.

NCCER STANDARDIZED CRAFT TRAINING PROGRAM

The National Center for Construction Education and Research (NCCER) provides a standardized national program of accredited craft training. Key features of the program include instructor certification, competency-based training, and performance testing. The program provides trainees, instructors, and companies with a standard form of recognition through a National Craft Training Registry. The program is described in full in the *Guidelines for Accreditation*, published by the NCCER. For more information on standardized craft training, contact the NCCER by writing us at P.O. Box 141104, Gainesville, FL 32614-1104; calling 352-334-0911; or e-mailing info@nccer.org. More information may be found at our Web site, www.nccer.org.

HOW TO USE THIS ANNOTATED INSTRUCTOR'S GUIDE

Each page presents two sections of information. The larger section displays each page exactly as it appears in the Trainee Module. The narrow column ties suggested trainee and instructor actions to each page and provides icons to call your attention to material, safety, audiovisual, or testing requirements. The bottom of each page includes space for your notes.

Teaching Tip

If you see the Teaching Tip icon, that means there is a teaching tip associated with this section. Also refer to any suggested teaching tips at the end of the module.

SAFETY CONSIDERATIONS

Ensure that the trainees are equipped with appropriate personal protective equipment. Emphasize the importance of following all safety precautions and procedures when working with power tools.

PREPARATION

Before teaching this module, you should review the Module Outline, Objectives, Performance Task, and the Materials and Equipment List. Be sure to allow ample time to prepare your own training or lesson plan and gather all required equipment and materials.

MATERIALS AND EQUIPMENT LIST

- Transparencies
- Markers/chalk
- Blank acetate sheets
- Transparency pens
- Pencils and scratch paper
- Module Examinations*
- Performance Profile Sheets*
- Overhead projector and screen
- Whiteboard/chalkboard
- Appropriate personal protective equipment
- Copies of *49 CFR Part 192* (Gas) and/or *49 CFR Part 195* (Liquid)
- Copies of your company policy and procedures manual
- A variety of transducers, including:
  - Velocity transducers
  - Accelerometers
  - Displacement transducers
- A variety of vibration test equipment, including:
  - Meters
  - Oscilloscopes
  - Spectrum analyzers
  - Electronic filters
  - Stroboscopes
- A variety of vibration recording instruments, including strip chart recorders and data collectors
- Sample vibration monitoring forms

* Located in the Test Booklet.

TEACHING TIME FOR THIS MODULE

An outline for use in developing your lesson plan is presented below. Note that each Roman numeral in the outline equates to one session of instruction. Each session has a suggested time period of 2½ hours. This includes 10 minutes at the beginning of each session for administrative tasks and one 10-minute break during the session. Approximately 5 hours are suggested to cover *Vibration Analysis.* You will need to adjust the time required for hands-on activity and testing based on your class size and resources.

Topic	Planned Time
Session I. Vibration Causes, Analysis, and Test Equipment	
A. Causes of Vibration	____________
B. Vibration Analysis	____________
C. Vibration Test Equipment	____________
1. Performance Testing	____________
Session II. Vibration Monitoring, Machinery, and Resonance	
A. Vibration Monitoring	____________
1. Laboratory	____________
B. Field-Balancing Machines	____________
C. Controlling Natural Resonance	____________
D. Summary	
1. Summarize module	____________
2. Answer questions	____________
E. Module Examination	____________
1. Trainees must score 70% or higher to receive recognition from the NCCER.	
2. Record the testing results on Craft Training Report Form 200 and submit the results to the Training Program Sponsor.	
F. Performance Testing	____________
1. Trainees must perform each task to the satisfaction of the instructor to receive recognition from the NCCER.	
2. Record the testing results on Craft Training Report Form 200 and submit the results to the Training Program Sponsor.	

Module 63303-02

Vibration Analysis

Instructor's Notes:

Instructor's Notes:

Introduction to the Pipeline Operator Qualification Rule, API, and NCCER

It could be said that petroleum products keep America running. Petroleum products like oil and natural gas are the principal sources of energy in the United States and are used in many manufacturing and chemical processes. Petroleum products are primarily transported by pipeline. There are more than a million miles of liquid and natural gas pipeline in this country, ranging in diameter from two inches to four feet. These pipelines bring crude oil and natural gas from production fields to processing plants and refineries—and ultimately to the consumer.

Pipeline Operator Qualification

Petroleum products are considered hazardous, so pipeline personnel must take extra precautions and follow specific procedures when working with them. To help ensure the safe handling and transportation of petroleum products, the United States Department of Transportation (DOT) mandates safe practices as federal regulations.

One such regulation, the Operator Qualification (OQ) rule, requires pipeline operators to develop a written qualification plan to evaluate their personnel and their contract personnel (see *49 CFR Part 192 Subpart N—Gas,* and *49 CFR Part 195 Subpart G—Liquid*). The evaluation hinges on the operator's ability to perform certain covered tasks and to be able to recognize and respond to abnormal operating conditions that may be encountered while performing these covered tasks.

Covered Tasks

A pipeline performance task is deemed *covered* if the activity passes the following four-part test:

1. Is it performed on a pipeline facility?
2. Is it an operations or maintenance task?
3. Is it performed as a requirement of *49 CFR Part 195 (Liquid)* or *49 CFR Part 192 (Gas)*?
4. Does it affect the operation or integrity of the pipeline facility?

If the answer to all four questions is *yes*, then the activity is considered *covered* and must figure into the evaluation process of pipeline personnel.

American Petroleum Institute

The OQ rule and its related covered tasks provided the catalyst to develop a unified pipeline training and assessment program—so that pipeline operators and owners have ready access to the assessment and training tools they need. This effort was the direct result of the vision and leadership provided by The American Petroleum Institute (API) and API members.

Efforts to unify the industry are nothing new to API. Indeed, the impetus for forming the Institute in 1919 was the need to standardize engineering specifications for drilling and production equipment. Today, the most pressing issues revolve around public perceptions and government policies toward the industry. Speaking with one voice on these issues has become as essential as having interchangeable parts in the field. API's members determine the petroleum industry's positions on public policy as well as the standards that govern its day-to-day operations worldwide.

Consortium on Operator Qualification

API assembled a focus group whose primary purpose was to identify covered tasks and provide compliance recommendations to pipeline operators.

The Consortium on Operator Qualification (COOQ) became the guiding force for the identification and development of covered tasks for the liquid pipeline industry. Their work product, API publication 1161, *Guidance Document for the Qualification of Liquid Pipeline Personnel,* formed the basis for the resulting Pipeliner Training and Assessment Program.

The covered task list created in API publication 1161 has now evolved to include gas covered tasks and is maintained by NCCER. The most current covered task list can be downloaded at www.nccer.org.

National Center for Construction Education and Research

With the guidance and support of API, its members, and the Consortium on Operator Qualification, the National Center for Construction Education and Research (NCCER) was able to design and implement the Pipeliner Training and Assessment Program.

NCCER is a not-for-profit 501(c)(3) education foundation established in 1995 by the world's largest and most progressive construction and maintenance contractors and national contractor associations. NCCER was founded to address the

severe workforce shortage facing the construction and maintenance industries and to develop a standardized training process and curricula. Today, NCCER is supported by hundreds of leading construction and maintenance companies, manufacturers, and national associations and developed more than 35 construction and maintenance titles.

NCCER also maintains a National Registry, which contains the records of trainees and instructors who have received NCCER training. All completion information for every trainee is sent to the NCCER and filed within the Registry. The Registry can then confirm training and skills for workers as they move from company to company, state to state, or even within their own company.

Standardized Curriculum (NCCER's Contren™ Learning Series)

NCCER curricula are developed using the expertise of industry subject matter experts (SMEs). The use of SMEs as an integral part of every new development or revision ensures that the curriculum is current, industry-driven, and indicative of industry practices. The development and/or revision process is facilitated from start to finish by NCCER project managers, who work closely with the SMEs and a select team of technical writer(s) on the development and presentation of the curriculum.

Pipeline: Five Training Paths

The pipeline curriculum focuses on essential skills and competencies within five skill areas:

- Corrosion Control
- Electrical and Instrumentation
- Maintenance
- Mechanical
- Operations (Field and Control Center)

SMEs from the pipeline industry have determined the skills and competencies needed based on API's covered tasks and the consensus of the group/industry. Specific company information is generally not incorporated into the curriculum unless the SMEs determine that the information is the industry standard.

Training Modules and Covered Tasks

The smallest unit of instruction in the curriculum is a *module*. Each curriculum is composed of modules (the equivalent of stand-alone chapters) that are logically sequenced into complete *levels* of instruction. Every module is task-oriented and requires the successful completion of a written (knowledge-based) examination and a performance (hands-on) test. Modules may be individually selected and/or combined for efficient task training. Modules in the pipeline curricula, to the extent possible, will align with API-developed covered tasks. While not all modules will relate to covered tasks, all covered tasks will relate to at least one module.

Benefits of Training to the Pipeline Marketplace

Applying NCCER's standardized training process allows pipeline owners and contractors in the United States to establish the benchmark for acceptable skill and competency levels within the pipeline industry. This process will assist training organizations in developing and/or improving standardized pipeline training programs that effectively match their resources and operating conditions while responding to regulatory demands and to the needs of the industry. The process will also recognize training organizations that meet the minimum standards of competence, warranting regulatory, public, and professional confidence. This in turn will foster national unity within the pipeline industry for the mutual benefit of individual technicians, owners, contractors, and ultimately pipeline-users and society.

Additional Resources

To review the Pipeliner Training and Assessment Program Implementation and Resource Guide and the latest version of pipeline covered tasks, visit NCCER's Web site at www.nccer.org, or contact:

National Center for Construction Education and Research
PO Box 141104
Gainesville, FL 32614-1104
(888) 622-3720
Web site: www.nccer.org

For more information on API, please contact:

American Petroleum Institute
1220 L Street, Northwest
Washington, D.C. 20005
(202) 682-8000
Web site: www.api.org

To order *API Pub 1161*, contact Global Engineering Documents by calling 1-800-854-7149 or visiting their Web site at www.global.his.com.

Additional sources of information on the pipeline operator qualification regulation include:

DOT – Office of Pipeline Safety
www.ops.dot.gov

Transportation Safety Institute
www.tsi.dot.gov

Instructor's Notes:

COURSE MAP

This course map shows all of the modules in the third level of the Pipeline Mechanical curriculum. The suggested training order begins at the bottom and proceeds up. Skill levels increase as you advance on the course map. The local Training Program Sponsor may adjust the training order.

PIPELINE MECHANICAL LEVEL THREE

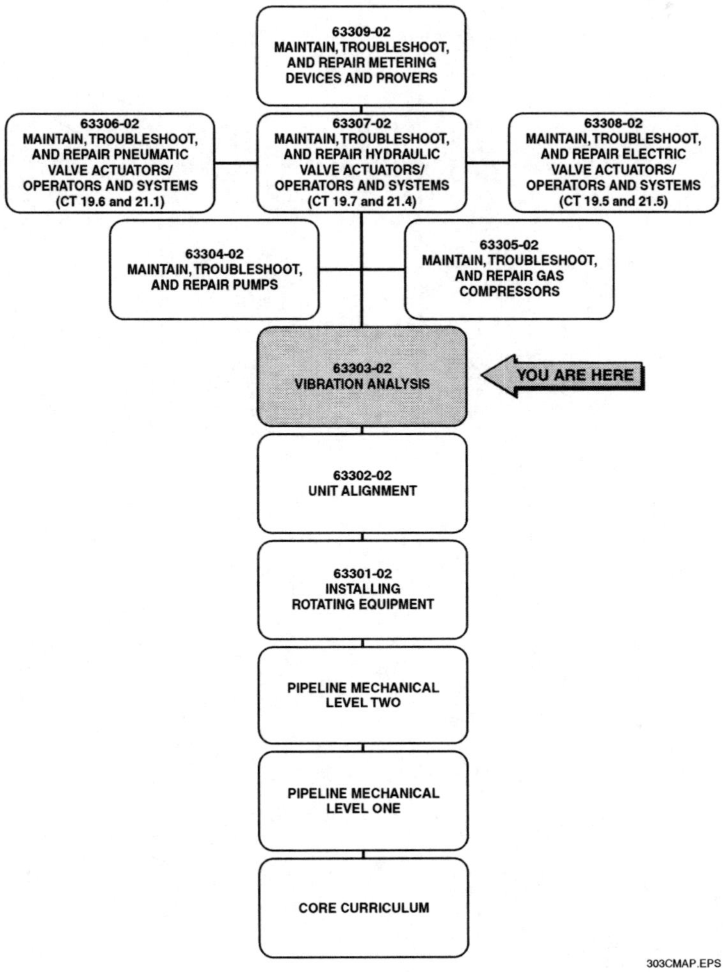

303CMAP.EPS

Homework

Assign reading of Module 63303-02.

Instructor's Notes:

MODULE 63303-02 CONTENTS

Instructor's Notes:

Figures

Instructor's Notes:

MODULE 63303-02

Vibration Analysis

Objectives

When you have completed this module, you will be able to do the following:

1. Explain the causes of vibration.
2. Explain vibration analysis.
3. Identify and explain the different kinds of basic vibration test equipment.
4. Explain vibration monitoring.
5. Explain field balancing of machines.

Prerequisites

Before you begin this module, it is recommended that you successfully complete the following: Core Curriculum; Pipeline Core, Modules 66101-02 and 66102-02; Pipeline Mechanical Level One; Pipeline Mechanical Level Two; Pipeline Mechanical Level Three, Modules 63301-02 and 63302-02.

Required Trainee Materials

1. Pencil and paper
2. Appropriate personal protective equipment

1.0.0 ◆ INTRODUCTION

Every rotating machine vibrates to some extent, even when it is new. However, as machines are used, bearings and other parts are subjected to wear and damage, resulting in an increase in vibration. Vibration monitoring and analysis allow you to pinpoint problems before they cause damage to the equipment or become safety hazards.

This module explains the causes of vibration, vibration analysis and monitoring techniques, vibration test equipment, and field balancing. Each of the vibration analysis and monitoring techniques explained in this module is a form of nondestructive evaluation (NDE). NDE is a method of examining an object or event without destroying it in the process.

2.0.0 ◆ CAUSES OF VIBRATION

Understanding the causes of vibration is the first step toward controlling vibration in rotating machinery. Early detection and control of vibration can greatly extend the life of the machinery. There are many possible causes of vibration in rotating machinery, including:

- Unbalance
- Misalignment
- Bent drive shafts
- Loose mounting bolts
- Worn or damaged bearings
- Improper gear meshing
- Loose drive belts
- Insufficient lubrication
- Electrical problems
- Destructive resonant frequencies
- Mechanical looseness or wear
- Cavitation/improper operation
- External vibration sources
- Flow turbulence
- Natural frequencies

Materials

Ensure that you have everything required to teach the course. Check the Materials and Equipment List at the front of this module.

Classroom

Explain that terms shown in bold are defined in the Glossary at the back of this module.

Audiovisual

Show Transparency 1, Course Objectives.

Show Transparency 2, Performance Task.

Teaching Tip

See the general Teaching Tip at the end of this module.

Classroom

Discuss the possible causes of vibration in rotating machinery.

Audiovisual

Show Transparency 3 (Figure 1). Discuss the causes of unbalance.

Classroom

Refer to Figure 2. Review the types of misalignment that affect vibration.

Discuss damage caused by bent drive shafts and loose mounting bolts.

Audiovisual

Show Transparency 4 (Figure 3). Explain what causes vibration in sleeve bearings.

2.1.0 Unbalance

Unbalance is a common cause of vibration. It is caused by unequal weight or stress at two opposite points around the axis or shaft of a rotating machine. Unbalance can be caused by a manufacturing problem, such as a hidden void in a casting (see *Figure 1[A]*), or by an operational problem, such as the full or partial loss of a fan blade (see *Figure 1[B]*).

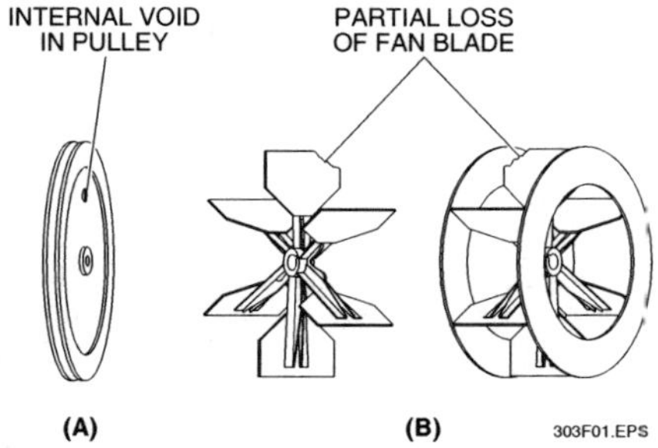

Figure 1 ◆ Two causes of unbalance.

2.2.0 Misalignment

Misaligned couplings, bearings, or V-belt pulleys can cause severe vibration. Misalignment can be parallel offset, angular offset, or a combination of the two (see *Figure 2*).

2.3.0 Bent Drive Shafts

Bent drive shafts can also cause severe vibration. If not detected and corrected early, a bent drive shaft can damage bearings, equipment mountings, and gears.

2.4.0 Loose Mounting Bolts

Loose mounting bolts result in a pounding action that is detected as vibration. The severity of the vibration depends on how loose the bolts are. As vibration continues, the bolts become progressively looser. Complete destruction of the related equipment can result if the problem is not corrected.

2.5.0 Worn or Damaged Bearings

Worn or damaged antifriction bearings create a high-**frequency** vibration. The high frequency results from the size differences between the balls or rollers and the larger inner and outer races. Bearing vibration can sometimes be detected as a high-pitched noise.

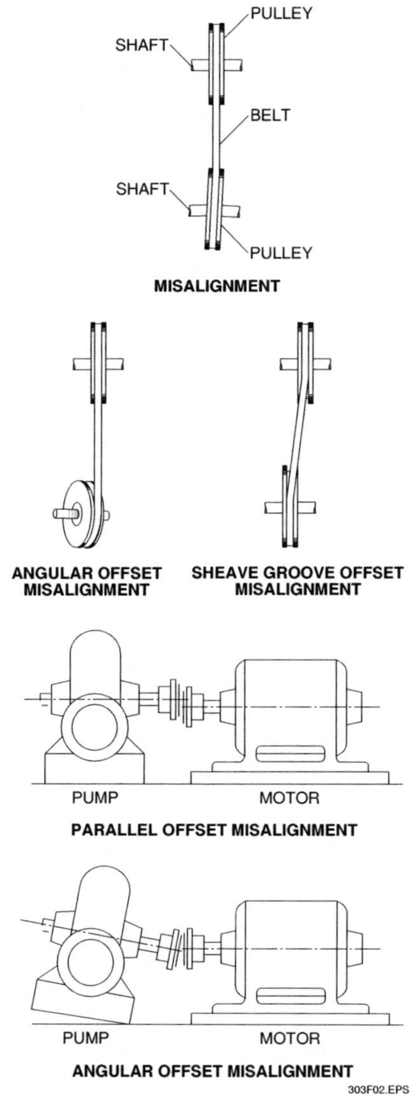

Figure 2 ◆ Misalignment.

Vibration in sleeve bearings can be caused by incorrect clearances or oil whirl (see *Figure 3*). Too little clearance causes low-amplitude, high-frequency vibration. Too much clearance due to wear causes high-amplitude, low-frequency vibration.

Instructor's Notes:

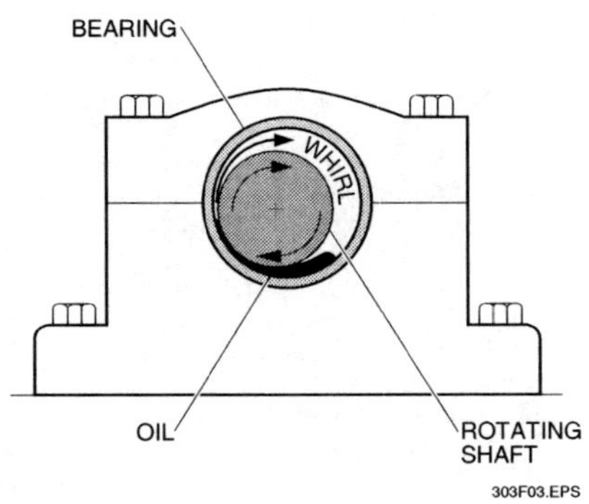

Figure 3 ◆ Oil whirl in a sleeve bearing.

Oil whirl can occur in loose-fitting, pressure-lubricated bearings in high-speed machinery. If any outside influence, such as imbalance or a sudden shock, displaces the shaft from its normal position in the bearing, pressurized oil rushes in to fill the gap left by the displaced shaft. This further displaces the shaft, causing the shaft to whirl in the direction of rotation.

2.6.0 Improper Gear Meshing

Improper gear meshing produces vibration between gears and can be the result of several different factors. Improper gear alignment, broken or severely worn gear teeth, and scored gear teeth are the most common causes. Improper gear meshing usually does not result in immediate destruction of the equipment, but the condition slowly worsens if it is not corrected.

2.7.0 Loose Drive Belts

Loose drive belts can usually be easily detected, except when one or more belts are loose on a multiple-belt drive. Use a tensiometer to check the tension of the belts, and adjust them as necessary. Refer to the equipment or belt manufacturer's instructions for the proper tension of the belt.

2.8.0 Insufficient Lubrication

Insufficient lubrication between a shaft and a sleeve bearing can cause vibration. Normally, a lubricating grease or heavy oil fills the void between the shaft and the sleeve bearing. If the lubricant dries out or is completely used, any imbalance in the load carried by the shaft will cause the shaft to vibrate in the bearing.

2.9.0 Electrical Problems

Electrically caused vibration in a motor or generator is the result of unequal magnetic forces between the rotor and stator. Electrical vibration can be easily identified. If the vibration stops immediately after removing electrical power, the cause is electrical. If the vibration level decreases slowly as the motor or generator slows to a stop, the problem is mechanical. Usually, the only mechanical problems in a motor are bad bearings, a bent rotor shaft, or an off-balance cooling vane in the back of the motor.

2.10.0 Mechanical Looseness

Mechanical looseness involves housing joints, flanges, fittings, shaft components, other parts that gradually loosen over time, and excessive bearing-to-housing clearance. It may be the result of improper torque applied to the bolts or nuts, or it can result from the normal running vibration of the machine. Loosening may occur slowly and gradually and may not be noticed until it becomes excessive and destructive. If not detected in time, mechanical looseness results in major damage to the unit involved.

2.11.0 External Vibration Sources

Vibration from external sources involves frequencies that are transmitted to a machine through a connection, such as a power coupling, shaft, or belt. Some vibration can be transmitted through mutual framing or a floor system. This type of vibration can be destructive and may be difficult to trace and repair. External vibration sources, if not repaired, can result in damage to more than one machine. For example, vibration in a driver can damage a pump bearing and shaft.

2.12.0 Cavitation

Cavitation is a problem common to pumps. The sudden formation and collapse of low-pressure bubbles in liquids caused by mechanical forces will cause the pump to vibrate. The frequency of cavitation is usually random, but it can become intermixed with or covered by the frequency of the impeller blade pass frequency. The condition can be accompanied by loud noise that imitates the sound of gravel inside the pump casing. The condition suggests that the pump is being starved on low suction. If the cavitation problem is not corrected soon, the pump impeller will become eroded, pitted, and possibly crack or disintegrate (see *Figure 4*).

Classroom

Review problems caused by improper gear meshing. Answer any questions trainees have.

Explain how to check for loose drive belts and electrical problems.

Discuss problems caused by mechanical looseness and external vibration sources.

Refer to Figure 4. Explain how to correct cavitation.

Classroom

Explain how changes in pressure and velocity affect flow turbulence.

Ensure that trainees understand how natural frequencies are activated. Explain when and how resonance occurs.

Discuss the relationship between critical speed and peak resonance. Go over any questions trainees have.

Describe several causes of vibration. Ask trainees to identify each cause and to discuss the consequences of each.

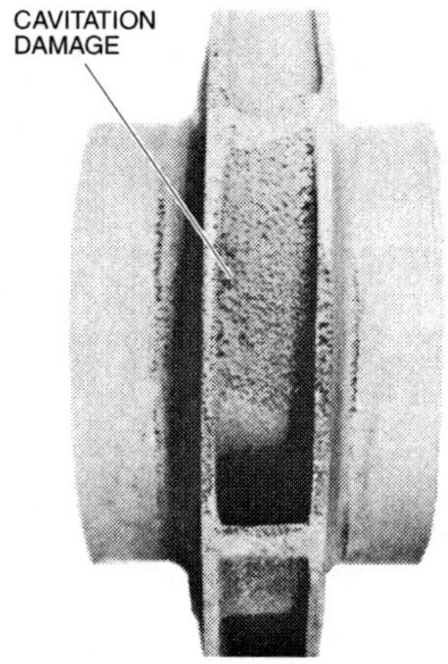

303F04.EPS

Figure 4 ◆ Impeller damaged by pump cavitation.

2.13.0 Flow Turbulence

Flow turbulence is associated with equipment such as industrial blowers and heavy air-handling equipment. Changes in the pressure and **velocity** of the air flowing into and out of the fan and the flow interference in the associated passageways creates turbulence that may cause random, low-frequency vibration.

2.14.0 Natural Frequencies

Every machine has natural frequencies that are often undetectable. These frequencies are invariably a natural part of each component in the machine and its structural elements. The natural frequency is the frequency at which the component vibrates if there are no restraints. This frequency does not normally manifest itself unless it is affected by some outside influence.

For example, a rotating shaft has a natural frequency that may not result in any vibration at all. However, when an outside force appears and matches or nearly matches the natural frequency of the shaft, the natural frequency of the shaft is activated and often multiplied in force.

Once these natural frequencies are activated, the amount of strain on machine components can be multiplied by 100 times or more. However, even though a machine may have the potential to experience multiple frequency problems, means can be provided to ensure that they are never activated.

2.14.1 Resonance

Natural frequencies are activated by an outside force called resonance. Resonance occurs when the frequency of one component matches or nearly matches that of another. In some cases, more than one natural frequency can be activated at one time, creating a greatly increased amount of vibration. Such vibration from resonance can greatly accelerate wear, metal fatigue, and other damage. As instruments and methods of detection have improved, resonant vibration has been found to be a more prominent cause of machine problems than originally thought.

The natural frequencies involved in the resonant vibration of a particular machine might include frequencies from structural elements, such as the floor, foundation, or other supporting structure. A machine might be receiving a frequency from another location being transmitted through a support component. The resonance of a machine is defined by a set of parameters, such as mounting type, mass, and shaft elasticity.

Resonance can be the result of some common problems, occurring in other machinery at or near the site, that include but are not limited to:

- Unbalance
- Misalignment
- Machine looseness
- Bearing problems
- Defective gears
- Worn or defective belts

2.14.2 Critical Speed

Critical speed is the rate of rotation that exactly corresponds to a harsh or violent vibration. Critical speed corresponds to the point of peak resonance. An example of critical speed is the vibration experienced in an automobile with a tire not properly balanced, which causes noticeable vibration at a particular speed.

The turning element might be a shaft, rotor, flywheel, or some other type of unit. Vibration of the element may be generated at the normal turning speed, or it may be generated in multiples of the normal speed, such as a frequency equivalent to two, three, or more times each revolution of the turning element. Machinery designers, through complex testing and analysis, try to determine the critical speed of a machine and then compare it to the optimum operating speed of the machine. Sometimes, either the critical speed or the operating speed must be adjusted to ensure that the critical speed of the machine is far enough from the optimum operating speed of the machine to avoid

Instructor's Notes:

problems. If the operating speed cannot be changed, the designers can alter the critical speed through various adjustments and alterations to the parameters of the machine.

3.0.0 ◆ VIBRATION ANALYSIS

Each vibration has a set of distinctive characteristics known as its **vibration signature**. The process of gathering vibration signatures, determining their causes, and comparing them with stored baseline data is known as vibration analysis.

Vibration of rotating machinery can be separated into individual vibration **cycles**. Each cycle begins with the shaft centered in the housing. As the shaft rotates, it bends because of loads and loose bearings or as compensation for a bent shaft. As the revolutions per minute (rpm) increase, the shaft moves away from the center line until it reaches a peak. The shaft then moves back through the center line to the peak in the opposite direction. The shaft deflection reverses back to the center line, and this is considered one cycle of vibration (see *Figure 5*). As the load or rpm increases, this creates a more noticeable wobble that is measured as increased frequency and amplitude. The vibration is caused by the reversal of movement at each peak of the cycle.

A vibration cycle consists of four measurable characteristics that make up the vibration signature, or fingerprint. In the vibration analysis process, each characteristic is monitored and recorded. Analyzing the characteristics helps identify the source of the vibration and the defect that caused it. Each type of defect in a rotating piece of equipment has its own specific vibration signature. For example, a bent shaft causes axial vibration at a frequency of one or two times the rotational speed of the shaft. The following are the four characteristics of a vibration cycle:

- Frequency
- Velocity
- Acceleration
- Displacement

3.1.0 Frequency

The frequency of a vibration is the number of individual vibration cycles that occur in a given time period. In vibration analysis, frequency is measured in cycles per minute (cpm). If the vibration cycle occurs 60 times in 1 minute, the frequency is 60 cpm.

3.2.0 Velocity

Vibration velocity is the speed at which the vibrating shaft completes one cycle. The velocity is not constant. When the shaft reaches the maximum distance in each direction from the center line, the velocity is zero because the shaft stops and changes direction. The velocity is at its maximum as the shaft travels through the center line. This highest, or peak, velocity is used to analyze the velocity and is recorded in inches per second. *Figure 6* shows zero and peak velocity.

Classroom

Discuss vibration signatures and the characteristics of a vibration cycle.

Audiovisual

Show Transparency 5 (Figure 5). Review what is meant by one cycle of vibration.

Classroom

Explain how frequency and velocity are measured.

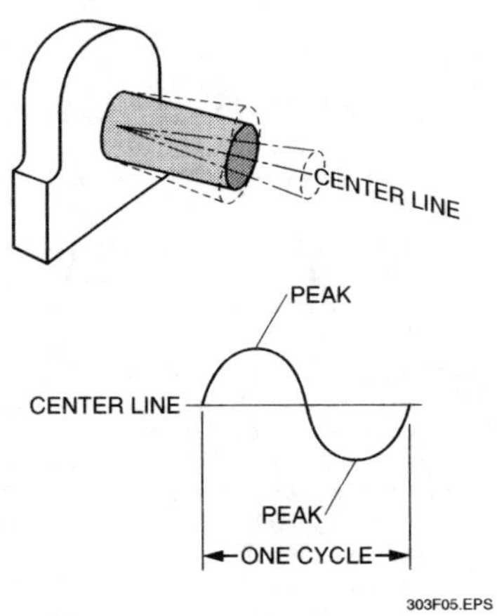

Figure 5 ◆ One cycle of vibration.

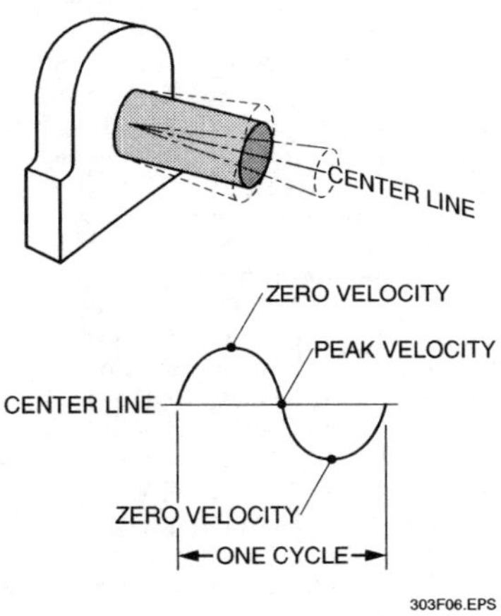

Figure 6 ◆ Zero and peak velocity.

Classroom

Discuss the effect of acceleration and displacement on vibration.

Refer to Figure 9. Ask a trainee to explain how a transducer records waveforms.

Audiovisual

Show Transparency 6 (Figure 10). Explain that the amplitude of a wave is the representation of all vibrations in one waveform.

3.3.0 Acceleration

In vibration analysis, acceleration is the increasing rate of change in velocity. Acceleration is measured in Gs. One G is equal to 9.8 meters per second or 32.2 feet per second. One G is also equal to the acceleration produced by the earth's gravity at sea level. Vibration acceleration refers to the force of Gs at each cycle peak. The acceleration of a vibration cycle is zero at the center line and reaches peak acceleration at each peak (see *Figure 7*).

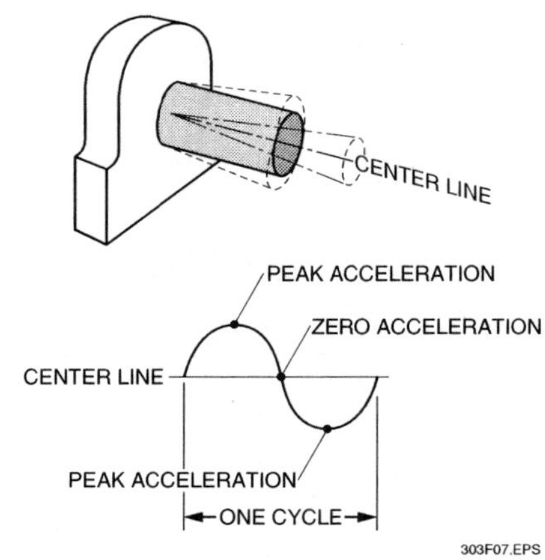

Figure 7 ◆ Zero and peak acceleration.

3.4.0 Displacement

Displacement, also known as amplitude, is the total distance that the shaft travels from one peak to the other peak during one cycle of vibration. This distance is measured in mils. One mil is $\frac{1}{1{,}000}$ inch. *Figure 8* shows vibration displacement.

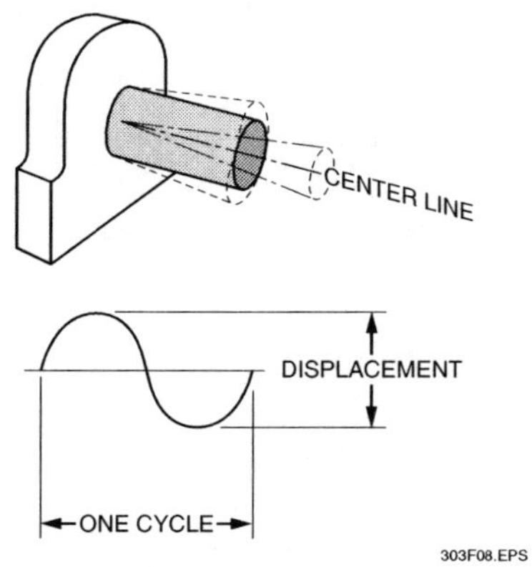

Figure 8 ◆ Vibration displacement.

3.5.0 Measuring Movement

The movement of a vibrating machine is demonstrated by the creation of sinusoidal waves. To understand a sinusoidal wave created on a strip-type recording instrument, consider a marking pen attached to a weight. The weight is suspended by a spring from a vibrating component. If the point of the pen is touching the strip of paper as it advances at a controlled speed and the machine is then stopped, the pen will create a straight line on the advancing paper. However, if the machine is running, the regular vibration of the machine will cause the weight to ride up and down on the spring in a movement that coincides with that of the machine. The movement of the weight creates a wave line as the paper advances that is representative of the actual vibration of the machine. In *Figure 9*, the weight suspended on the spring is equal to the action of the **transducer** installed in the bearing support housing on the left. The line that demonstrates the vibration cycle is the equivalent of the time waveform.

Each sinusoidal wave on the paper demonstrates one complete vibration cycle. As the movement of the machine begins, the recorded cycle begins with the weight and pen ascending from the central position to the maximum point of travel in one direction, descending back past the central position, and then descending further to the maximum point of travel in the other direction, creating one complete wave on the advancing paper. If allowed to continue, a set of waves would be drawn on the advancing paper, and all the waves would be identical time waveforms. Knowing the speed of the paper, such as inches per minute, a frequency can be calculated from the number of waves on a measured length of paper at the given speed of advancement. The upward and downward wave on the paper can then be measured to gain a reference for the amplitude of the wave, which is the representation of all vibrations together in one time waveform (see *Figure 10*).

Data received from transducers is in the form of a time waveform. Computers and microprocessors transform the signals from the transducers. A fast fourier transform (FFT) is created and displayed on a screen so that the various vibrations can be viewed and understood separately. This display on the screen is called the signature.

It is imperative that you understand the use of a time waveform and its conversion to an FFT, which demonstrates the complete vibration

Instructor's Notes:

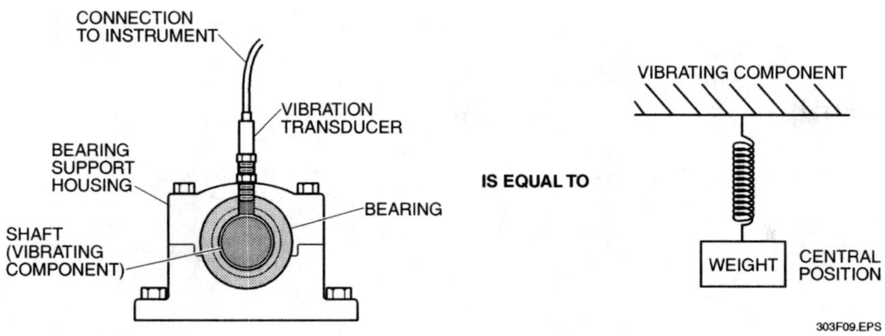

Figure 9 ◆ Vibration transducer.

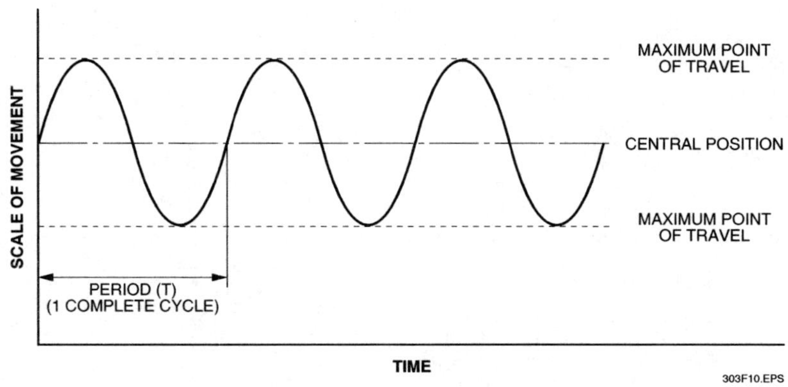

Figure 10 ◆ Frequency demonstrated on a time waveform.

movement of the machine or other entity to which the transducers are attached. The FFT is the vibration signature; it is somewhat like a fingerprint that can be used in solving a crime. The FFT transforms the time domain data into frequency domain data. An FFT can be very helpful in diagnosing vibration problems.

3.6.0 Time and Frequency Domains

The time domain data is the amplitude of the frequency compared to the specified time (minute or second). Frequency is normally articulated in either cycles per minute (cpm) or cycles per second (cps). Cycles per second are normally stated in hertz, where 1 hertz (Hz) is equal to 60 cps. The frequency is the number of times the component moves during the specified time. The amplitude of the vibration is the measurement of the distance the vibrating component moves (mils, inches per second, or Gs). The frequency domain is the amplitude of the frequency compared to the frequency. *Figure 11* is a typical spectrum as demonstrated by the FFT.

3.6.1 Vibration Velocity

Velocity is the speed at which a component moves as it passes through its vibration cycle. Velocity is highest at the midpoint of the vibration cycle and is zero at the maximum points of the cycle. At the maximum point, the movement of the vibrating component must stop in order to change directions. At the midpoint in the cycle, the change of direction is over, and the component has accelerated completely before slowing and reversing direction again. Velocity is usually articulated as inches per second (in/sec) or millimeters per second (mm/sec). *Figure 12* illustrates the points of acceleration and velocity changes.

Audiovisual

Show Transparency 7 (Figure 11). Explain how time and frequency domains are displayed on a fast fourier transform (FFT).

Classroom

Discuss the changes in velocity that occur at various points in one vibration cycle. Ask whether trainees have any questions.

Classroom

Discuss the relationship between the amounts of stress and damage and the amounts of change in direction and acceleration that a machine experiences.

Audiovisual

Show Transparency 8 (Figure 13). Explain how multiple phases of vibration cause more damage than a single phase.

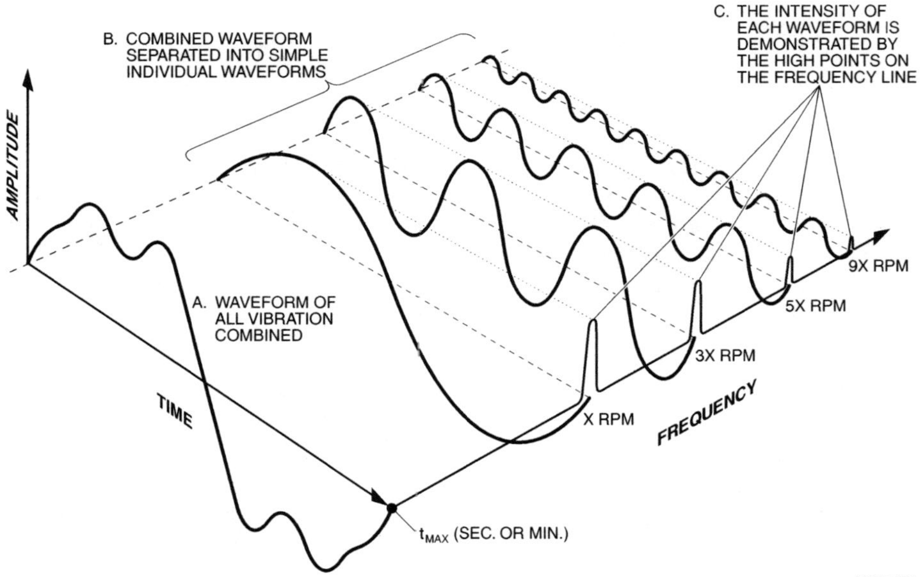

Figure 11 ◆ Fast fourier transform (FFT).

3.6.2 Vibration Acceleration

Every machine component that vibrates experiences repeated accelerations and decelerations because of the speed changes inherent to the process of vibration. The greatest acceleration occurs at the point at which the component has stopped movement in one direction and begins to pick up speed as it rebounds in the other direction.

The amount of stress and the resultant damage to the machine is directly proportional to the amount of change in direction and acceleration being experienced. It is imperative that you are able to understand acceleration to be able to comprehend its affects and the importance of using an accelerometer on the job. Most of the effects of vibration are demonstrated and measured from the standpoint of acceleration (see *Figure 12*).

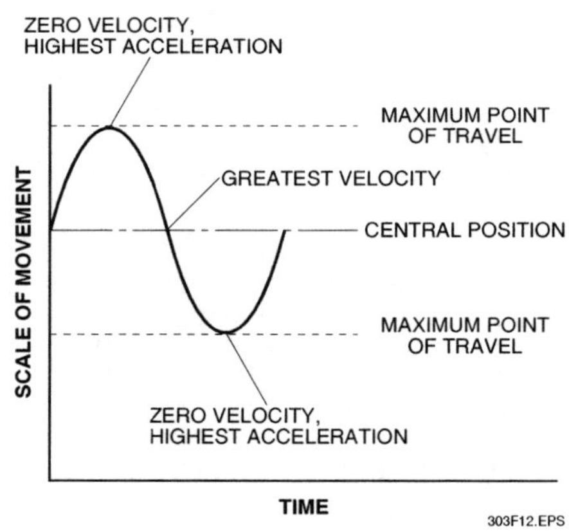

Figure 12 ◆ Waveform illustrating velocity.

3.7.0 Phase Differences

Machines or components that vibrate in different phases probably experience more destructive effects from vibration than they would if the vibration were only single phase. The reason for this is that after one phase reaches its maximum travel and changes direction, a few degrees later, another reaches its travel limit with a hard jolt and changes direction. As this procedure continues at high speed, the conditions of vibration occurs. If there are multiple phases of vibration (as is often the case), the machine is receiving a continual beating from several directions (see *Figure 13*).

Instructor's Notes:

Consider a shaft running through a machine with two vibration points, one on each end of the shaft. However, the vibration points are not synchronized; they are offset by 90 degrees. The jolt of the shaft moving, stopping, and accelerating again from the second point follows the same experience from the first. What makes matters worse is that that the jolt from the first point tries to lift one end of the shaft while the jolt from the other point tries to lift the other end of the shaft, ultimately trying to rock the machine from end-to-end if it were possible (see *Figure 14*). It is not possible, so the destructive forces are imparted to the shaft, bearings, and other components.

In the field, you might have a hard time discerning exactly what the real causes of vibration are without the use of electronic instruments. Without the use of spectrum analysis, some vibration problems might never be solved. However, with the use of computerized instruments and spectrum analysis equipment, it is possible to understand the realistic causes of the vibration and where it originates and to quickly form a plan to remedy the problem.

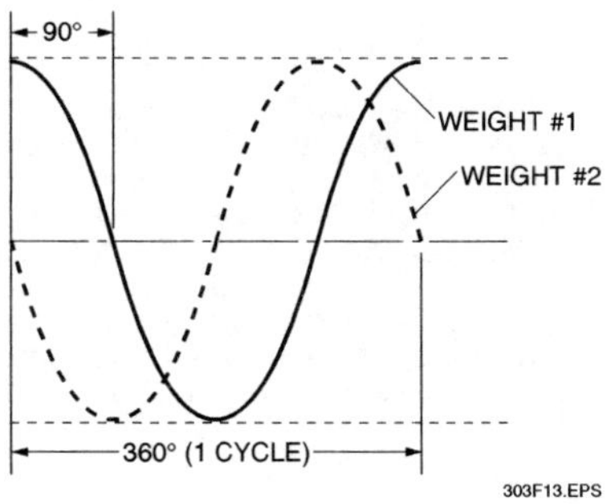

Figure 13 ◆ Different phases of vibration.

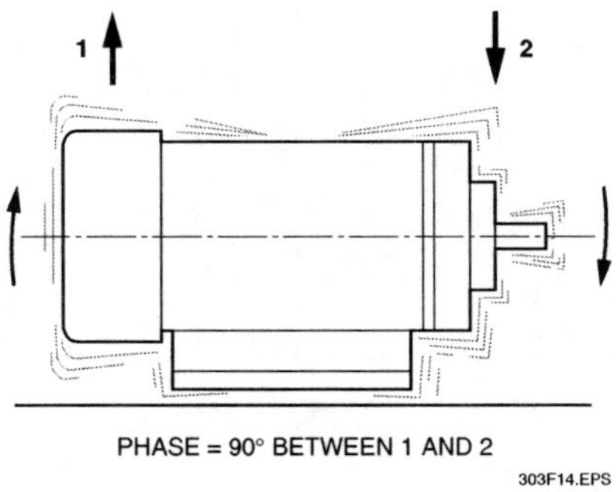

Figure 14 ◆ Phase offset of 90°.

4.0.0 ◆ VIBRATION TEST EQUIPMENT

Knowing the frequency, velocity, acceleration, and displacement of a vibration cycle will enable you to determine the location and cause of the vibration. Vibration analysis allows you to pinpoint the problem in the machinery in order to make the necessary repairs. Vibration test equipment is used to analyze equipment and determine the type of vibration defect. Basic vibration test equipment includes transducers, vibration analysis equipment, and vibration recording instruments. This equipment is used to monitor and analyze machine vibrations.

4.1.0 Transducers

Transducers convert mechanical vibrations into electrical signals that are reproduced as sounds or graphic displays. A transducer provides the link between the equipment being analyzed and the vibration analysis equipment. The three types of transducers are:

- Velocity transducers
- Accelerometers
- Displacement transducers

4.1.1 Velocity Transducers

Velocity transducers are known as contact transducers because they are mounted directly on the vibrating surface. These transducers pick up frequency signals in the intermediate range, which is the frequency range of most industrial equipment. These transducers can be used with portable, hand-held test equipment, or they can be permanently mounted to the vibrating surface for use with a continuous monitoring system.

A velocity transducer consists of a coil of fine wire wrapped around a metal core that is supported by soft springs (see *Figure 15*). The coil assembly is surrounded by a magnet attached to the inside of the transducer case. When the transducer is in contact with the equipment being analyzed, the case and magnet move back and forth with the vibrating equipment, while the coil assembly inside the transducer remains in one place. As the transducer moves back and forth around the coil, the magnetic field surrounding the coil induces voltage in the coil. This voltage is proportional to the vibration velocity. The faster the transducer case and magnet vibrate, the higher the output voltage of the coil is.

Classroom

Refer to Figure 14. Explain problems that result when vibration points are not synchronized.

Ask trainees to list the characteristics of a vibration cycle. Have trainees explain how each characteristic affects vibration and to explain how each is measured.

Review the types and uses of vibration test equipment.

Compare the three types of transducers. Answer any questions trainees have.

Audiovisual

Show Transparency 9 (Figure 15). Review the components of a velocity transducer.

Classroom

Explain why accelerometers are able to take the most accurate measurement of vibration acceleration.

Refer to Figure 16. Discuss the role of the crystal disc in an accelerometer.

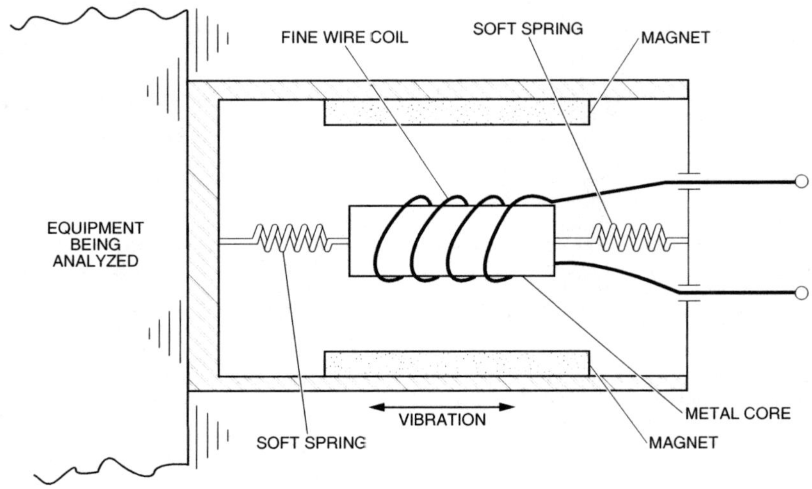

Figure 15 ◆ Velocity transducer.

4.1.2 Accelerometers

The most accurate measurement of vibration acceleration is obtained with an accelerometer. Accelerometers are contact transducers with wider frequency ranges than velocity transducers. Accelerometers are used to analyze various forces, loads, and stresses that emit low- to high-frequency sounds. Some types of accelerometers require an amplifier to strengthen low-frequency signals.

Accelerometers can be permanently mounted to the equipment being monitored. The three ways to mount an accelerometer to a machine are to attach it with a heavy magnet, screw it into a tapped hole, or screw it into an attached adapter.

An accelerometer contains crystal discs that, when the crystal is compressed, generate a voltage proportional to the vibration acceleration (see *Figure 16*). When the machine accelerates, **G forces**

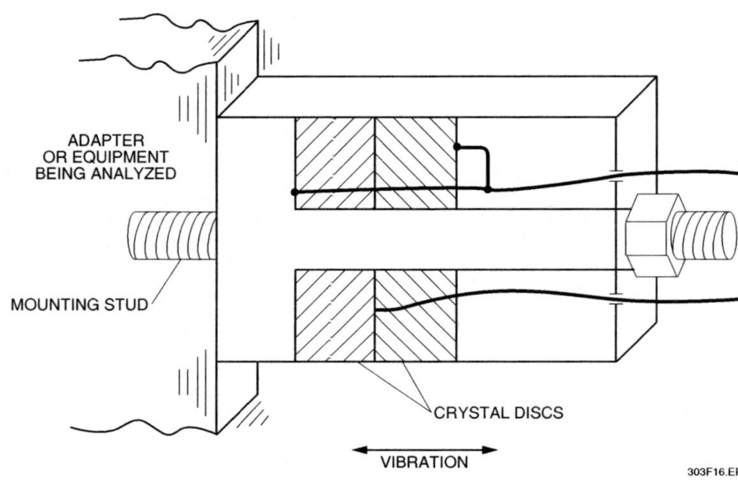

Figure 16 ◆ Accelerometer.

Instructor's Notes:

cause the crystal discs to compress. The induced voltage is amplified and fed to the analyzer to be displayed as vibration acceleration.

4.1.3 Displacement Transducers

Displacement transducers are noncontact devices that are mounted a specific distance from the rotating shaft of the machine being analyzed. They are used to detect the small amount of vibration produced by lightweight, high-speed shafts in large, heavy casings. Noncontact displacement transducers are more accurate than contact transducers for this application because they sense the movement of the shaft through a magnetic field.

The displacement transducer sensor is placed 0.020" to 0.100" from the shaft surface. The displacement transducer sensor produces a magnetic field that is distorted by the movement of the shaft. As the shaft vibrates, the gap size and the amount of magnetic field distortion change, generating varying voltages. A **pickup coil** in the transducer senses the changes in voltage and transmits the information to the vibration recorder or analyzer (see *Figure 17*). Displacement transducers can provide accurate vibration measurements on turbines, compressors, centrifugal pumps, and similar machinery.

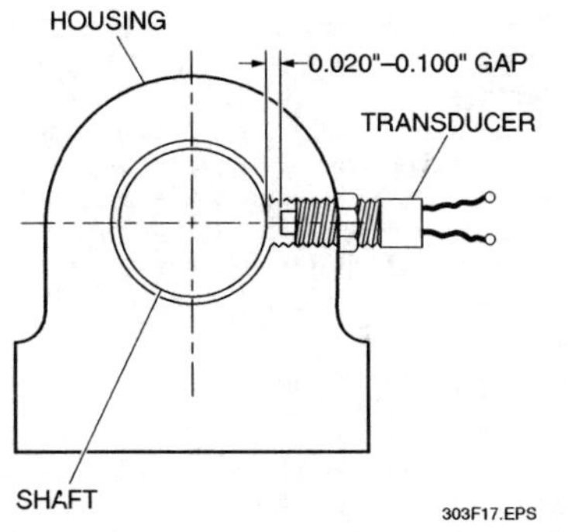

Figure 17 ◆ Displacement transducer.

4.2.0 Vibration Analysis Equipment

Vibration analysis equipment is used to determine the cause of vibration. Machinery malfunctions can be identified by comparing the frequency of the vibration with the running speed of the machine. These comparisons are used to determine the sources of the vibration. The electronic equipment used for vibration analysis includes:

- Meters
- Oscilloscopes
- Spectrum analyzers
- Electronic filters
- Stroboscopes

4.2.1 Meters

Vibration meters (see *Figure 18*) are portable instruments used to measure vibrations that occur within a selected frequency. Most portable meters give the amplitude of the vibration displacement in mils, the acceleration of the vibration in Gs, and the velocity of the vibration in inches per second. Vibration meters are generally battery powered and use contact transducers.

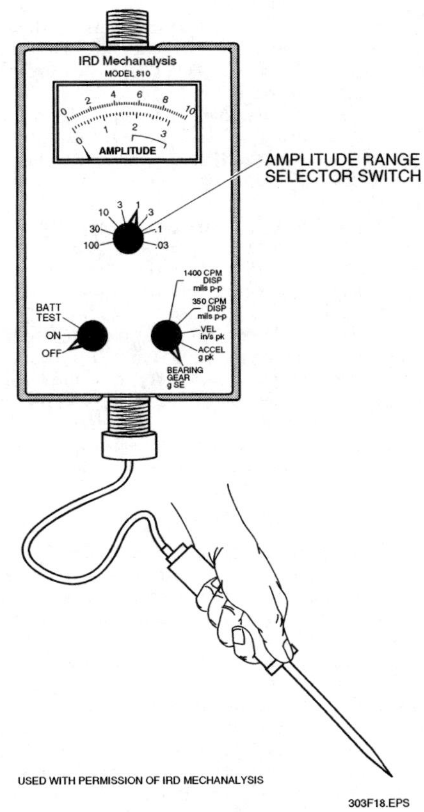

Figure 18 ◆ Vibration meter.

Audiovisual

Show Transparency 10 (Figure 17). Explain how the pickup coil in a displacement transducer senses the change in voltage.

Classroom

Review the types of electronic equipment use for vibration analysis.

Refer to Figure 18. Explain how meters are used to measure vibration.

Audiovisual

Show Transparency 11 (Figure 20). Explain how oscilloscopes measure and record vibration sine waves.

Classroom

Refer to Figure 21. Discuss the two amplitude scales used by spectrum analyzers and explain the factors that determine which to use.

Some vibration meters are made with the meter and sensor pickup in one housing. However, vibration meters with the sensor connected to a flexible cable are easier to use in a confined space and around obstructions. The three controls on the vibration meter are the displacement/velocity/acceleration selector switch, the amplitude range selector switch, and the battery test/on/off selector switch.

4.2.2 Oscilloscopes

Cathode-ray oscilloscopes are used to test electronic circuits. Oscilloscopes display graphs of rapidly changing voltage and current, but they are also capable of giving information concerning frequency values, phase differences, and voltage amplitude. The principle components of a basic oscilloscope include a cathode-ray tube (CRT), a sweep oscillator, horizontal and vertical deflection amplifiers, controls, switches, and input terminals (see *Figure 19*). A CRT is a vacuum tube that projects a beam onto a fluorescent screen. The deflection amplifier changes the direction of the beam both vertically and horizontally on the screen. The sweep oscillator, or generator, moves the electronic beam across the screen to form a wave pattern.

The oscilloscope is used as a vibration analysis instrument because it provides a signal that can be displayed on the oscilloscope screen. Once it appears on the screen, this signal can be recorded, photographed, and analyzed.

Small, portable oscilloscope units are used to monitor signals in the field. Laboratory oscilloscopes are useful when several people must view the screen simultaneously. A third type of oscilloscope is equipped with a camera adapter and is used to photograph certain signals. *Figure 20* shows a vibration sine wave measurement.

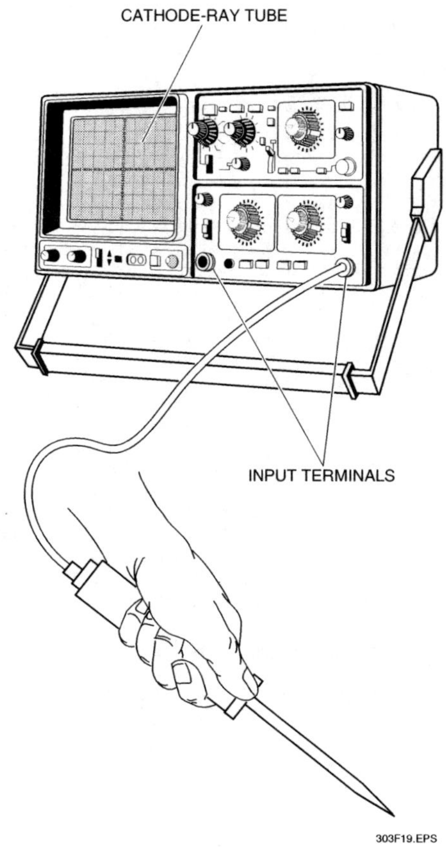

Figure 19 ◆ Oscilloscope.

4.2.3 Spectrum Analyzers

Vibration spectrum analyzers (see *Figure 21*) measure and compare the amplitude and frequency characteristics of a vibration. Machinery vibration and noise are generally complex, and the spectrum analyzer is capable of separating each vibration frequency for individual monitoring.

The spectrum analyzer has two amplitude scales. The top scale measures amplitude from zero to one, and the bottom scale measures amplitude from one to three. The severity of the vibration determines which scale to use. A displacement/velocity selector switch allows you to measure either vibration displacement in mils or vibration velocity in inches per second.

The frequency meter capability of the spectrum analyzer measures the frequency of the indicated vibration in cycles per minute (cpm), and a frequency range selector switch allows you to select the desired frequency range of the frequency meter.

A turntable filter selector switch allows you to select from different filter bandwidths, ranging from sharp to broad. The broad filter bandwidth is used for rapid scanning to quickly determine the vibration frequencies present. The sharp filter bandwidth is used to analyze a particular vibration frequency.

Instructor's Notes:

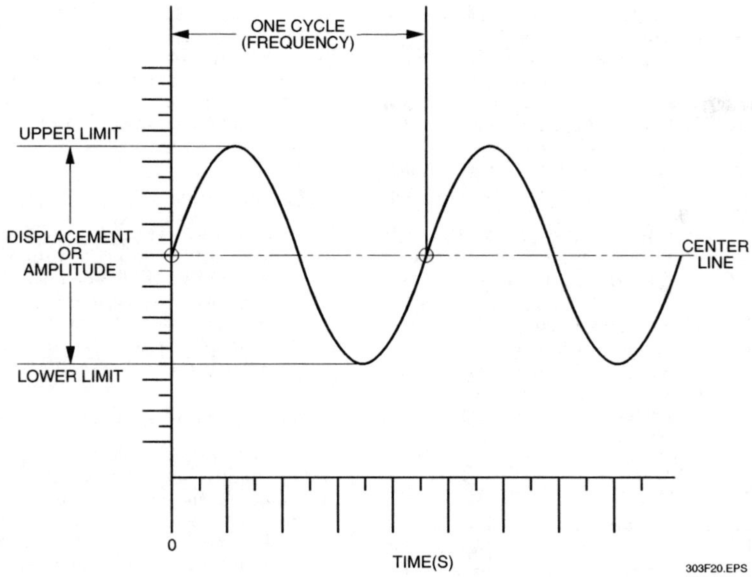

Figure 20 ◆ Vibration sine wave measurement.

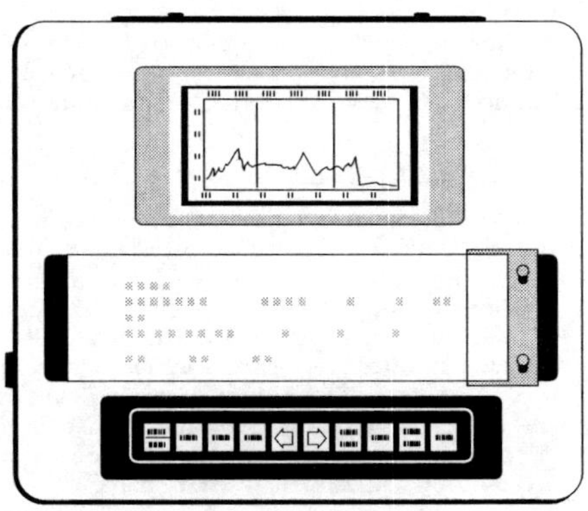

USED WITH PERMISSION OF IRD MECHANALYSIS

303F21.EPS

Figure 21 ◆ Vibration spectrum analyzer.

4.2.4 Electronic Filters

When a machine vibrates, a signal composed of many frequencies is generated. Basic vibration meters cannot pinpoint specific problems. To determine the frequencies present in the signal, electronic filters are used. They can detect a vibration frequency within a specific, narrow band range. Electronic filters can be separate components connected to the vibration analyzer, or they can be built into the analyzer. Filters can be classified in four ways:

- *Passband frequency* – Passband electronic filters are the most basic types of filters. A passband filter detects a specified band of frequencies and eliminates frequencies above or below the desired passband. High-pass filters detect all frequencies above a specified frequency. Low-pass filters detect frequencies below a specified frequency.
- *Method of tuning* – Fixed filters, tunable filters, and automatic filters are used to tune vibration analyzers to specific frequencies. Fixed filters are commonly used for monitoring where known or predictable frequencies occur. Tunable filters allow the filter frequency band to be adjusted so that specific frequencies can be isolated. This type of filter saves time in searching for specific frequencies and can eliminate a cluttered frequency signal. The tuning signal in an automatic, or tracking, filter is generated by and synchronized with the part of the machine being analyzed. The automatic filter is used in shaft-balancing applications and for tracking specific frequencies during startup. It is also used when the system is coasting to a stop.

Classroom

Review the classifications for electronic filters. Go over any questions trainees have.

Audiovisual

Show Transparency 12 (Figure 22). Explain how to set the stroboscope timing.

Classroom

Review the types of vibration recording instruments and explain how they are used to collect and transfer vibration data.

Teaching Tip

See the Teaching Tip for Section 4.0.0 at the end of this module.

Performance Testing

Have each trainee perform Performance Task 1 to your satisfaction. Fill out a Performance Profile Sheet for each trainee.

Homework

Have trainees review Sections 5.0.0–7.3.0 and answer the Review Questions for the next session.

- *Component type* – Electronic filters can also be classified as active or passive, depending on the type of components used in their construction. Active filters are commonly used in machinery analysis because they are small and stable and provide a voltage increase. These filters use operational amplifiers to strengthen the vibration signal. Passive filters are much larger and heavier than active filters, limiting their use and portability. However, the vibration signals from passive filters are much sharper because amplifiers are not required.
- *Roll-off characteristic* – An electronic filter receives frequency signals within a certain frequency band. It does not immediately cut off all frequencies outside the passband. The farther the frequencies are outside the passband, the weaker the frequency signals. This effect is called the filter roll-off. The frequency attenuation varies among filters.

4.2.5 Stroboscopes

Stroboscopes (see *Figure 22*) shine a short flash of intense light when triggered by a transducer. The transducer is attached to a piece of equipment. A target on the rotating part of the equipment being analyzed is lighted each time the vibration causes the transducer to trigger the stroboscope.

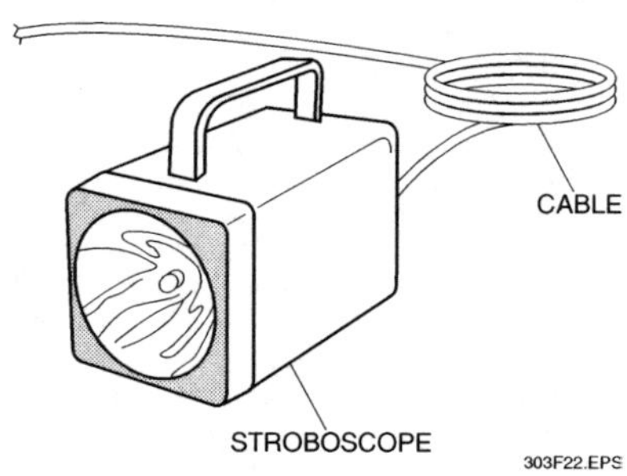

Figure 22 ◆ Stroboscope.

The stroboscope timing can be set to correspond with the rpm of the element being analyzed in order to locate the vibration, or it can be set to match the vibration to determine the frequency of the vibration. When it is set to match the turning speed of the element, a balanced element appears stable in the light. If the element appears to move, it is unbalanced and vibrating. For example, the light can be moved along a shaft; where the shaft appears to move, there is vibration. Where the shaft appears to be still, or flat, there is no vibration. This flat spot, called a node, is a good place to locate a bearing or some type of meter so that it is not affected by vibration. When the stroboscope is set to the frequency of the vibration but the part still appears to be moving, the vibration is coming from another source.

4.3.0 Vibration Recording Instruments

Recording instruments provide either a printed or a magnetic tape recording of vibration data for a permanent record of vibration. Recording instruments are also used to transfer vibration data from one analytical device to another. The major types of recording instruments used in vibration analysis are strip chart recorders and data collectors.

4.3.1 Strip Chart Recorders

Strip chart recorders print the machinery vibration signature on chart paper. The machinery vibration signature is the frequency and amplitude of machine vibration. A pen records the vibration motion on chart paper that moves between two rollers at a set rate of speed. The chart paper is divided at regular intervals and can be used to record the strength and frequency of the vibration. This provides a hard copy of the machine's signature, which can be reviewed and analyzed. *Figure 23* shows a strip chart recorder built into a vibration spectrum analyzer and an example of the information recorded on the strip.

4.3.2 Data Collectors

Data collectors (see *Figure 24*) gather complete spectral analyses, time-waveform analyses, and phase measurements for on-site analysis of machinery conditions. This information is stored in internal computer memory. The data collector can later be interfaced with vibration data-collection computer software to be studied, analyzed, and compared to baseline vibration. The data collector can record displacement, velocity, and acceleration values in its memory bank.

5.0.0 ◆ VIBRATION MONITORING

Vibration characteristics should be checked and logged when a machine is first installed and then on a regular schedule to monitor increased vibration. Checking vibration regularly and entering the vibration readings on a data sheet or in a data recorder are known as vibration monitoring. A vibration-monitoring program can greatly reduce

Materials

Ensure that you have everything required for the laboratory and testing during this session.

Classroom

Explain when vibration characteristics should be checked and logged.

Instructor's Notes:

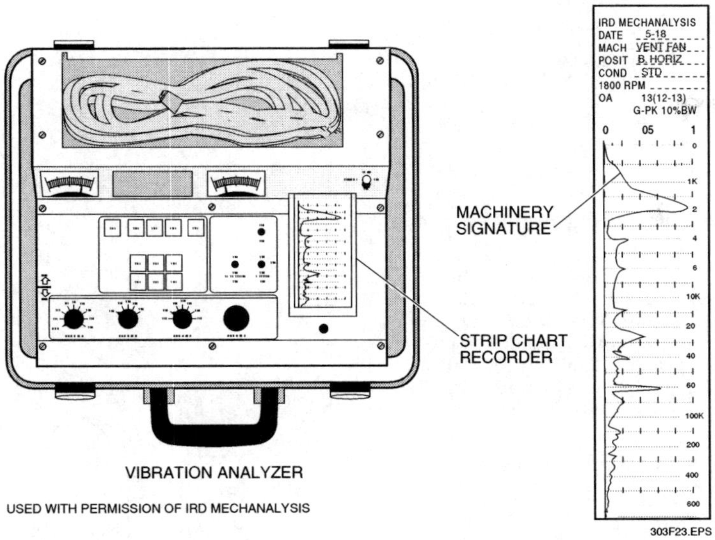

Figure 23 ◆ Strip chart recorder and strip example.

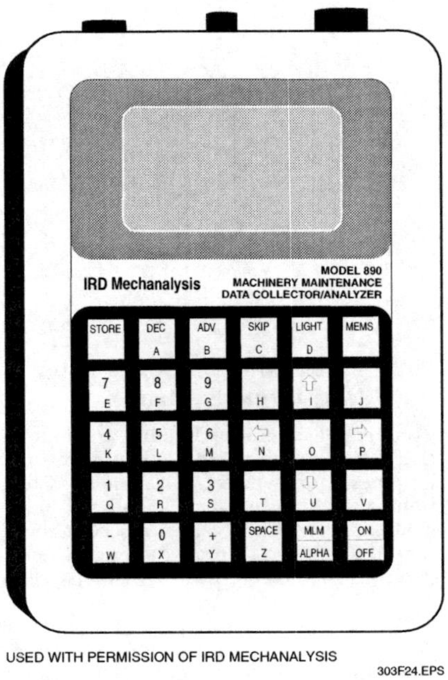

Figure 24 ◆ Data collector.

production downtime, equipment failure, and safety hazards. There are several purposes for implementing a vibration monitoring program:

- To identify the equipment to be monitored
- To establish a schedule for monitoring each piece of equipment
- To determine the locations of monitoring points on each piece of equipment
- To develop a continuous monitoring system for trending
- To develop and maintain vibration monitoring records for each piece of equipment being monitored

5.1.0 Identifying Equipment to Be Monitored

The first step in setting up a vibration monitoring program is to identify the equipment to be included in the program. The factors to consider include:

- Cost of replacement or repair
- Importance of the equipment to operations
- Possibility of safety hazards
- Lead time for replacement parts

Classroom

Review the considerations for implementing a vibration monitoring program.

Discuss the equipment selection criteria for a vibration monitoring program.

Classroom

Emphasize that the maximum interval for checking important machinery is three months.

Audiovisual

Show Transparency 13 (Figure 25). Review the critical points to check for vibration.

Show Transparency 14 (Figure 26). Discuss the elements of a typical continuous vibration monitoring system.

Classroom

Review the classes of continuous vibration monitors. Ask whether trainees have any questions.

5.1.1 Cost of Replacement or Repair

Equipment is often selected for monitoring based on replacement or repair costs. Monitoring vibration can detect problems early and prevent damage to the equipment, which reduces costs. Because of this, expensive equipment is often included in a vibration monitoring program.

5.1.2 Importance of Equipment to Operations

Failure of some large machines can shut down a complete line and even affect the whole facility. A carefully performed vibration monitoring program can often prevent a disruption by predicting an impending failure. Replacement of the part or parts causing the increased vibration can then be performed during routine shutdown.

5.1.3 Possibility of Safety Hazards

Bearing failure is a possible safety hazard. Bearing seizure in a large, high-speed machine can cause disintegration of the shaft and anything connected to it. Pieces of flying metal are a safety hazard to anyone in the area. Vibration monitoring can detect the early stages of bearing failure and considerably reduce the safety hazard.

5.1.4 Lead Time for Replacement Parts

It can take weeks or months to receive machine replacement parts after the order is placed. A good vibration monitoring program can help predict when a part will need to be replaced. Early placement of the order helps ensure that the part is on hand when replacement is needed.

5.2.0 Establishing a Schedule for Monitoring

Many factors must be considered in establishing a schedule for vibration monitoring, including the machine speed, number of bearing points, reliability history of similar equipment, and importance of the machine. In a vibration monitoring program, the maximum interval for checking important machinery should be three months.

5.3.0 Determining Locations of Monitoring Points

Vibration should be checked at all critical points on equipment. To make the observation more useful, the vibration of each critical point, such as bearings and shaft couplings, should be measured in three axes: horizontal, vertical, and axial (see *Figure 25*).

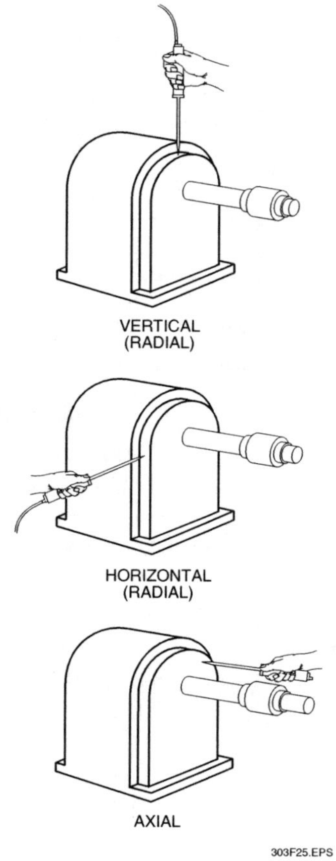

Figure 25 ◆ Vibration check in three axes.

5.4.0 Establishing a Continuous Monitoring System

Continuous monitoring requires an instrumentation system that monitors equipment operation at all times. Data collectors, computers with vibration-analysis software packages, and vibration analyzers are used to gather data from the monitors so that it can be more thoroughly analyzed. *Figure 26* shows a typical continuous vibration monitoring system.

The three classes of continuous vibration monitors are standard monitors for general-purpose applications, monitors with basic pre-engineered modifications, and fully engineered monitors. The first two types are used in most continuous

Instructor's Notes:

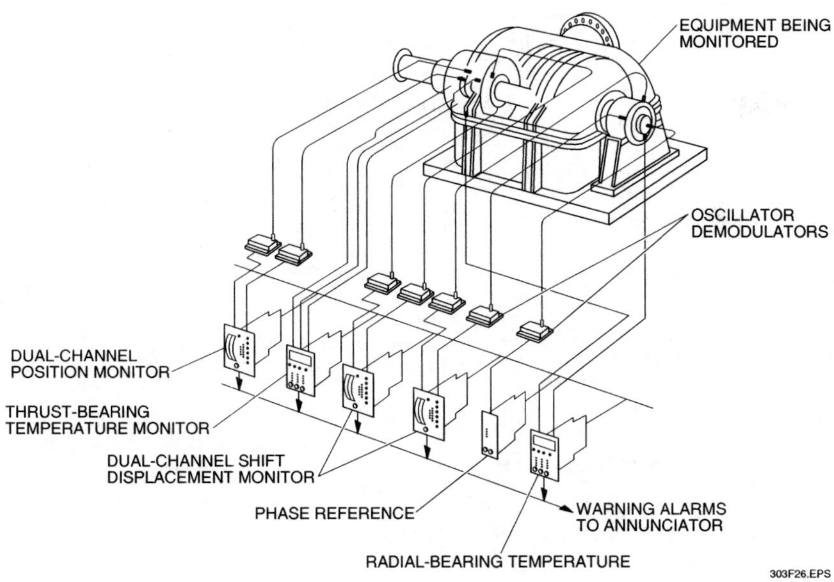

Figure 26 ◆ Continuous vibration monitoring system.

monitoring systems. Standard monitors are available as either single-channel monitors or multiple interchangeable-monitor modules. Single-channel monitors record vibrations from a single, fixed transducer. Multiple-monitor modules are attached to several fixed transducers and record several types of vibrations simultaneously. Each channel monitor is able to perform the following functions:

- Monitor vibration
- Monitor axial shaft position
- Monitor thrust-bearing wear
- Incorporate warning and shutdown alarms
- Incorporate time delays and test alarms

5.5.0 Developing and Maintaining Vibration Monitoring Records

When vibration is monitored manually, all vibration readings are entered on a data sheet or hand-held data recorder. The recorded data is then compared to the machine baseline vibration data. Baseline data should be established during initial startup when a machine is new or after a machine has been repaired. This data is used as a basis for analyzing the vibration signatures gathered periodically throughout the life of a machine. The more data that can be collected makes it easier to identify problem areas. Some of the parameters that need to be trended for proper analysis include temperature, vibration amplitude, frequencies, flow rates, products involved, and any other pertinent information. *Figure 27* shows a standard vibration monitoring form.

6.0.0 ◆ FIELD-BALANCING MACHINES

Unbalance in a rotating machine is usually caused by a concentration of weight in one spot. Two heavy spots of the same weight exactly opposite each other and equally distanced from the shaft cancel out unbalance. Balancing a machine involves:

- Determining the causes of unbalance
- Calculating the unbalance force
- Determining the required corrective action

Classroom

Explain the functions that a channel monitor can perform.

Explain how to record data and maintain a vibration monitoring record.

Laboratory

Distribute sample vibration monitoring forms. Review the parameters that are required for proper vibration analysis. Ask trainees to record data that you specify.

Teaching Tip

See the Teaching Tip for Section 5.0.0 at the end of this module.

Classroom

Review the considerations for balancing a machine.

VIBRATION MONITORING DATA SHEET

☐ NOISE ☒ VIBRATION

FOR: ____________________

____

____

EQUIPMENT USED:

____

____

PERFORMED BY: ____________________

LEGEND:
- → PICKUP POINT
- X PLAIN BEARING
- ☒ ANTIFRICTION BEARING
- ⊣⊢ COUPLING

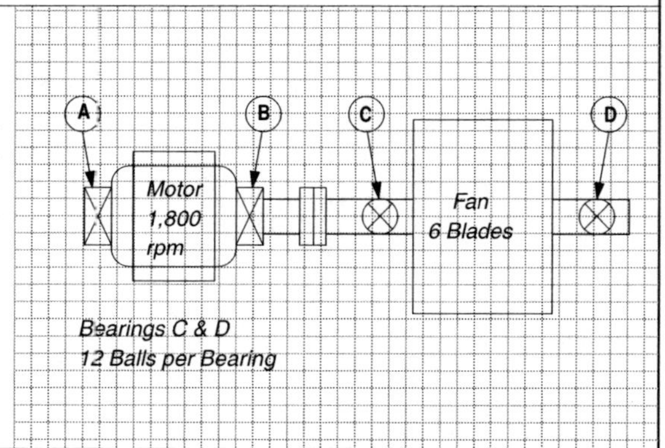

MEAS.		FILTER OUT VIBRATION						SPIKE ENERGY		FILTERED VIBRATION						
		DISP.		VEL.		ACCEL.										
POINTS	POS.	MIL.	CPM	IN/SEC	CPM	G	CPM	G/SE								
	H															
	V															
	A															
	H															
	V															
	A															
	H															
	V															
	A															
	H															
	V															
	A															
	H															
	V															
	A															
	H															
	V															
	A															

UCC DATA SHEET

303F27.EPS

Figure 27 ◆ Standard vibration monitoring form.

Instructor's Notes:

6.1.0 Determining Causes of Unbalance

Unbalance can be immediate or time related. Immediate causes of unbalance are related to manufacturing problems and result in unbalance when the machine or equipment is first started up. The time-related causes develop over a period of time as the equipment is used.

6.1.1 Immediate Causes of Unbalance

Flaws in casting, machining, and assembly show up as unbalance immediately upon startup. These flaws include casting blowholes, eccentricity, keys and keyways, and clearance tolerance buildup.

Casting blowholes, or sand traps, in castings are hidden flaws that are undetectable by visual inspection. The lack of metal caused by a hole in a casting results in weight loss in that area. This weight loss causes a heavy spot directly across the shaft from the blowhole. The result is significant unbalance. *Figure 28* shows a blowhole affecting balance in a cast **sheave**.

The displacement of a shaft from true center is called eccentricity. No matter how perfectly round and balanced a sheave, rotor, fan blade, or other component is, it rotates unbalanced if its shaft is not located exactly at the center of rotation. *Figure 29* shows an eccentric shaft.

Keys and keyways can also cause immediate unbalance. Motor manufacturers usually balance motors with a half key in the keyway. Sheave manufacturers usually check sheave balance with a half key in the keyway as well. When the motor and sheave are mated and a full key, which weighs twice as much as the half key does, is used, unbalance can result.

The buildup of clearance tolerances from assembling a machine can also result in unbalance. For example, the center hole in a sheave or other item has to have a larger inside diameter than the outside diameter of its mating shaft. This difference is required to permit a slip fit. When a setscrew is tightened to secure the item to the shaft, the center line of the item is displaced from the shaft center line, causing unbalance. *Figure 30* shows clearance tolerance unbalance.

6.1.2 Time-Related Causes of Unbalance

Rotating equipment, such as blowers, compressor rotors, pump impellers, and other equipment used in material handling, is subject to corrosion or abrasion over time. If the item does not wear evenly, unbalance results.

Another time-related cause of unbalance is the buildup of deposits on rotating equipment. Although deposits tend to form evenly, some unbalance still results. The problem becomes worse as pieces of the deposits begin to break off, increasing the unbalance and causing more pieces to break off. A serious unbalance usually results.

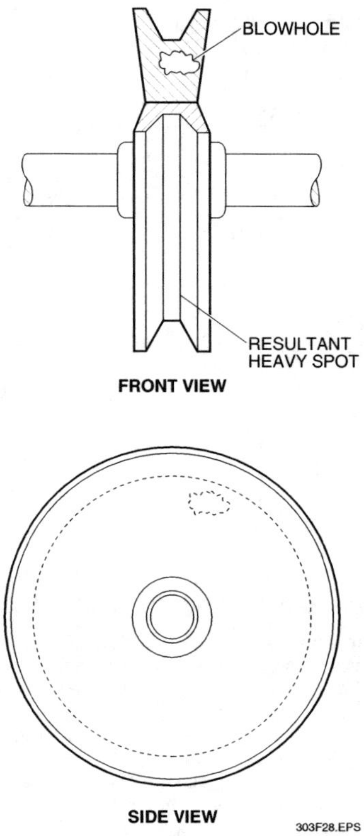

Figure 28 ◆ Blowhole affecting balance in cast sheave.

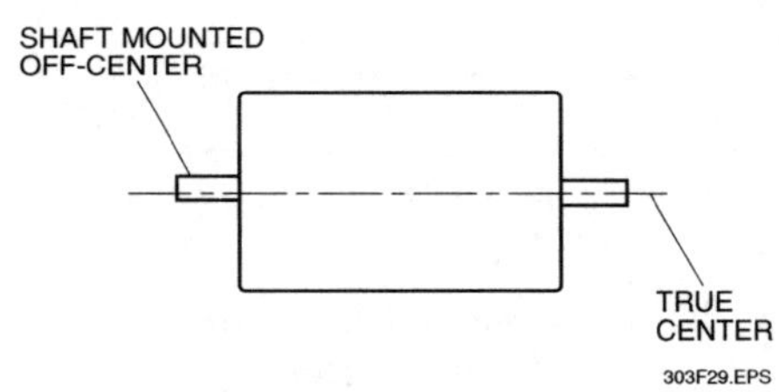

Figure 29 ◆ Eccentric shaft.

Classroom

Explain how to determine the immediate and time-related causes of unbalance.

Audiovisual

Show Transparency 15 (Figure 28). Explain how to detect casting blowholes.

Classroom

Refer to Figure 29. Discuss the concept of eccentricity.

Audiovisual

Show Transparency 16 (Figure 30). Review the causes of clearance tolerance unbalance. Answer any questions trainees have.

Classroom

Provide examples of time-related causes of unbalance.

Demonstration

On the whiteboard/chalkboard, demonstrate how to calculate the unbalance force.

Audiovisual

Show Transparency 17 (Figure 31). Review the basic components of a balancing machine.

Teaching Tip

See the Teaching Tip for Section 6.0.0 at the end of this module.

Classroom

Review the causes of resonance and discuss the solutions for vibration problems.

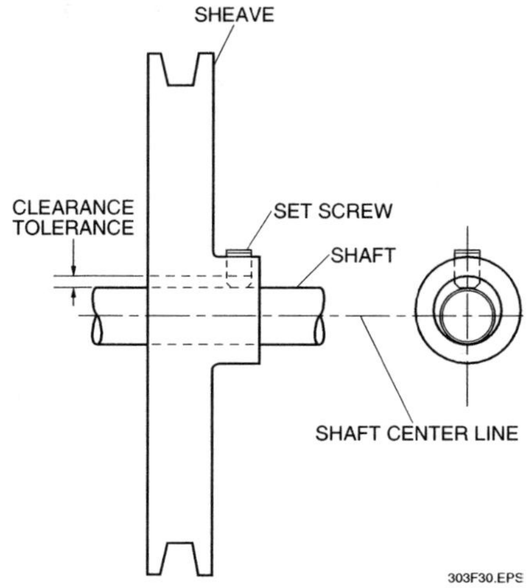

Figure 30 ◆ Clearance tolerance unbalance.

6.2.0 Calculating Unbalance Force

Unbalance in a rotating object generates a force that causes the object to vibrate. The amount of unbalance is the weight of the heavy spot multiplied by the distance of the heavy spot from the center of rotation. For example, a 2-ounce heavy spot located 3 inches from the center of rotation causes an unbalance of 6 ounce-inches (oz.-in.). The unbalance force depends on the amount of unbalance and the speed of rotation. The unbalance force can be calculated by using the following formula:

$$F = 1.77 \times (rpm/1{,}000)2 \times oz.\text{-}in.$$

F is equal to the unbalance force in pounds, rpm is the rotating speed of the part, oz.-in. is the amount of unbalance in ounce-inches, and 1.77 is the constant required to make the formula dimensionally correct.

6.3.0 Determining Required Corrective Action

Whenever possible, equipment balancing is performed without dismantling the machine. When the part requiring balancing must be removed, it is placed on a balancing machine capable of supporting it and driving it at its normal speed.

A balancing machine (see *Figure 31*) consists of three basic components: a stand with two bearing pedestals, two support assemblies with bearings or V-blocks to support the rotor, and a vibration instrument to determine the unbalance. It must also have the capability of rotating the rotor.

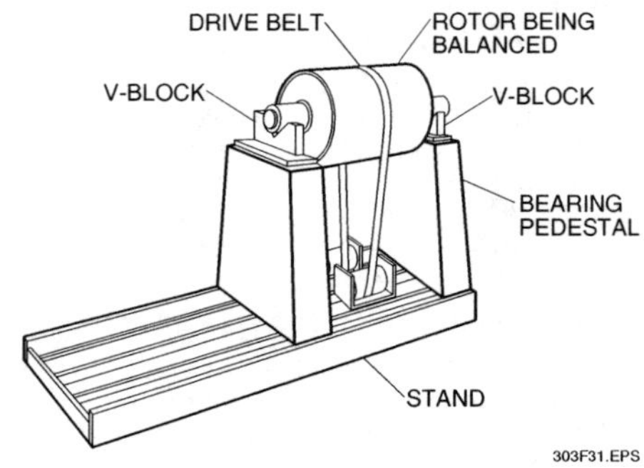

Figure 31 ◆ Typical balancing machine.

A strobe light is often used to detect the position of the heavy spot, and a vibration tester displays the vibration amplitude. From this data, you can calculate how much weight must be added to or removed from a given spot to correct the unbalance.

Other instruments are available that automatically compute the unbalance of a rotor and provide a numerical display for the technician to use in balancing the rotor. With these types of instruments, you do not need to perform any calculations because balance weights can be placed at the locations indicated by the test instrument.

7.0.0 ◆ CONTROLLING NATURAL RESONANCE

As was discussed earlier in this module, the resonance from the natural frequency of equipment can be destructive. This damaging resonance may be caused by:

- Unbalance
- Misalignment
- Machine looseness
- Bearing problems
- Defective gears
- Worn or defective belts

There are many solutions to a wide variety of vibration problem, including:

- Mass
- Stiffness
- Damping

Instructor's Notes:

7.1.0 Mass

Vibration can sometimes be controlled by adding mass (weight) to or removing mass (weight) from the rotating unit at the appropriate point on the unit. Adding mass to a rotating unit lowers the natural frequency.

Although it is far more complicated with a rotating machine part, it is the same principle used to balance a tire on an automobile. By adding a small amount of weight (mass) at exactly the right point on the tire rim, the vibration of the tire is controlled to the point of rolling smoothly.

7.2.0 Stiffness

Vibration can sometimes be controlled by adding stiffness to the machine or installation. Increasing stiffness raises the natural frequency.

One of the basic ways to add stiffness is to add supporting members. Adding extra material to frames and supports that tend to distort or bend under load often aids in controlling vibration. Fastening the feet of the machine to the floor or foundation has some of the same benefits as adding mass to the frames and supports.

Another technique is the removal of spacers or shims. Some spacers and shims may have been incorrectly placed under the machine when it was installed and aligned. Others might have been placed later because of warping of the machine frame.

Stiffness may also be increased by reducing the running clearances of bearings. The bearing clearance can be reduced by tightening some bearings (removing shims), or it may be necessary to install new bearings with less clearance.

7.3.0 Damping

Damping is the addition to or subtraction of material from the bulk of the vibrating component, which changes the natural frequency. However, such operations are highly technical and must be accomplished by vibration specialists. When properly accomplished, this technique can be very effective in preventing complete machine failure. In some cases, a tuned damper can be fitted to the component.

Summary

Vibration resulting from the natural forces of a moving mechanism is inherent to most machinery. Controlling vibration before it becomes destructive to the unit involved is a necessary part of mechanical work. Through extensive training, you will be able to understand the basic causes of vibration and ways to reduce and control the problem.

Machine balance, alignment, bearings, and a host of other component problems must be understood in order to diagnose vibration problems. The ability to comprehend the causes of vibration is gained through extensive training and experience in repairing vibration problems in mechanical equipment. A good understanding of the use of equipment documentation and historical data will greatly aid you in controlling and repairing vibration problems.

Training in the use of electronic diagnostic equipment is essential for troubleshooting vibration problems in machinery. As some vibration is caused by unbalance in rotating or moving parts, other types of vibration can be caused by unbalance in electrical or magnetic fields inside equipment. It is your responsibility to investigate and isolate the problem, identify the cause, and then repair it quickly and efficiently so that downtime is reduced and destructive problems are controlled. As you gain more training, knowledge, and skill in vibration control and repair, your troubleshooting skills will become stronger.

This module explained how to approach mechanical vibration problems, investigate and diagnose the problems, and make efficient repairs to the equipment.

Classroom

Explain how to add mass and stiffness to a machine or installation. Go over any questions trainees have.

Explain how damping changes the natural frequency of a vibration component.

On the whiteboard/ chalkboard, write a list of basic problems that cause vibration. Have trainees identify whether the problem can be solved by changing mass, stiffening, and/or damping.

Classroom

Go over the Review Questions for this module, and answer any questions trainees have.

Review Questions

1. Vibration occurs to some extent in _____ rotating machine(s).
   a. very few
   b. every
   c. most
   d. some

2. Early _____ of vibration can greatly extend the life of the machinery.
   a. balancing and tightening
   b. analysis and testing
   c. detection and control
   d. alignment and correction

3. _____ is a common cause of vibration.
   a. Laminar flow
   b. Oil whirl
   c. Alignment
   d. Unbalance

4. _____ can occur in loose-fitting, pressure-lubricated bearings in high-speed machinery.
   a. Oil whirl
   b. Gear meshing
   c. Turbulence
   d. Offset

5. Use a(n) _____ to check if belts are loose, and adjust them as necessary.
   a. stroboscope
   b. analyzer
   c. tensiometer
   d. transducer

6. Electrically caused vibration in a motor or generator is the result of unequal _____.
   a. current surges
   b. magnetic forces
   c. centrifugal forces
   d. balancing

7. When one of a machine's vibration frequencies coincides with a natural frequency, it creates a _____ frequency.
   a. digital
   b. low
   c. high
   d. resonant

8. The frequency of cavitation is usually _____.
   a. random
   b. predictable
   c. harmonic
   d. high

9. Each vibration has a set of distinctive characteristics known as the vibration _____.
   a. cycle
   b. velocity
   c. signature
   d. spectrum

10. All of the following are measurable characteristics of the vibration cycle *except* _____.
    a. frequency
    b. velocity
    c. volume
    d. acceleration

11. The movement of a vibrating machine is demonstrated by the creation of _____.
    a. turbulence
    b. sinusoidal waves
    c. vibration cycles
    d. cycle peaks

12. Data received from transducers is in the form of a(n) _____.
    a. time waveform
    b. FFT
    c. sinusoidal wave
    d. signature

13. If there are multiple phases of vibration, the machine is receiving a continual _____ from several directions.
    a. acceleration
    b. signal
    c. beating
    d. waveform

14. Once natural frequencies are activated, the amount of strain on the machine components can be multiplied by _____ times or more.
    a. 25
    b. 50
    c. 75
    d. 100

Instructor's Notes:

15. A(n) _____ is that rate of rotation that exactly corresponds to a harsh or violent vibration.
   a. natural frequency
   b. FFT
   c. critical speed
   d. cruising speed

16. All of the following methods are commonly used to control vibration *except* _____.
   a. adding stiffness
   b. reducing lubrication
   c. adding damping
   d. removing mass

17. _____ convert mechanical vibrations into electrical signals which are reproduced as sounds or graphic displays.
   a. Transducers
   b. Transformers
   c. Accelerometers
   d. Adaptors

18. Cathode-ray oscilloscopes are used to test _____.
   a. shaft movement
   b. selected frequencies
   c. vibration velocity
   d. electronic circuits

19. A _____ is capable of separating each vibration frequency for individual monitoring.
   a. cathode-ray oscilloscope
   b. vibration meter
   c. spectrum analyzer
   d. stroboscope

20. _____ electronic filters are the most basic type of filters.
   a. Fixed
   b. Passband
   c. Tunable
   d. Automatic

21. Electronic filters can be classified in each of the following ways *except* _____.
   a. passband frequency
   b. active or passive
   c. tuning method
   d. heavy or light

22. You should consider all of the following factors *except* _____ when identifying which equipment should be included in a vibration monitoring program.
   a. size of equipment
   b. repair or replacement cost
   c. lead time for parts
   d. importance of equipment to operations

23. _____ data should be established during initial startup.
   a. Frequency
   b. Repair
   c. Baseline
   d. Location

24. Unbalance can be immediate or _____.
   a. delayed
   b. time-related
   c. nonexistent
   d. sporadic

25. The displacement of a shaft from true center is called _____.
   a. unbalance
   b. off-centered
   c. angularity
   d. eccentricity

Examination

Administer the Module Examination. Be sure to record the results of the Exam on Craft Training Report Form 200 and submit the results to the Training Program Sponsor.

Performance Testing

Ensure that the Performance Test has been completed and Performance Profile Sheets for each trainee are filled out. Be sure to record the results of the testing on Craft Training Report Form 200 and submit the results to the Training Program Sponsor.

Instructor's Notes:

GLOSSARY

Trade Terms Introduced in This Module

Cycle: A segment of the continuing periodic change in displacement with respect to a fixed location, such as the periodic, repeated shaking of a piece of moving machinery.

Frequency: The number of cycles occurring in a giving period of time.

G force: A unit of measurement for bodies going through the stress of acceleration.

Pickup coil: A device that converts measurable quantities into corresponding electric signals.

Sheave: A grooved wheel or pulley.

Transducer: A device that converts an input signal into an output signal of a different form. Examples include microphones, loudspeakers and ultrasonic test probes. An ultrasonic transducer converts alternating current energy above 20 kilohertz into mechanical vibrations of the same frequency and uses the vibrations to measure the thickness of pipeline walls.

Velocity: Speed, expressed in distance over time, such as ft/sec or ft/min.

Vibration signature: The specific characteristics of a vibration measured in frequency, acceleration, displacement, and velocity.

MODULE 63303-02 — TEACHING TIPS

The following are suggested activities or instructional methods to help you teach the material in this AIG.

General

When you call on someone to answer a question, the rest of the class relaxes or even tunes out because they expect that the question and answer will take place only between you and the trainee you called on. Instead, use this technique to involve more trainees in answering questions and to keep them on their toes.

1. Ask trainees to define a term or explain a concept.
2. After one trainee has answered, ask a trainee seated nearby if the answer is right. Then ask whether a trainee in the back of the room agrees.
3. Ask trainees to explain why they think an answer is right or wrong.
4. Use the session to clear up incorrect ideas and encourage trainees to learn from their mistakes.

Section 4.0.0 *Vibration Test Equipment*

Trainees have reviewed the types of equipment used to monitor and analyze machine vibrations. This exercise will familiarize trainees with this equipment and its uses. For this exercise, you will need a variety of transducers, vibration test equipment, and vibration recording instruments. Trainees will need their Trainee Guides and pencils and paper to take notes for discussion. Allow 30 minutes for this exercise.

1. Set up several stations with a variety of equipment.
2. Ask trainees to visit each station and identify the type of equipment and how and why it is used. Discuss safety considerations for using the equipment.
3. Review and discuss trainees' answers. Go over any questions trainees have about vibration test equipment.

Section 5.0.0 *Vibration Monitoring*

This exercise will familiarize trainees with vibration monitoring programs. For this exercise, trainees will need their Trainee Guides and pencils and paper to take notes for discussion. Allow 30 minutes for this exercise.

1. On the whiteboard/chalkboard, write the criteria for equipment to be included in several hypothetical vibration monitoring programs.
2. Divide the class into small groups according to the number of scenarios you provide. Have each group review and discuss the provided information.
3. Ask each group to outline the elements of a vibration monitoring program and to consider the following:
   - Equipment to be monitored
   - A schedule for monitoring
   - Location of monitoring points
   - A continuous monitoring system
4. Review the programs developed by each group. Make recommendations as warranted. Answer any questions trainees have about vibration monitoring.

Section 6.0.0 *Field-Balancing Machines*

In this exercise, trainees will practice identifying causes of unbalance. This exercise will require an overhead projector, sheets of acetate, and transparency markers. Trainees will need their Trainee Guides, pencils, and paper to take notes for discussion. Allow 20 minutes for this exercise.

1. On a sheet of acetate, write a list of problems that cause unbalance. Potential problems could include the following:
   - Casting blowholes
   - Eccentricity
   - Keys and keyways
   - Buildup of clearance tolerances
   - Corrosion and abrasion
   - Buildup of deposits
2. On a sheet of paper, have trainees identify whether the cause is immediate or time related, explain the problem, and provide potential solutions.
3. Review and discuss trainees' answers. Go over any questions trainees have about field-balancing machines.

MODULE 63303-02 — ANSWERS TO REVIEW QUESTIONS

Answer	Section
1. b	1.0.0
2. c	2.0.0
3. d	2.1.0
4. a	2.5.0
5. c	2.7.0
6. b	2.9.0
7. d	2.14.1
8. a	2.12.0
9. c	3.0.0
10. c	3.0.0
11. b	3.5.0
12. a	3.5.0
13. c	3.7.0
14. d	2.14.0
15. c	2.14.2
16. b	7.0.0
17. a	4.1.0
18. d	4.2.2
19. c	4.2.3
20. b	4.2.4
21. d	4.2.4
22. a	5.1.0
23. c	5.5.0
24. b	6.1.0
25. d	6.1.1

CONTREN™ LEARNING SERIES — USER UPDATES

The NCCER makes every effort to keep these textbooks up-to-date and free of technical errors. We appreciate your help in this process. If you have an idea for improving this textbook, or if you find an error, a typographical mistake, or an inaccuracy in NCCER's Contren™ textbooks, please write us, using this form or a photocopy. Be sure to include the exact module number, page number, a detailed description, and the correction, if applicable. Your input will be brought to the attention of the Technical Review Committee. Thank you for your assistance.

Instructors – If you found that additional materials were necessary in order to teach this module effectively, please let us know so that we may include them in the Equipment/Materials list in the Instructor's Guide.

Write: Curriculum Revision and Development Department
National Center for Construction Education and Research
P.O. Box 141104, Gainesville, FL 32614-1104

Fax: 352-334-0932

E-mail: curriculum@nccer.org

Craft ____________________ Module Name ____________________

Copyright Date __________ Module Number __________ Page Number(s) __________

Description

____

____

____

____

(Optional) Correction

____

____

____

(Optional) Your Name and Address

____

____

____

Module 63304-02

Maintain, Troubleshoot, and Repair Pumps

MODULE OVERVIEW

This module introduces common pump troubleshooting and maintenance procedures. Trainees will learn how to prepare a pump for shutdown, as well as how to remove, disassemble, reassemble, and install a pump.

PREREQUISITES

Please refer to the Course Map in the Trainee Module. Prior to training with this module, it is recommended that the trainee shall have successfully completed the following:

Core Curriculum; Pipeline Core, Modules 66101-02 and 66102-02; Pipeline Mechanical Levels One and Two; Pipeline Mechanical Level Three, Modules 63301-02 through 63303-02

OBJECTIVES

Upon completion of this module, the trainee will be able to:

1. Describe the preventive maintenance requirements for a pump.
2. Describe the inspection requirements for a pump.
3. Identify common troubleshooting techniques and problems for a pump.
4. Identify the common steps required to prepare a pump for shutdown for maintenance or repair.
5. Identify the common steps required to remove a pump from a pipeline system for maintenance or repair.
6. Identify the common steps to disassemble and reassemble a pump.
7. Identify the common steps required to install the pump after the pump has been reassembled.
8. Identify the common steps to prepare the pump for startup and operational check after maintenance or repair has been completed.

PERFORMANCE TASKS

Under the supervision of the instructor, the trainee should be able to:

1. Describe the preventive maintenance requirements for a pump.
2. Describe the inspection requirements for a pump.
3. Demonstrate common troubleshooting techniques and identify common problems for a pump.
4. Demonstrate the common steps to perform each of the following:
   - Prepare a pump for shutdown for maintenance or repair.
   - Remove a pump from a pipeline system for maintenance or repair.
   - Disassemble and reassemble a pump.
   - Install a pump after it has been reassembled.
   - Prepare a pump for startup and operation check after maintenance or repair has been completed.

NCCER STANDARDIZED CRAFT TRAINING PROGRAM

The National Center for Construction Education and Research (NCCER) provides a standardized national program of accredited craft training. Key features of the program include instructor certification, competency-based training, and performance testing. The program provides trainees, instructors, and companies with a standard form of recognition through a National Craft Training Registry. The program is described in full in the *Guidelines for Accreditation*, published by the NCCER. For more information on standardized craft training, contact the NCCER by writing us at P.O. Box 141104, Gainesville, FL 32614-1104; calling 352-334-0911; or e-mailing info@nccer.org. More information may be found at our Web site, www.nccer.org.

HOW TO USE THIS ANNOTATED INSTRUCTOR'S GUIDE

Each page presents two sections of information. The larger section displays each page exactly as it appears in the Trainee Module. The narrow column ties suggested trainee and instructor actions to each page and provides icons to call your attention to material, safety, audiovisual, or testing requirements. The bottom of each page includes space for your notes.

Teaching Tip

If you see the Teaching Tip icon, that means there is a teaching tip associated with this section. Also refer to any suggested teaching tips at the end of the module.

SAFETY CONSIDERATIONS

Ensure that the trainees are equipped with appropriate personal protective equipment. Emphasize the importance of following all safety precautions and procedures when working with power tools.

PREPARATION

Before teaching this module, you should review the Module Outline, Objectives, Performance Tasks, and the Materials and Equipment List. Be sure to allow ample time to prepare your own training or lesson plan and gather all required equipment and materials.

MATERIALS AND EQUIPMENT LIST

Transparencies

Markers/chalk

Blank acetate sheets

Transparency pens

Pencils and scratch paper

Module Examinations*

Performance Profile Sheets*

Overhead projector and screen

Whiteboard/chalkboard

Appropriate personal protective equipment

Copies of *49 CFR Part 192* (Gas) and/or *49 CFR Part 195* (Liquid)

Copies of your company policy and procedures manual

Completed inspection reports

Sample manufacturers' manuals, including procedures, schematics, and line drawings

Copies of *Figure 13* with the callouts covered

* Located in the Test Booklet.

TEACHING TIME FOR THIS MODULE

An outline for use in developing your lesson plan is presented below. Note that each Roman numeral in the outline equates to one session of instruction. Each session has a suggested time period of 2½ hours. This includes 10 minutes at the beginning of each session for administrative tasks and one 10-minute break during the session. Approximately 10 hours are suggested to cover *Maintain, Troubleshoot, and Repair Pumps*. You will need to adjust the time required for hands-on activity and testing based on your class size and resources.

Topic	Planned Time
Session I. Maintaining and Troubleshooting Pumps for Repair	
A. Performing Preventive Maintenance on Pumps	________
B. Troubleshooting Pumps	________
C. Performance Testing	________
Session II. Preparing, Removing, and Disassembling the Pump	
A. Preparing for Shutdown and Repair	________
B. Removing the Pump from the System	________
C. Disassembling the Pump	________
Session III. Reassembling the Pump	
A. Replacing Lock Washers and Locking Devices	________
B. Replacing O-Rings and Gaskets	________
C. Cleaning Bearing Surfaces	________
D. Assembling the Impeller	________
E. Pre-Startup Final Inspections	________
Session IV. Installing and Testing the Pump	
A. Installing the Pump	________
B. Pump Startup Procedure and Operational Tests	________
C. Summary	
1. Summarize module	________
2. Answer questions	________
D. Module Examination	________
1. Trainees must score 70% or higher to receive recognition from the NCCER.	
2. Record the testing results on Craft Training Report Form 200 and submit the results to the Training Program Sponsor.	
E. Performance Testing	________
1. Trainees must perform each task to the satisfaction of the instructor to receive recognition from the NCCER.	
2. Record the testing results on Craft Training Report Form 200 and submit the results to the Training Program Sponsor.	

Module 63304-02

Maintain, Troubleshoot, and Repair Pumps

Instructor's Notes:

Instructor's Notes:

Introduction to the Pipeline Operator Qualification Rule, API, and NCCER

It could be said that petroleum products keep America running. Petroleum products like oil and natural gas are the principal sources of energy in the United States and are used in many manufacturing and chemical processes. Petroleum products are primarily transported by pipeline. There are more than a million miles of liquid and natural gas pipeline in this country, ranging in diameter from two inches to four feet. These pipelines bring crude oil and natural gas from production fields to processing plants and refineries—and ultimately to the consumer.

Pipeline Operator Qualification

Petroleum products are considered hazardous, so pipeline personnel must take extra precautions and follow specific procedures when working with them. To help ensure the safe handling and transportation of petroleum products, the United States Department of Transportation (DOT) mandates safe practices as federal regulations.

One such regulation, the Operator Qualification (OQ) rule, requires pipeline operators to develop a written qualification plan to evaluate their personnel and their contract personnel (see *49 CFR Part 192 Subpart N—Gas,* and *49 CFR Part 195 Subpart G—Liquid*). The evaluation hinges on the operator's ability to perform certain covered tasks and to be able to recognize and respond to abnormal operating conditions that may be encountered while performing these covered tasks.

Covered Tasks

A pipeline performance task is deemed *covered* if the activity passes the following four-part test:

1. Is it performed on a pipeline facility?
2. Is it an operations or maintenance task?
3. Is it performed as a requirement of *49 CFR Part 195 (Liquid)* or *49 CFR Part 192 (Gas)*?
4. Does it affect the operation or integrity of the pipeline facility?

If the answer to all four questions is *yes,* then the activity is considered *covered* and must figure into the evaluation process of pipeline personnel.

American Petroleum Institute

The OQ rule and its related covered tasks provided the catalyst to develop a unified pipeline training and assessment program—so that pipeline operators and owners have ready access to the assessment and training tools they need. This effort was the direct result of the vision and leadership provided by The American Petroleum Institute (API) and API members.

Efforts to unify the industry are nothing new to API. Indeed, the impetus for forming the Institute in 1919 was the need to standardize engineering specifications for drilling and production equipment. Today, the most pressing issues revolve around public perceptions and government policies toward the industry. Speaking with one voice on these issues has become as essential as having interchangeable parts in the field. API's members determine the petroleum industry's positions on public policy as well as the standards that govern its day-to-day operations worldwide.

Consortium on Operator Qualification

API assembled a focus group whose primary purpose was to identify covered tasks and provide compliance recommendations to pipeline operators.

The Consortium on Operator Qualification (COOQ) became the guiding force for the identification and development of covered tasks for the liquid pipeline industry. Their work product, API publication 1161, *Guidance Document for the Qualification of Liquid Pipeline Personnel,* formed the basis for the resulting Pipeliner Training and Assessment Program.

The covered task list created in API publication 1161 has now evolved to include gas covered tasks and is maintained by NCCER. The most current covered task list can be downloaded at www.nccer.org.

National Center for Construction Education and Research

With the guidance and support of API, its members, and the Consortium on Operator Qualification, the National Center for Construction Education and Research (NCCER) was able to design and implement the Pipeliner Training and Assessment Program.

NCCER is a not-for-profit 501(c)(3) education foundation established in 1995 by the world's largest and most progressive construction and maintenance contractors and national contractor associations. NCCER was founded to address the

severe workforce shortage facing the construction and maintenance industries and to develop a standardized training process and curricula. Today, NCCER is supported by hundreds of leading construction and maintenance companies, manufacturers, and national associations and developed more than 35 construction and maintenance titles.

NCCER also maintains a National Registry, which contains the records of trainees and instructors who have received NCCER training. All completion information for every trainee is sent to the NCCER and filed within the Registry. The Registry can then confirm training and skills for workers as they move from company to company, state to state, or even within their own company.

Standardized Curriculum (NCCER's Contren™ Learning Series)

NCCER curricula are developed using the expertise of industry subject matter experts (SMEs). The use of SMEs as an integral part of every new development or revision ensures that the curriculum is current, industry-driven, and indicative of industry practices. The development and/or revision process is facilitated from start to finish by NCCER project managers, who work closely with the SMEs and a select team of technical writer(s) on the development and presentation of the curriculum.

Pipeline: Five Training Paths

The pipeline curriculum focuses on essential skills and competencies within five skill areas:

- Corrosion Control
- Electrical and Instrumentation
- Maintenance
- Mechanical
- Operations (Field and Control Center)

SMEs from the pipeline industry have determined the skills and competencies needed based on API's covered tasks and the consensus of the group/industry. Specific company information is generally not incorporated into the curriculum unless the SMEs determine that the information is the industry standard.

Training Modules and Covered Tasks

The smallest unit of instruction in the curriculum is a *module*. Each curriculum is composed of modules (the equivalent of stand-alone chapters) that are logically sequenced into complete *levels* of instruction. Every module is task-oriented and requires the successful completion of a written (knowledge-based) examination and a performance (hands-on) test. Modules may be individually selected and/or combined for efficient task training. Modules in the pipeline curricula, to the extent possible, will align with API-developed covered tasks. While not all modules will relate to covered tasks, all covered tasks will relate to at least one module.

Benefits of Training to the Pipeline Marketplace

Applying NCCER's standardized training process allows pipeline owners and contractors in the United States to establish the benchmark for acceptable skill and competency levels within the pipeline industry. This process will assist training organizations in developing and/or improving standardized pipeline training programs that effectively match their resources and operating conditions while responding to regulatory demands and to the needs of the industry. The process will also recognize training organizations that meet the minimum standards of competence, warranting regulatory, public, and professional confidence. This in turn will foster national unity within the pipeline industry for the mutual benefit of individual technicians, owners, contractors, and ultimately pipeline-users and society.

Additional Resources

To review the Pipeliner Training and Assessment Program Implementation and Resource Guide and the latest version of pipeline covered tasks, visit NCCER's Web site at www.nccer.org, or contact:

National Center for Construction Education and Research
PO Box 141104
Gainesville, FL 32614-1104
(888) 622-3720
Web site: www.nccer.org

For more information on API, please contact:

American Petroleum Institute
1220 L Street, Northwest
Washington, D.C. 20005
(202) 682-8000
Web site: www.api.org

To order *API Pub 1161*, contact Global Engineering Documents by calling 1-800-854-7149 or visiting their Web site at www.global.his.com.

Additional sources of information on the pipeline operator qualification regulation include:

DOT – Office of Pipeline Safety
www.ops.dot.gov

Transportation Safety Institute
www.tsi.dot.gov

Instructor's Notes:

COURSE MAP

This course map shows all of the modules in the third level of the Pipeline Mechanical curriculum. The suggested training order begins at the bottom and proceeds up. Skill levels increase as you advance on the course map. The local Training Program Sponsor may adjust the training order.

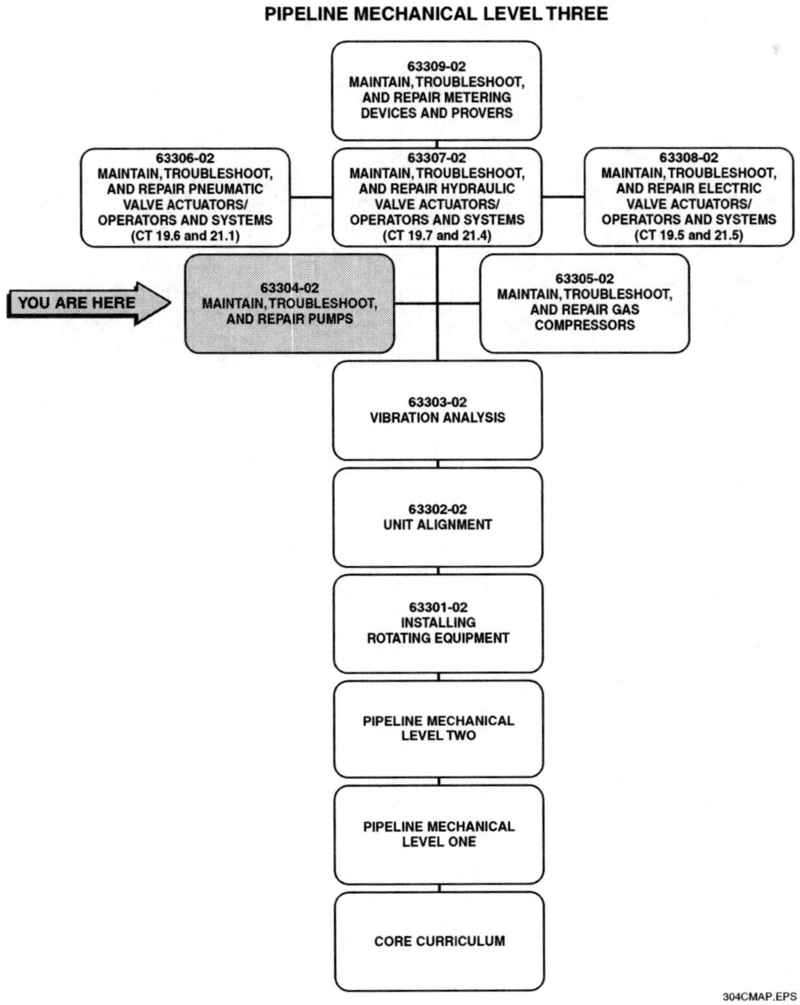

Homework

Assign reading of Module 63304-02.

Instructor's Notes:

MODULE 63304-02 CONTENTS

Figures

Tables

Instructor's Notes:

MODULE 63304-02

Maintain, Troubleshoot, and Repair Pumps

Objectives

When you have completed this module, you will be able to do the following:

1. Describe the preventive maintenance requirements for a pump.
2. Describe the inspection requirements for a pump.
3. Identify common troubleshooting techniques and problems for a pump.
4. Identify the common steps required to prepare a pump for shutdown for maintenance or repair.
5. Identify the common steps required to remove a pump from a pipeline system for maintenance or repair.
6. Identify the common steps to disassemble and reassemble a pump.
7. Identify the common steps required to install the pump after the pump has been reassembled.
8. Identify the common steps to prepare the pump for startup and operational check after maintenance or repair has been completed.

Prerequisites

Before you begin this module, it is recommended that you successfully complete the following: Core Curriculum; Pipeline Core, Modules 66101-02 and 66102-02; Pipeline Mechanical Level One; Pipeline Mechanical Level Two; Pipeline Mechanical Level Three, Modules 63301-02 through 63303-02.

Required Trainee Materials

1. Pencil and paper
2. Appropriate personal protective equipment

1.0.0 ◆ INTRODUCTION

Pumps are used to move many types of materials, from thin liquids to thick semisolids. There are many different types and designs of pumps. The type of pump used in a system depends on the system design, the types of materials to be moved, and the construction of the pump. A pump designed for one type of system and material may not work in a different system; therefore, a pump should be used only in the system for which it was designed. This module will introduce you to common pump troubleshooting and maintenance procedures. Always remember to follow the pump manufacturer's maintenance procedures when maintaining pumps and to make sure that the pump is not under any type of manufacturer's warranty before disassembling the pump. *Figure 1* shows a typical pipeline pumping installation.

2.0.0 ◆ PERFORMING PREVENTIVE MAINTENANCE ON PUMPS

There is a wide variety of pump types, sizes, designs, and materials used in pipeline systems. The manufacturer's instruction books should always be referred to for the recommended preventive maintenance (PM) for that pump. Your company will likely have specific preventive maintenance procedures for you to use. Always follow your company procedures when performing the

MAINTAIN, TROUBLESHOOT, AND REPAIR PUMPS — TRAINEE MODULE 63304-02 4.1

Materials

Ensure that you have everything required to teach the course. Check the Materials and Equipment List at the front of this module.

Classroom

Explain that terms shown in bold are defined in the Glossary at the back of this module.

Audiovisual

Show Transparency 1, Course Objectives.

Show Transparency 2, Performance Tasks.

Teaching Tip

See the general Teaching Tip at the end of this module.

Classroom

Review the factors that determine the type of pump used in a system.

Emphasize the importance of following manufacturers' instruction books for recommended preventive maintenance.

Review the pump parameters to monitor as part of preventive maintenance.

Discuss the general preventive maintenance procedures for common pumps used in a pipeline system.

Explain how to monitor a pump on a daily basis. Provide examples of abnormal operating conditions that trainees may encounter.

304F01.EPS

Figure 1 ◆ Typical pipeline pumping installation.

preventive maintenance procedures, and comply with your company's documentation requirements for scheduled inspections.

The key to an effective preventive maintenance program for pumps is to monitor and document the pump performance on a regular schedule.

There are numerous parameters that should be monitored to understand how a pump is performing. These include but are not limited to **suction** pressure (P_s), **discharge** pressure (P_d), flow (Q), pump speed (N_r), pumpage properties, and power. These indicators should be recorded per company policy and monitored regularly for any change that may indicate there is a problem with the pump efficiency. If a change is noticed in the performance of a pump, it should be inspected more closely to determine if there is a problem with the pump. Repair of the pump should be scheduled before it becomes inoperative.

Other keys to an effective pump preventive maintenance program are to monitor and record bearing temperatures and to check the motor and pump alignment periodically.

Following are the general preventive maintenance procedures for common types of pumps used in a pipeline system. Always follow the pump manufacturer's operation and maintenance manual along with company procedures when performing these inspections.

2.1.0 Daily Observations

The following steps may be taken to monitor a pump operation on a daily basis:

Step 1 Check and record at least daily the readings from the pressure gauges and flowmeters installed in the pipeline system.

> **NOTE**
> If there is a significant change in the operating parameters of the pipeline system from the normal limits, inspect the pump and the piping system within a reasonable time.

Step 2 Monitor the normal operational noise of the pump, and report any abnormal changes.

Step 3 Perform a walk-around inspection of the pump, and check for leaks and any abnormal conditions when the pump is accessible.

> **NOTE**
> Always report any abnormal conditions that are observed per company policy.

Instructor's Notes:

2.2.0 Semiannual PM Activities

The following steps may be taken to perform semiannual preventive maintenance activities:

Step 1 Review the pump daily operation readings since the last pump inspection.

Step 2 Compare the pump and system instrument readings on the pump with the operation reading transmitted to the control center to ensure that the instruments are transmitting the correct information.

Step 3 Inspect the stuffing box packing for excessive wear, damage, and sealing effect. Replace as required.

Step 4 Inspect the stuffing box glands for free movement.

Step 5 Clean the stuffing box gland bolts with an approved solvent, and lubricate them.

Step 6 Check the pump and pump motor for angular and parallel alignment, and adjust them as required. *Figure 2* shows angular and parallel alignment.

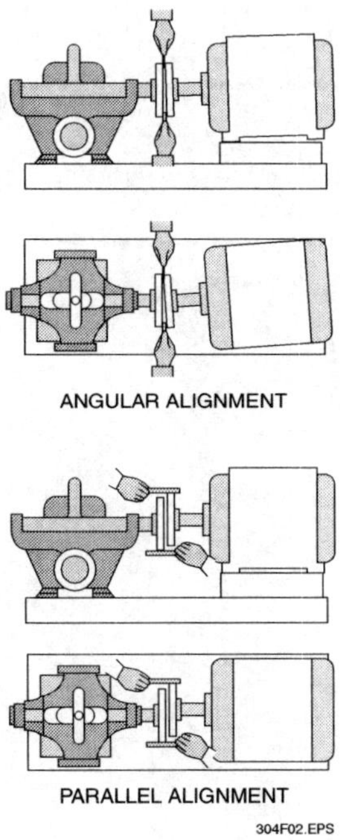

Figure 2 ◆ Angular and parallel alignment.

Step 7 Inspect and lubricate the pump bearings as required.

Step 8 Take a temperature reading of the pump bearings, and compare the temperature to the manufacturer's specifications.

Step 9 Clean and inspect the strainer if equipped.

Step 10 Inspect the pump and motor baseplate for damage, cracked welds, clear drain connection, levelness, loose grout, and corrosion.

Step 11 Take a pump oil sample for an oil analysis test if applicable.

Step 12 Record and report any deficiencies observed during the inspection.

> **NOTE**
> Immediate action should be taken on those items that are critical to the safe operation of the pipeline system.

Step 13 Complete all company-required documentation.

2.3.0 Annual PM Activities

The following steps may be taken to perform annual preventive maintenance activities on a pump:

Step 1 Review the previously completed inspection reports. Look for trends in the efficiency of the pump operation.

Step 2 Complete all the steps as listed in the semiannual PM activities.

Step 3 Recalibrate all the instruments and flowmetering devices.

Step 4 Remove the antifriction bearings, and inspect them for scratches and wear. Replace the bearings and sleeve as required.

Step 5 Clean and lubricate the bearings.

Step 6 Inspect the vertical shaft movement of a pump, with the sleeve bearings at both ends and the packing removed, for excessive movement and wear. Replace the bearings and sleeve as required.

Step 7 Measure the end play of the bearings, and compare them to the manufacturer's specifications. Replace as required.

Classroom

Review the tasks to perform semiannual preventive maintenance activities. Answer any questions trainees have.

Audiovisual

Show Transparency 3 (Figure 2). Explain how to check for and adjust angular and parallel alignment.

Classroom

Distribute completed inspection reports. Ask trainees to review the reports and identify trends in the pump operation efficiency. Ensure that trainees understand how to complete all required documentation.

Discuss annual preventive maintenance activities. Provide examples from your own experience maintaining pumps.

See the Teaching Tip for Section 2.0.0 at the end of this module.

Classroom

Review basic trouble-shooting techniques used by mechanical technicians.

Safety

Explain why belt-driven units must be locked out and tagged before inspecting belts and pulleys.

Step 8 Inspect and flush all auxiliary piping, such as drains, sealing water piping, and cooling water piping.

Step 9 Inspect and flush auxiliary coolers if installed.

Step 10 Record and report any deficiencies observed during the inspection.

Step 11 Complete all company-required documentation.

3.0.0 ◆ TROUBLESHOOTING PUMPS

Troubleshooting is used by mechanical technicians to determine what is wrong with a piece of equipment, such as a pump, and the most efficient way to repair the problem. Maintenance technicians must have a strong understanding of the system to which they are assigned and must know how the pumps are integrated into the system to perform troubleshooting responsibilities. The operating principles of the pumps must be completely understood by the technicians to be able to troubleshoot problems quickly.

If the cause of the problem is not immediately understood, the process of elimination is often the procedure of choice. After the problem is identified, the technician makes a mental or written list of the possible causes. The technician then begins with the simplest and easiest possibility and advances in order until the problem is identified. Possible causes that would require disassembly and extensive downtime should not be considered until all other possible causes have been eliminated.

3.1.0 General Troubleshooting Procedure

In order to troubleshoot a pump, a mechanical technician might follow these very general steps:

Step 1 Inspect the pump installation for any leaks from the casing, lines, or any other connection.

Step 2 Examine all the joints and connections for the possibility of air leaking into the pump.

Step 3 Listen for any unusual noise that might reveal a problem.

Step 4 Use temperature-sensing equipment to detect signs of excessive temperature in the pump or driver.

Step 5 Verify that all the mounting bolts are tight.

Step 6 Observe the installation closely for excessive vibration.

> **NOTE**
> Vibration may indicate a problem with pump bearings, misalignment, or restricted suction.

Step 7 Inspect for excessive buildup of dirt and debris that might restrict cooling.

Step 8 Load test the driver to determine if the power being used exceeds the maximum on the nameplate. *Figure 3* shows a clamp-on meter used for load testing.

> **NOTE**
> Load readings that exceed the load stamped on the nameplate of the pump indicate a bearing problem, alignment problem, overloaded pump, or worn parts.

Step 9 Inspect the belts and pulleys on the belt-driven pumps for their condition, alignment, and proper tension adjustment.

> **CAUTION**
> Belt-driven units must be locked out and tagged before attempting to inspect the condition of belts and pulleys.

304F03.EPS

Figure 3 ◆ Clamp-on ammeter.

Instructor's Notes:

3.2.0 Troubleshooting Charts

The need for mechanical technicians to have a strong understanding of pumps and their integration into the system cannot be overstressed. However, troubleshooting charts, especially those provided by pump manufacturers, often save time and frustration for many technicians. When the problem has been diagnosed, a troubleshooting chart can quickly suggest a possible cause and solution. The following charts are for rotary, reciprocating, and centrifugal pumps, respectively. *Table 1* shows a troubleshooting chart for a rotary pump. *Table 2* shows a troubleshooting chart for a reciprocating pump. *Table 3* shows a troubleshooting chart for a centrifugal pump.

Table 1 Troubleshooting Chart for a Rotary Pump

Pump Symptom	Possible Cause
Pump fails to discharge	• Pump not properly primed • Suction pipe not submerged • Strainer clogged • Foot valve clogged • Foot valve leaking • Suction lift too high • Wrong direction of rotation • Low speed • Pump worn
Pump is noisy	• Air leaking into suction • Insufficient liquid supply • Excessive pressure • Drive shaft bent • Coupling out of balance or alignment • Relief valve chatter
Pump wears rapidly	• Excessive pressure • Grit or dirt in liquid • Pump runs dry • Pipe strain on pump casing • Corrosion
Pump not up to capacity	• Strainer clogged • Insufficient suction • Air leaking into suction • Suction pipe too small • Low speed • Pump worn • Air leak at packing • Relief valve improperly seated
Pump starts, then loses suction	• Pump not properly primed • Suction pipe not submerged • Air leaking into suction • Suction pipe too small • Insufficient liquid supply
Pump takes excessive power	• Viscosity higher than specified • Obstruction in discharge line • Drive shaft bent • Pipe strain on pump casing • Packing too tight

304T01.EPS

Table 2 Troubleshooting Chart for a Reciprocating Pump

Pump Symptom	Possible Cause
Liquid end noise	• Insufficient suction pressure • Partial loss of prime • Poorly supported piping, abrupt turns in piping, pipe too small, piping misalignment • Air in liquid • Overpressure or overspeed • Valves broken or badly worn • Packing worn • Obstruction under valve
Power end noise	• Main bearing loose or worn • Connecting rod bearing and pinion shaft bearings worn • Oil level low • Plunger loose
Overheated power end	• Overpressure or overspeed • Oil level low • Main bearings tight • Ventilation inadequate • Belts too tight • Driver misaligned
Water in crankcase	• Condensation • Leaking atmospheric vent
Oil leak from crankcase	• Seals worn • Oil level too high
Rapid packing or plunger wear	• Dirty liquid • Dirty environment • Pump not level and rigid • Packing too tight • Worn rod
Pitted valves or seats	• Cavitation • Dirty liquid • Corrosion • Foreign matter
Valves hanging up	• Valve binding • Valve spring broken • Foreign matter
Leak at cylinder valve hole plugs	• Overpressure and overspeed • Water hammer • Cylinder plug loose • O-ring seal damaged
Loss of prime	• Insufficient suction pressure • Insufficient suction • Leaking suction at foot valve • Acceleration head requirement too high

304T02.EPS

Classroom

Stress the importance of accurately diagnosing problems when working on pumps.

Refer to Tables 1 through 3. Review symptoms and possible causes of problems for rotary, reciprocating, and centrifugal pumps. Go over any questions trainees have.

Table 3 Troubleshooting Chart for a Centrifugal Pump (1 of 3)

Pump Symptom	Possible Cause
Pump does not deliver fluid	• Pump not primed • Pump or suction pipe not completely filled with liquid • Insufficient suction • Insufficient margin between suction pressure and vapor pressure • Air pocket in suction line • Inlet of suction pipe insufficiently submerged • Speed too low • Direction of rotation wrong • Total head of system higher than design head of pump • Parallel operation of pumps unsuitable for such operation • Foreign matter in impeller • Strainers dirty or clogged • Suction or discharge valve in incorrect position
Insufficient capacity delivered	• Pump or suction pipe not completely filled with liquid • Insufficient suction • Insufficient margin between suction pressure and vapor pressure • Excessive amount of air or gas in liquid • Air leaks in suction line • Air leaking into pump through stuffing boxes • Foot valve too small • Foot valve partially clogged • Inlet of suction pipe insufficiently submerged • Speed too low • Total head of system higher than design head of pump • Viscosity of liquid different from design • Parallel operation of pumps unsuitable for such operation • Foreign mater in impeller • Wear rings worn • Impeller damaged • Casing gasket defective, permitting internal leakage
Insufficient pressure developed	• Excessive amount of air or gas in liquid • Speed too low • Direction of rotation wrong • Total head of system higher than design head of pump • Viscosity of liquid different from design • Parallel operation of pumps unsuitable for such operation • Wear rings worn • Impeller damaged • Casing gasket defective, permitting internal leakage
Pump loses prime after starting	• Pump or suction pipe not completely filled with liquid • Insufficient suction • Excessive amount of air or gas in liquid • Air pocket in suction line • Air leaks in suction line • Air leaking into pump through stuffing boxes • Inlet of suction pipe insufficiently submerged • Seal coolant line plugged • Seal cage improperly located in stuffing box, preventing sealing fluid from entering space to form seal
Pump requires excessive power	• Speed too high • Direction of rotation wrong • Total head of system higher than design head of pump • Total head of system lower than design head of pump • Specific gravity of liquid different from design • Viscosity of liquid different from design

304T03A.EPS

Instructor's Notes:

Table 3 Troubleshooting Chart for a Centrifugal Pump (2 of 3)

Pump Symptom	Possible Cause
Pump requires excessive power (continued)	• Foreign matter in impeller • Misalignment • Shaft bent • Rotating part rubbing on stationary part • Wear rings worn • Packing improperly installed • Type of packing incorrect for operating conditions • Gland too tight, resulting in no flow of liquid to lubricate packing
Stuffing box leaks excessively	• Seal cage improperly located in stuffing box, preventing sealing fluid from entering space to form seal • Mechanical seal failure • Misalignment • Shaft bent • Shaft or shaft sleeves worn or scored at packing • Packing improperly installed • Type of packing incorrect for operating conditions • Shaft running off center because of worn bearings or misalignment • Rotor out of balance, causing vibration • Cooling liquid not being provided to water-cooled stuffing boxes • Excessive clearance at bottom of stuffing box between shaft and casing, causing packing to be forced into pump interior • Dirt or grit in sealing liquid, leading to scoring of shaft or shaft sleeve
Packing has short life	• Seal cage improperly located in stuffing box, preventing sealing fluid from entering space to form seal • Misalignment • Shaft bent • Bearings worn • Shaft or shaft sleeves worn or scored at packing • Packing improperly installed • Type of packing incorrect for operating conditions • Shaft running off center because of worn bearings or misalignment • Rotor out of balance, causing vibration • Gland too tight, resulting in no flow of liquid to lubricate packing • Cooling liquid not being provided to water-cooled stuffing boxes • Excessive clearance at bottom of stuffing box between shaft and casing, causing packing to be forced into pump interior • Dirt or grit in sealing liquid, leading to scoring of shaft or shaft sleeve
Pump vibrates or is noisy	• Pump or suction pipe not completely filled with liquid • Insufficient suction • Insufficient margin between suction pressure and vapor pressure • Foot valve too small • Foot valve partially clogged • Inlet of suction pipe insufficiently submerged • Operation at very low capacity • Foreign matter in impeller • Misalignment • Foundations not rigid • Shaft bent • Rotating part rubbing on stationary part • Bearings worn • Impeller damaged • Shaft running off center because of worn bearings or misalignment • Internal clearance allowing recirculation • Rotor out of balance, causing vibration • Pump running outside of operating parameters/pump curve

304T03B.EPS

Teaching Tip

See the Teaching Tip for Section 3.0.0 at the end of this module.

Performance Testing

Have each trainee perform Performance Tasks 1 through 3 to your satisfaction. Fill out a Performance Profile Sheet for each trainee.

Homework

Have trainees review Sections 4.0.0–6.0.0 for the next session.

Materials

Ensure that you have everything required to teach this session.

Classroom

Emphasize the importance of relieving all pressure from a system before performing any maintenance activities.

Review the general recommendations to prepare a system for shutdown.

Table 3 Troubleshooting Chart for a Centrifugal Pump (3 of 3)

Pump Symptom	Possible Cause
Pump vibrates or is noisy (continued)	• Excessive internal clearances allowing recirculation • Excessive thrust caused by mechanical failure inside pump or by failure of hydraulic balancing device if any • Excessive grease or oil in antifriction bearing housing or lack of cooling, causing excessive bearing temperature • Lack of lubrication • Improper installation of antifriction bearings (damaged during assembly, incorrect assembly of stacked bearings, use of unmatched bearings as a pair, etc.) • Dirt in bearings • Rusting of bearings from water in housing • Excessive cooling of water-cooled bearing, resulting in condensation of atmospheric moisture in bearing housing
Bearing has short life	• Misalignment • Shaft bent • Rotating part rubbing on stationary part • Bearings worn • Shaft running off center because of worn bearings or misalignment • Rotor out of balance, causing vibration • Excessive thrust caused by mechanical failure inside pump or by failure of hydraulic balancing device if any • Excessive grease or oil in antifriction bearing housing or lack of cooling, causing excessive bearing temperature • Lack of lubrication • Improper installation of antifriction bearings (damaged during assembly, incorrect assembly of stacked bearings, use of unmatched bearings as a pair, etc.) • Dirt in bearings • Rusting of bearings from water in housing • Excessive cooling of water-cooled bearing, resulting in condensation of atmospheric moisture in bearing housing
Pump overheats and seizes	• Pump not primed • Insufficient margin between suction pressure and vapor pressure • Operation at very low capacity • Parallel operation of pumps unsuitable for such operation • Misalignment • Rotating part rubbing on stationary part • Bearing worn • Shaft running off center because of worn bearing or misalignment • Rotor out of balance, causing vibration • Improper installation of antifriction bearings (damaged during assembly, incorrect assembly of stacked bearings, use of unmatched bearings as a pair, etc.) • Pump running against a closed valve

304T03C.EPS

4.0.0 ◆ PREPARING FOR SHUTDOWN AND REPAIR

Before repairing or replacing any pump, steps must be taken to prepare for this maintenance activity. It is critical to relieve all pressure from the system before performing any maintenance on any component in the system. Always follow company policy and procedure and the pump manufacturer's service manual for specific instructions regarding the pump being repaired.

Follow these general steps to prepare for the system to be shut down before repairs are started:

Step 1 Obtain and review the past service and inspection reports for the pump. Look for any unusual conditions that might need to be addressed during the repair process.

Step 2 Obtain the manufacturer's service manual for the pump that is to be serviced. Review the repair and installation procedures described in it, paying particular attention to component specifications and tolerances.

Instructor's Notes:

Step 3 Obtain all the tools and instruments required for the repair process.

Step 4 Inspect the tools and instruments for serviceability, and replace those tools or instruments that are not serviceable.

Step 5 Obtain any personal protective equipment (PPE) that is required for the repair process per company policy.

Step 6 Obtain all the general equipment or supplies required for the repair process, such as a fire extinguisher, waste containers, cleaning solvents, shop cloths, lubricants, replacement bearings, and seals.

Step 7 Verify the location of the pump to be repaired.

Step 8 Survey the area around the pump for any unusual conditions, such as a spill that presents a fire hazard or requires additional assistance. Ensure that the area around the pump is safe to enter and work in.

> **NOTE**
> If the pump area is not safe to work in, contact the control center, and take safe and appropriate action as directed by the control center.

Step 9 Verify the accessibility of the pump by removing any obstructions.

Step 10 Verify the pump number and nameplate data to ensure that the correct pump is being repaired.

> **NOTE**
> If the nameplate is missing, contact the control center to verify the pump, and replace the nameplate according to company policy and procedures.

Step 11 Notify the operations control center and/or affected personnel that you are beginning the repair.

Step 12 Isolate all energy sources.

> **WARNING!**
> All energy sources must be isolated to prevent equipment damage or personal injury when the pressure is relieved.

Step 13 Shut down the power to the pump system.

Step 14 Relieve the pressure throughout the pump system.

Step 15 Lock out and tag out the pump motor switches and all other energy sources.

> **NOTE**
> Always follow company policy and procedure for lock out and tag out.

Step 16 Obtain a line-opening permit, and purge the pump and lines as required by company policy.

> **NOTE**
> Always follow company policy and procedure for opening a pipeline system.

Step 17 Remove any auxiliary piping and seal flush lines as required.

Step 18 Remove the coupling guard from the motor to the pump as applicable.

Step 19 Remove the coupling components, and match-mark them to aid in reassembly.

Step 20 Clean and inspect the coupling components.

> **NOTE**
> If the coupling components show signs of excessive wear or damage, order a new coupling with all components so that it will be available when it is needed for reassembly.

5.0.0 ◆ REMOVING THE PUMP FROM THE SYSTEM

When removing a pump from the system, company policy and procedure must be followed. The manufacturer's manual should be consulted.

> **WARNING!**
> Ensure that the system has been prepared for maintenance and that it is locked out and tagged before removing a pump. Ensure that all pressure has been relieved in the system before disconnecting piping from the pump. Check the bleeder valves to ensure that there is zero pressure. Pressure left in the system when the pipe connection is broken could cause personal injury.

Classroom

Distribute sample manufacturers' manuals. Have trainees review the repair and installation procedures. Discuss the manufacturers' specifications and requirements.

Safety

Discuss problems that can result if the energy sources are not isolated.

Ask a trainee to explain why a system must be locked out and tagged before removing a pump. Have another trainee explain why it's important to relieve all pressure in the system.

Classroom

Review the steps to remove a pump from the pipeline system. Ask whether trainees have any questions.

Refer to Figure 4. Explain that the pump must be reinstalled in the same position after repairs are completed.

Refer to Figure 5. Explain how to remove the shims from under the pump foot.

Follow these general steps to remove a pump from the pipeline system:

Step 1 Clean the pump thoroughly around each pipe connection to ensure that no dirt or contaminants enter the interior of the pump or piping system.

Step 2 Disconnect the coupling between the motor and the pump.

Step 3 Disconnect the piping from the pump, and cap or seal the ends of each pipe that is disconnected to prevent dirt and/or contaminants from entering the piping system.

Step 4 Take note of the pump mounting arrangement so that it can be reinstalled in the same position after all repairs are completed. *Figure 4* shows a typical mounting arrangement for a pipeline pump.

Step 5 Remove the pump mounting bolts from the pump and baseplate.

Step 6 Remove the shims from under the pump foot. *Figure 5* shows a typical shim installed between the pump foot and base.

NOTE
Note the number and location of shims at each mounting bolt so that the same number of shims can be reinstalled at each point later. Measure the thickness of each shim pack, and number the shims according to each pump foot. These shims are required for reassembly.

Step 7 Collect the rigging hardware that will be used to lift the pump from the baseplate.

Step 8 Inspect each piece of the rigging hardware for excessive wear, damage, and cracks.

NOTE
Replace any rigging hardware that is not serviceable.

Step 9 Rig the pump assembly for removal using appropriate rigging based on the size of the pump.

NOTE
Some manufacturer's maintenance manuals provide pump rigging instructions for their pumps. When these instructions are available, follow the recommended pump rigging procedures to prevent damage to the pump and to lift the pump safely.

Step 10 Loosen and remove the frame foot hold-down bolts.

Steps 11 Slowly lift the pump assembly from the base, and place the pump assembly in a safe, clean, and stable work area.

Step 12 Clean and inspect the pump assembly mounting platform.

Figure 4 ◆ Typical mounting arrangement for a pipeline pump.

Instructor's Notes:

Figure 5 ◆ Typical precut shim.

> **NOTE**
> The mounting platform should be clean and level with no signs of corrosion, wear, or cracks in the metal frame.

Step 13 Repair any damage to the mounting platform before installing a replacement pump.

6.0.0 ◆ DISASSEMBLING THE PUMP

Mechanical technicians who are preparing to disassemble a pump should ensure that all initial preparations are made according to company policy and procedure. Depending upon whether the pump is in service at the time of disassembly, initial preparations might include some or all the following procedures.

- Securing the appropriate work orders and permits
- Securing the appropriate manufacturer's manual for the pump being disassembled
- Notifying the control center and other required personnel
- Performing lockout and tagout procedures
- Closing all valves and lines connected to the pump
- Draining all liquid from the pump
- Removing all connections that interfere with disassembly
- Disconnecting the driver from the pump
- Using the appropriate pulling tool to remove the drive coupling from the pump shaft

> **CAUTION**
> Care should be used to avoid disturbing the alignment of the pump casing and driver.

6.1.0 Beginning Disassembly

The end covers, bearing housings, and other associated parts must be removed before the pump shafts are exposed (see *Figure 6*). Mechanical drawings in the manufacturer's manual should be studied before beginning any procedures. Initial procedures involve removing the stuffing box, packing gland and packing, or the mechanical seals, depending on the pump model (see *Figure 7*).

6.1.1 Gear-Type Oiling Systems

Some pump bearings are equipped with oiling systems that use gear-type oil pumps. These units require that the reservoir and pump be removed, sometimes as a unit, and set aside. Manufacturer's instructions and recommendations should be consulted and followed to ensure that they are removed correctly.

Classroom

Discuss the procedures for preparing a pump for disassembly.

Safety

Discuss the consequences of disturbing the alignment of the pump casing and driver.

Audiovisual

Show Transparency 4 (Figure 7). Review the components in a pump bearing arrangement.

Classroom

Distribute sample manufacturer's manuals. Have trainees locate and study the mechanical drawings. Emphasize the importance of reviewing the drawings before beginning disassembly procedures.

Classroom

Explain how to remove bearings. Stress that the location of the bearings affects the removal procedure.

Safety

Explain how to avoid damaging shaft surfaces when removing pump bearings.

6.1.2 Removing Pump Bearings

There are several different types of large bearings. The manufacturer's manual and instructions should be reviewed to determine exactly which type of bearing is involved and how it is mounted before attempting to dismantle the assembly. Bearings mounted directly on the shaft require a procedure that is different from those mounted on a sleeve or in the housing. Bearing oil deflectors and protective parts must be removed in the proper sequence. Thrust bearings are often mounted directly on the shaft (refer to *Figure 7*). Appropriate pullers must be used to remove thrust bearings, taking care not to lose or damage adjustment shims. Once sealing assemblies and bearings are removed, technicians can proceed with the disassembly of the pump and its internal parts.

Figure 6 ◆ Pipeline pump in service.

> **CAUTION**
> When removing pump bearings, care should be exercised to avoid damaging shaft surfaces.

Some bearings are installed with shims, locating pins, and other small components. Technicians should arrange to place such parts in a clean container until they are needed. It is a good practice to label the individual collections of small parts. Technicians should make good notes on the sequence and number of parts, keys, setscrews, deflectors, and other items as they are removed, for later reference. Rough hand-drawn illustrations are often valuable in the reassembly of mechanical parts.

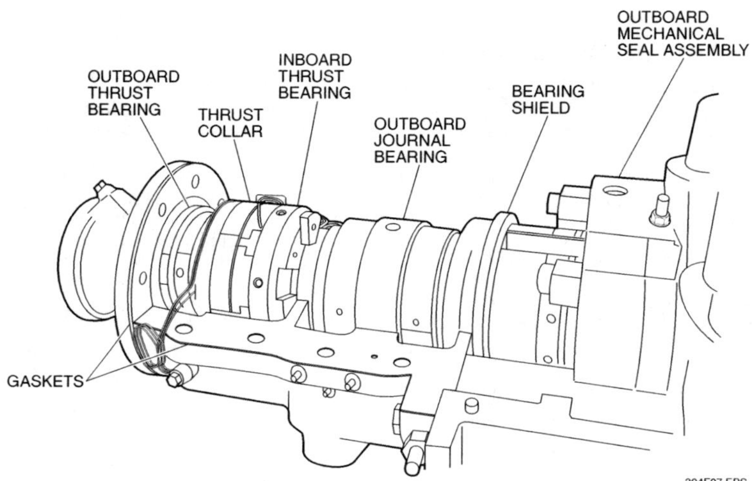

Figure 7 ◆ Pump bearing arrangement.

Instructor's Notes:

6.2.0 Disassembling Internal Components

Appropriate hoisting and rigging equipment should be on site before repairs are begun. To avoid serious injury or damage to components, manufacturer's instructions and recommendations for appropriate rigging and hoisting procedures should be consulted in advance and followed carefully.

> **CAUTION**
> Pump components can be extremely heavy and hard to handle.

6.2.1 Removing the Rotating Element

Pump casings must be separated in order to remove the rotating element or impeller (see *Figure 8*). Normally, the shaft and impeller must be removed as a unit after the casing is separated. In order to remove the shaft and impeller from a pump with a horizontally separated case, follow these general steps:

Step 1 Remove the nuts from all casing studs.

Step 2 Using jackscrews in the manufacturer's designated casing holes, tighten all screws evenly in a sequence until the casing separates.

Step 3 Remove the jackscrews from the casing.

Step 4 Use the manufacturer's recommended attachments, rigging, and a hoist to lift the upper half of the case off and away from the lower casing.

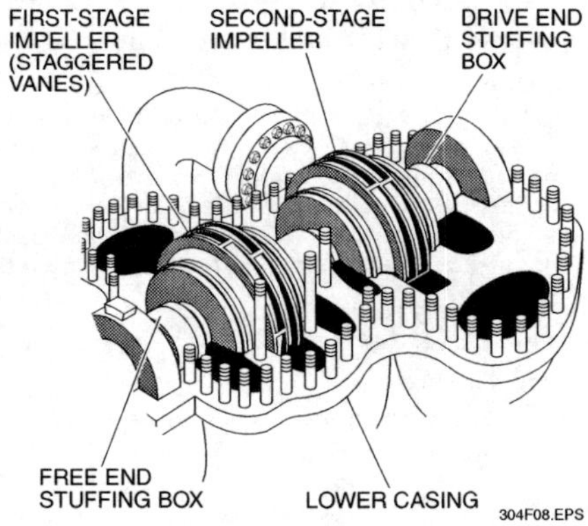

Figure 8 ◆ Centrifugal pump with top casing removed.

Step 5 Attach a lifting sling as close to the impeller as possible, and using a hoist with slight pressure, use a pry bar to unseat shaft rings and bushings from the lower casing.

Step 6 Hoist the shaft and impeller carefully from the inside of the lower casing.

Step 7 Carefully hoist the shaft and impeller assembly to a preselected and prepared location (see *Figure 9*).

> **WARNING!**
> Personnel must not allow any body part or extremity to pass under the shaft and impeller while they are being hoisted and moved.

Step 8 Extract any loose parts from the pump shaft, noting location and number.

304F09.EPS

Figure 9 ◆ Shaft and impeller assembly properly stored.

Safety

Emphasize that pump components can be difficult to handle.

Audiovisual

Show Transparency 5 (Figure 8). Explain that the pump casing must be separated to remove the rotating element.

Classroom

Review the procedures to remove the shaft and impeller from the pump. Answer any questions trainees have.

Safety

Discuss possible consequences of allowing any body parts to pass under the shaft and impeller.

Classroom

Explain how to remove the impeller from the shaft.

Safety

Ask a trainee to explain why the shaft should be wrapped in wet cloths.

Emphasize that trainees must use the appropriate personal protective equipment when handling the hot impeller.

Classroom

Explain that it is useful to clean the pump components before conducting an inspection. Review the procedures for inspecting the components of a pump.

Discuss the consequences of excessive shaft runout. Go over any questions trainees have.

6.2.2 Removing the Impeller

After the shaft and impeller has been removed as an assembly, the impeller can be removed from the shaft using the following general steps:

Step 1 Remove the setscrews or other locking mechanisms from the large nuts or threaded sleeves on both sides of the impeller.

Step 2 Using appropriate wrenches, remove the nuts or sleeves from both sides of the impeller.

> **NOTE**
> One of the nuts on one side of the impeller usually has left-handed threads so that the rotation of the shaft helps to keep it tight.

Step 3 Take off any impeller spacers, sleeves, or rings, noting the number and sequence.

Step 4 Place index marks on the impeller and shaft so that the impeller can be reinstalled in exactly the same location.

Step 5 Apply heat to the impeller as prescribed by the manufacturer's recommendations. Begin applying heat from the outside edge of the impeller and move inward to the center. After removal, cool the impeller slowly.

> **CAUTION**
> The shaft should be wrapped in wet cloths to protect it from the heat being applied to the impeller.

Step 6 When the manufacturer's specified temperature is reached, bump the impeller firmly with a soft mallet to separate it from the shaft.

> **CAUTION**
> Use appropriate PPE to handle the hot impeller. Use appropriate hoisting equipment as required.

> **NOTE**
> Clean pump components that will not be assembled soon should have a coat of light oil applied to the machined surfaces and should be covered for protection from dust and dirt for a few days. If components will not be used for a longer duration, proper storage should be arranged.

6.3.0 Inspecting Pump Components

Once the pump shaft and impeller have been disassembled, the various parts of the unit should be cleaned. When possible, use the solvents and cleaning procedures described in the manufacturer's instructions and recommendations. When the parts are properly cleaned, wear, damage, and imperfections can be more readily seen. The following general procedures might be followed to properly inspect the components of a pump:

Step 1 Closely examine the pump shaft for burrs and rough places, and repair it with a file and an emery cloth.

Step 2 Inspect the shaft for runout, following the manufacturer's instructions and recommendations.

Step 3 Measure the running clearances inside the pump; compare them to the manufacturer's instructions and recommendations.

> **NOTE**
> Clearance must be within the tolerances published in *API Standard 610.*

6.4.0 Measuring Shaft Runout Using a Dial Indicator and Base

After the pump shaft and impeller have been separated, the shaft should be placed in the appropriate supports, and the shaft runout should be measured with a dial indicator (see *Figure 10*). Excessive shaft runout can lead to a number of serious problems and eventual pump failure. Manifestations of excessive shaft runout include:

- Premature bearing failure
- Loss of power
- Seal failure
- Excessive operating temperatures

Instructor's Notes:

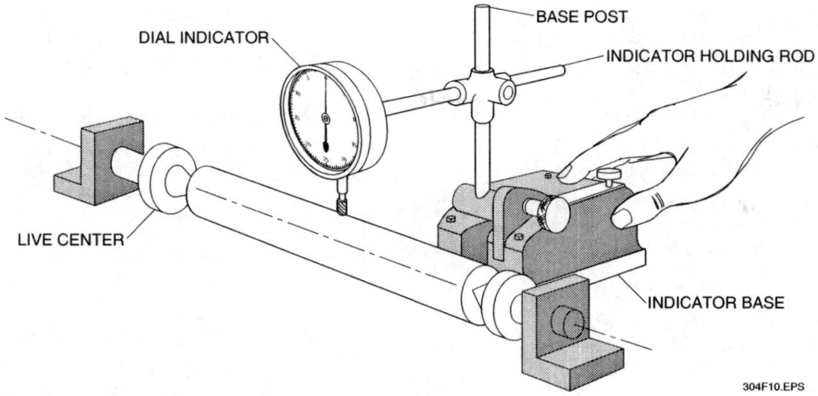

Figure 10 ◆ Measuring shaft runout with a dial indicator setup.

To effectively measure shaft runout on each particular model of pump, the manufacturer's instructions and recommendations must be followed for the locations and procedures required. After the runout has been measured, the manufacturer's specifications can be consulted for the maximum amount of allowable runout and for the recommended procedures for correcting the problem. In general, to measure runout:

Step 1 Clean all surfaces that the dial indicator plunger will contact.

Step 2 Determine the appropriate location, and attach the indicator base.

Step 3 Mount the dial indicator (see *Figure 11*) on the indicator base as appropriate, and tighten the locking device.

Step 4 Preload the indicator by moving the plunger against the shaft area so that about half of the plunger is pressed into the sleeve.

Step 5 Holding the indicator plunger in that position, tighten the locking mechanism to maintain position.

Step 6 Zero the indicator by rotating the face of the dial to read absolute zero.

Step 7 Carefully tighten the dial face locking mechanism, maintaining the absolute zero position.

CAUTION

Take care not to bottom the indicator plunger, which will damage the instrument.

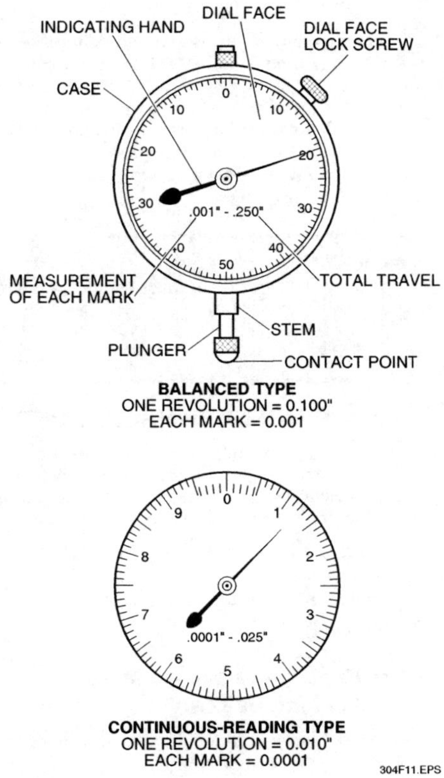

Figure 11 ◆ Typical dial indicators.

Audiovisual

Show Transparency 6 (Figure 10). Discuss the steps to measure the shaft runout using a dial indicator.

Show Transparency 7 (Figure 11). Explain how to mount the dial indicator.

Safety

Discuss the consequences of bottoming the indicator plunger.

Classroom

Explain when the impeller and case rings should be replaced.

Teaching Tip

See the Teaching Tip for Sections 4.0.0–6.0.0 at the end of this module.

Homework

Have trainees review Section 7.0.0 for the next session.

Materials

Ensure that you have everything required to teach this session.

Classroom

Review the basic procedures to reassemble a pump.

Audiovisual

Show Transparency 8 (Figure 12). Explain how to verify that the impellers are installed in the proper direction.

Step 8 Verify that all adjustments on the instrument mounts have been tightened.

Step 9 Observe the dial while slowly rotating the shaft to read the total runout (the highest number reached on the dial).

Step 10 Record the highest numbers while rotating the shaft in both directions

Step 11 Reposition the indicator base to two other points of measurement, and repeat the procedure so that all the measurements can be averaged to obtain the average runout for the shaft.

Step 12 Calculate the average of all measurements, and use this figure for assessing the shaft condition.

6.5.0 Replacing Impeller and Case Rings

The impeller and case rings should be replaced if wear is indicated by substandard performance levels. As a matter of good practice, some company policies and procedures require the rings to be replaced any time the pump is dismantled. Follow the manufacturer's instructions and recommendations for removal and replacement of the impeller and case rings.

> **NOTE**
> Installing new impeller rings necessitates the installation of new case rings. New impeller rings would prevent the older case rings from seating properly if they were not replaced at the same time.

7.0.0 ◆ REASSEMBLING THE PUMP

Reassembling the pump requires that some procedures be performed as a matter of good practice and professionalism. Reassembling the pump should include the following basic procedures:

- Replacing all internal lock washers and locking devices
- Installing new O-rings and gaskets
- Cleaning the bearing surfaces

7.1.0 Replacing Lock Washers and Locking Devices

New lock washers and locking devices should be installed on all fastenings within the pump case while it is disassembled and the work can be easily done. If a weakened lock washer or locking device fails after startup, serious damage could occur, and the entire unit would have to be disassembled again to repair it.

7.2.0 Replacing O-Rings and Gaskets

It is considered good practice to install new O-rings and gaskets (manually cut or factory made) when any unit is being reassembled. Most old gaskets and O-rings are damaged during removal and are at best hardened with age and exposure. Many problems with leakage are eliminated after new O-rings and gaskets are properly installed.

7.3.0 Cleaning Bearing Surfaces

All parts, especially the bearing surfaces, should be cleaned meticulously before installation. Such cleaning prevents small particles of dust, grit, grime, and residues from causing problems later. Manufacturer's instructions and recommendations should be followed for selecting solvents and cleaning bearing surfaces before assembly.

7.4.0 Assembling the Impeller

Impellers are designed to turn in one direction only (see *Figure 12*). Technicians must verify that the impeller is being installed in the proper direction on the shaft.

In order to properly assemble a pump impeller and shaft, the manufacturer's instructions and recommendations should be consulted. Line drawings and schematics of the pump should be studied along with written procedures for the exact model of pump being assembled. *Figure 13* shows a general layout of pipeline pump components.

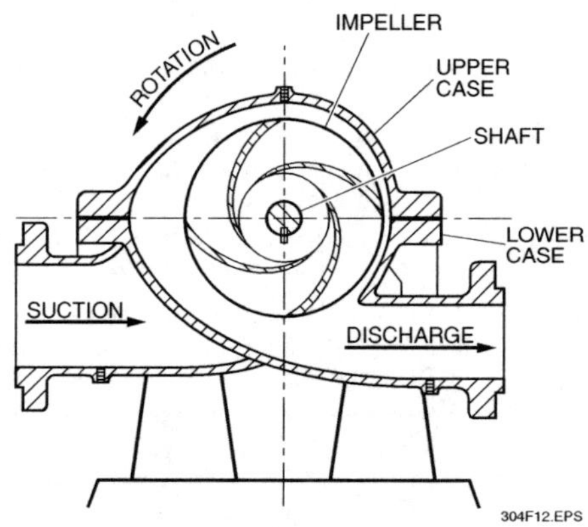

Figure 12 ◆ Pump showing impeller rotational direction.

Instructor's Notes:

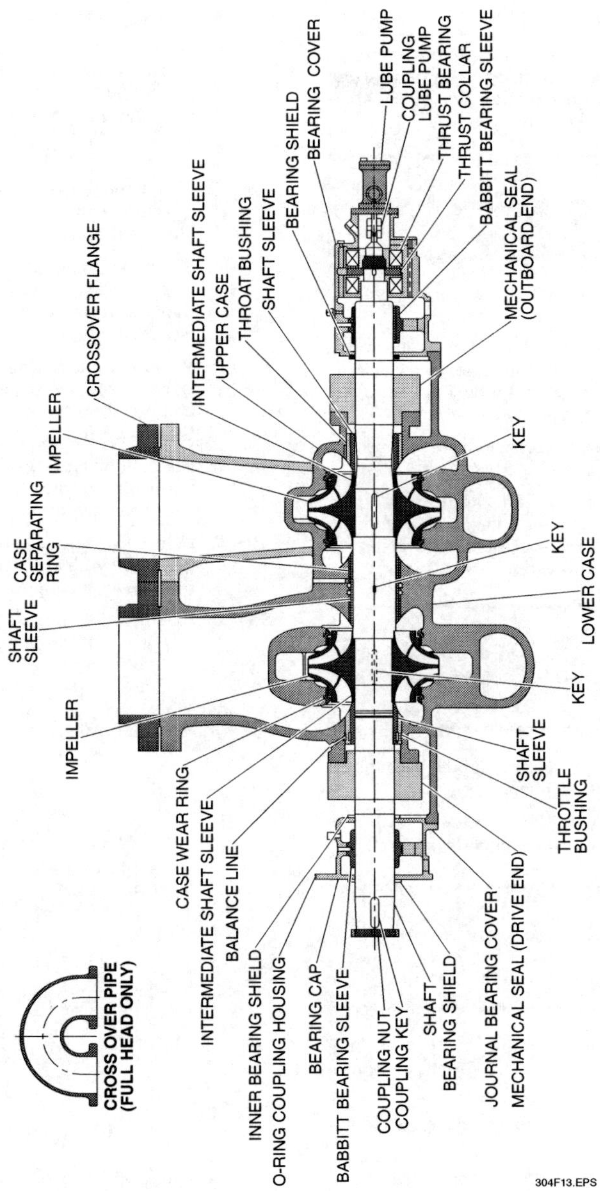

Figure 13 ◆ General layout of pipeline pump components.

Classroom

Distribute copies of Figure 13 with the callouts covered. Ask trainees to identify the components and explain how to install each when reassembling a pump.

Classroom

Review the general steps for assembling a pump impeller and shaft.

Safety

Discuss the consequences of failing to wear the appropriate personal protective equipment.

Audiovisual

Show Transparency 9 (Figure 14). Explain how to correctly place the dowels into the indentations.

To assemble a pump impeller and shaft, follow these general steps:

Step 1 Evenly heat the impeller to the manufacturer's temperature recommendation, preferably in an oven if available. Otherwise, use a low torch flame.

> **CAUTION**
> Use the appropriate PPE to handle the heated impeller.

Step 2 Verify that the shaft key is in the proper position.

Step 3 Locate the impeller between the index marks made at disassembly, or calculate the proper location for the impeller using the manufacturer's instructions.

Step 4 Mount the spacer sleeves in the proper locations on each side of the impeller.

Step 5 Install the impeller retaining nuts or threaded sleeves on the shaft, and using the appropriate tool, tighten them according to manufacturer's instructions.

> **NOTE**
> One of the nuts or sleeves will normally have a left-handed thread to aid in staying tight while in operation.

Step 6 To lock the setscrews, tighten them in the retaining nuts or rings.

Step 7 Mount any additional sleeves with new O-rings on the proper ends of the shaft.

> **NOTE**
> Most sleeve-type components are not interchangeable and must be installed on the correct end of the shaft.

Step 8 If the pump is equipped with a stuffing box or sleeve assembly, slip these components onto the shaft.

Step 9 Install keys in the keyways on each end of the shaft.

Step 10 Place the retaining rings onto the shaft.

Step 11 Mount the casing rings and any shaft bushings on the shaft.

Step 12 Using rigging and a hoist according to manufacturer's instructions, slowly lower the shaft and impeller assembly into the bottom casing, stopping just before contact with the seats of the casing rings and throat bushings.

Step 13 Adjust the position of the rings and bushings until the machined indentations align with the dowels in the casing.

Step 14 Slowly and carefully lower the impeller and shaft assembly into place, ensuring that the dowels enter the indentations (see *Figure 14*).

Step 15 Verify that the assembly has seated properly in the lower casing and that the rings and bushings were not displaced during the procedure.

Step 16 Place the casing gasket on the mating surface, verifying that all holes are cut and that other areas align properly.

Step 17 Using rigging and a hoist, align the casing bolt holes with the casing studs, and carefully lower the top casing onto the studs.

Step 18 Slowly lower the casing into place on the lower half of the pump casing, verifying that the case rings and bushings have remained in position.

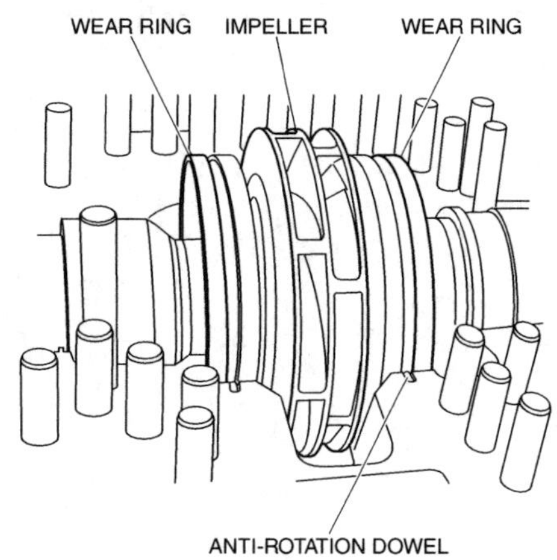

304F14.EPS

Figure 14 ◆ Impeller with wear rings.

Instructor's Notes:

Step 19 Place all the nuts onto the studs, and tighten them according to the sequence and torque specifications laid out in the manufacturer's instructions and recommendations.

> **NOTE**
> Some pump casings require a particular sequence for tightening the casing nuts to prevent damage and help ensure a good seal.

Step 20 Install the pump seal assembly (stuffing box, packing, or mechanical seals as applicable) while referring to and following the manufacturer's instructions and recommendations for the appropriate assembly.

Step 21 Install the pump bearings according to the manufacturer's instructions and recommendations, verifying that the correct bearings are being used.

Because pump manufacturers sometimes vary the type of bearings used in a particular model, technicians should study the procedures, schematics, and line drawings in the manual to ensure that the correct parts and installation procedure are being used. *Figure 15* shows a typical pump main bearing.

> **CAUTION**
> It is important that technicians verify and adjust all clearances according to the manufacturer's instructions and recommendations.

7.5.0 Pre-Startup Final Inspections

Final inspections of a pump that has been repaired can seem redundant to some people, but considering the cost, labor, and downtime that might result from an error, it is definitely necessary. Before any attempt to start a newly assembled pump is made, it is essential that technicians make the following final inspections:

- If mechanical seals were installed, verify that their setting is at the manufacturer's recommended clearances and that the drive collar setscrews have been tightened (refer to the manual schematics and line drawings).
- Verify that all piping and other connections that were removed during disassembly are in place and tight.
- Examine the shaft alignment closely and adjust it according to the manufacturer's instructions before assembling the coupling.
- Ensure that lubrication has been properly applied and that all systems are properly filled and prepared for startup.
- Verify that all fasteners have been tightened evenly to the manufacturer's torque specifications.

8.0.0 ◆ INSTALLING THE PUMP

After it has been verified that the pump installation site is cleaned, repaired, and ready for installation, follow these general steps to install the pump:

Step 1 Using hoisting and rigging equipment as directed in the manufacturer's instructions and recommendations, carefully and slowly lower the pump onto the mounting base.

Step 2 Partially install the pump mounting bolts, leaving them loose to enable adjustment on the base.

Step 3 Refer to the notes from the removal of the previous pump, and install the exact number and sizes of shims between the pump and base points, respectively.

304F15.EPS

Figure 15 ◆ Typical pump main bearing.

Teaching Tip

See the Teaching Tip for Section 7.0.0 at the end of this module.

Safety

Emphasize that is it important to verify and adjust all clearances according to the manufacturer's instructions.

Classroom

Discuss the elements and importance of a final inspection. Ask whether trainees have any questions.

Homework

Have trainees review Sections 8.0.0–9.0.0 and answer the Review Questions for the next session.

Materials

Ensure that you have everything required for the testing during this session.

Classroom

Discuss the general steps for installing a pump. Review problems you may have encountered when installing a pump.

Classroom

Review the procedures to start and test a new or repaired pump. Go over any questions trainees have.

Safety

Stress the importance of shutting down the pump in the event of excessive noise, vibration, or temperature.

Teaching Tip

See the Teaching Tip for Sections 8.0.0–9.0.0 at the end of this module.

Step 4 Verify that the suction and discharge of the pump are clean of any foreign material.

Step 5 Connect the pipeline connections according to company policy and procedure.

Step 6 Tighten the base mounting bolts just enough to be snug.

Step 7 Align the pump and driver shafts according to the manufacturer's instructions and recommendations.

Step 8 Install the pump coupling according to manufacturer's instructions and recommendations.

> **NOTE**
> The pump and driver must be perfectly aligned to prevent damage and the premature breakdown of the pump and/or driver.

Step 9 Ensure that all lubrication points and systems are properly filled and prepared.

Step 10 Verify that all fastenings are tightened according to manufacturer's instructions.

Step 11 Remove all locks and tags from the system, and restore power.

Step 12 Complete all required documentation.

9.0.0 ◆ PUMP STARTUP PROCEDURE AND OPERATIONAL TESTS

When starting up a pump that has been repaired or installed, manufacturer's instructions and recommendations must be followed for the pump to create flow and to avoid pump damage. The following general steps should be followed when attempting to start a newly installed or repaired pump:

Step 1 Open the pump inlet valve to fill the pump with liquid.

Step 2 Bleed all the air from the pump.

Step 3 Examine the pump and seals for leaks.

Step 4 Slightly open the pump discharge valve.

Step 5 Verify that the lubrication levels at all points of the system are correct.

Step 6 Remove all tools, equipment, and debris from the pump area.

Step 7 Secure permission from the control center to perform the operational tests, and follow the control center's directions.

Step 8 Start the pump and observe the operation for excessive noise, vibration, or temperature.

> **CAUTION**
> In the case of unusual noise, vibration, or temperature, shut down the pump immediately and investigate.

Step 9 When flow has begun, slowly increase the setting of the throttle valve until the appropriate pressure level is reached.

Step 10 Examine the motor and pump for excessive heat buildup.

Step 11 Observe the pump and motor to see if there is any type of vibration.

Step 12 Inspect all the seals and connections for leaks.

> **NOTE**
> Some pumps are equipped with a type of seal or packing that is expected to drip up to 60 drops per minute.

Step 13 Observe and verify that all gauges remain within a normal range for the system.

Step 14 Notify the control center that the testing is completed.

Step 15 Complete all required documentation.

Instructor's Notes:

Summary

Pipelines depend on powerful hydraulic pumps for the movement of all products and petroleum. In order to maintain shipping schedules and deliveries, the pumps must be maintained in good working order. They must operate in a safe, efficient, and economic manner in order to supply the pressure required by the pipeline. Mechanical technicians are responsible for maintaining the pumps in a safe and efficient condition.

Trained and qualified mechanical technicians must have a strong understanding of the pumps and the hydraulics involved in the pipeline system in order to carry out an organized schedule of inspection and preventive maintenance that supports the efficient and economical operation of the pumps. Maintaining documented records of the equipment enables technicians to trace the problems associated with keeping each pump up to an efficient level of operation. By studying records, technicians may gain helpful insight into the development of a problem.

Review Questions

1. The following are parameters that should be monitored to understand how a pump is performing *except* _____.
   a. suction pressure
   b. flow
   c. pump speed
   d. lucidity rate

2. Beginning with the simplest possibility is the first step in the process of _____.
   a. trial and error
   b. troubleshooting
   c. disassembly
   d. elimination

3. It is critical to _____ from the system before performing any maintenance on any component.
   a. relieve all pressure
   b. purge all fluids
   c. remove the parts
   d. shut off the flow

4. All energy sources must be _____ to prevent equipment damage or personal injury.
   a. tagged
   b. reduced
   c. isolated
   d. identified

5. Check the bleeder valves before breaking the connection to ensure that there is _____ pressure.
   a. maximum
   b. zero
   c. steady
   d. sufficient

6. The _____ should be clean and level with no signs of corrosion, wear, or cracks in the metal frame.
   a. mounting platform
   b. shims
   c. pump casing
   d. rigging

7. Care should be used to avoid disturbing the _____ of the pump casing and driver.
   a. coupling
   b. covers
   c. connections
   d. alignment

8. The end covers, bearing housings, and other associated parts must be removed before the _____ are exposed.
   a. bearings
   b. valves
   c. pump shafts
   d. packing glands

9. _____ must be separated in order to remove the rotating element.
   a. Couplings
   b. Pump casings
   c. Pump bearings
   d. Mechanical seals

10. Shaft runout should be measured with a _____.
    a. dial indicator
    b. caliper
    c. machinist's rule
    d. scale

11. Installing new impeller rings necessitates the installation of new _____.
    a. lock washers
    b. bearings
    c. case rings
    d. gaskets

Classroom

Stress the importance of completing the proper documentation when maintaining, troubleshooting, and repairing pumps.

Go over the Review Questions for this module, and answer any questions trainees have.

Examination

Administer the Module Examination. Be sure to record the results of the Exam on Craft Training Report Form 200 and submit the results to the Training Program Sponsor.

Performance Testing

Have each trainee perform Performance Task 4 to your satisfaction. Fill out a Performance Profile Sheet for each trainee.

Ensure that all Performance Tests have been completed and Performance Profile Sheets for each trainee are filled out. Be sure to record the results of the testing on Craft Training Report Form 200 and submit the results to the Training Program Sponsor.

12. It is commonly considered good practice to install new _____ and _____ as any unit is being reassembled.

a. impeller rings, case rings
b. O-rings, gaskets
c. bearings, seals
d. sleeves, retaining nuts

13. To ensure that the correct bearings are used, you should study all of the following manual components *except* _____.

a. schematics
b. line drawings
c. procedures
d. sequential plots

14. When reinstalling the pump, install the exact number and sizes of _____ between the pump and the base parts.

a. shims
b. nuts
c. mounting bolts
d. rings

15. Some pumps are equipped with a type of seal or packing that is expected to drip up to _____ drops per minute.

a. 60
b. 40
c. 30
d. 20

Instructor's Notes:

GLOSSARY

Trade Terms Introduced in This Module

Discharge: The exit side of the pump from which the fluid leaves the pump.

Suction: The inlet side of a pump where the fluid is supplied to the pump.

MODULE 63304-02 — TEACHING TIPS

The following are suggested activities or instructional methods to help you teach the material in this AIG.

General

When you call on someone to answer a question, the rest of the class relaxes or even tunes out because they expect that the question and answer will take place only between you and the trainee you called on. Instead, use this technique to involve more trainees in answering questions and to keep them on their toes.

1. Ask trainees to define a term or explain a concept.
2. After one trainee has answered, ask a trainee seated nearby if the answer is right. Then ask whether a trainee in the back of the room agrees.
3. Ask trainees to explain why they think an answer is right or wrong.
4. Use the session to clear up incorrect ideas and encourage trainees to learn from their mistakes.

Section 2.0.0 *Performing Preventive Maintenance on Pumps*

This exercise emphasizes the importance of following company requirements for scheduled inspections. For this exercise, trainees will need copies of their company's policies and procedures. Allow 30 minutes for this exercise.

1. Review the general preventive maintenance requirements for a pump.
2. Ask trainees to locate and review their company's requirements for daily, semiannual, and annual preventive maintenance activities.
3. Compare trainees' company requirements with the general guidelines covered to this point. Review and discuss any major differences.
4. Answer any questions trainees have about performing preventive maintenance on pumps.

Section 3.0.0 *Troubleshooting Pumps*

This exercise will give trainees the opportunity to identify common troubleshooting problems for pumps. For this exercise, you will need an overhead projector, sheets of acetate, and transparency markers. Trainees will need their Trainee Guides and pencils and paper to take notes for discussion. Allow 30 minutes for this exercise.

1. On a sheet of acetate, write a list of symptoms that may indicate problems with rotary, reciprocal, and centrifugal pumps.
2. Ask trainees to review and diagnose each symptom. Emphasize that there may be more than one reason for each problem.
3. Review and discuss trainees' answers. Discuss potential solutions for each of the problems.
4. Go over any questions trainees have about troubleshooting pumps.

Sections 4.0.0–6.0.0 *Preparing, Removing, and Disassembling the Pump*

This exercise will familiarize trainees with the common steps for preparing, removing, and disassembling a pump. For this exercise, trainees will need their Trainee Guides and pencils and paper to take notes for discussion. Allow 30 minutes for this exercise.

1. Divide the class into small groups. Assign each group with the task of preparing, removing, or disassembling a pump.
2. Ask each group to identify the procedures to perform its assigned task.
3. Have each group write the list of procedures on the whiteboard/chalkboard.
4. Ask trainees to review the procedures and make any necessary corrections, additions, and/or changes.
5. Answer any questions trainees have about preparing, removing, or disassembling a pump.

Section 7.0.0 *Reassembling the Pump*

This exercise will familiarize trainees with techniques for reassembling a pump. For this exercise, you will need copies of manufacturers' manuals with procedures, schematics, and line drawings. Trainees will need their Trainee Guides and pencils and paper to take notes for discussion. Allow 20 minutes for this exercise.

1. Divide the class into small groups according to the number of manufacturers' manuals.
2. Ask trainees to locate and review the procedures, schematics, and line drawings.
3. Have trainees compare the information in the manual with the information in the Trainee Guide. Ask trainees to identify the following information for the pump featured in the manual:
   - Specific parts required
   - Additional steps or procedures
   - Safety considerations
4. Review and discuss trainees' answers. Go over any questions trainees have about reassembling a pump.

Sections 8.0.0–9.0.0 *Installing and Testing the Pump*

This exercise will give trainees the opportunity to witness a pump installation. For this exercise, you will need to locate a nearby site where an installation is scheduled. Trainees will need their Trainee Guides and pencils and paper to take notes for discussion. Allow at least 60 minutes for this exercise.

1. Arrange a visit to a local site where a pump installation is scheduled.
2. Ask the operations personnel to explain the scheduled procedure to the trainees.
3. Ask trainees to observe the installation and testing and to assist with the procedures, as appropriate.
4. Following the site visit, review trainees' observations.
5. Answer any questions trainees have about installing and testing a pump.

MODULE 63304-02 — ANSWERS TO REVIEW QUESTIONS

Answer	Section
1. d	2.0.0
2. d	3.0.0
3. a	4.0.0
4. c	4.0.0
5. b	5.0.0
6. a	5.0.0
7. d	6.0.0
8. c	6.1.0
9. b	6.2.1
10. a	6.4.0
11. c	6.5.0
12. b	7.2.0
13. d	7.4.0
14. a	8.0.0
15. a	9.0.0

CONTREN™ LEARNING SERIES — USER UPDATES

The NCCER makes every effort to keep these textbooks up-to-date and free of technical errors. We appreciate your help in this process. If you have an idea for improving this textbook, or if you find an error, a typographical mistake, or an inaccuracy in NCCER's Contren™ textbooks, please write us, using this form or a photocopy. Be sure to include the exact module number, page number, a detailed description, and the correction, if applicable. Your input will be brought to the attention of the Technical Review Committee. Thank you for your assistance.

Instructors – If you found that additional materials were necessary in order to teach this module effectively, please let us know so that we may include them in the Equipment/Materials list in the Instructor's Guide.

Write: Curriculum Revision and Development Department
National Center for Construction Education and Research
P.O. Box 141104, Gainesville, FL 32614-1104

Fax: 352-334-0932

E-mail: curriculum@nccer.org

Craft ______________________ Module Name ______________________

Copyright Date ____________ Module Number ____________ Page Number(s) ____________

Description

____

____

____

(Optional) Correction

____

____

(Optional) Your Name and Address

____

____

Module 63305-02

Maintain, Troubleshoot, and Repair Gas Compressors

MODULE OVERVIEW

This module reviews the types and designs of gas compressors. Trainees will learn to identify and demonstrate the procedures to maintain, troubleshoot, and repair gas compressors.

PREREQUISITES

Please refer to the Course Map in the Trainee Module. Prior to training with this module, it is recommended that the trainee shall have successfully completed the following:

Core Curriculum; Pipeline Core, Modules 66101-02 and 66102-02; Pipeline Mechanical Levels One and Two; Pipeline Mechanical Level Three, Modules 63301-02 through 63304-02

OBJECTIVES

Upon completion of this module, the trainee will be able to:

1. Identify the typical lubrication system components of a gas compressor.
2. Describe the preventive maintenance requirements for a gas compressor.
3. Identify the common troubleshooting techniques for a gas compressor.
4. Identify the common steps required to prepare for shutdown and repair of a gas compressor.
5. Identify the common steps required to isolate a gas compressor from a pipeline system.
6. Identify the common steps required to repair a rotary and reciprocating gas compressor.
7. Identify the common steps required to prepare the gas compressor for start-up and operational check after maintenance has been completed.

PERFORMANCE TASKS

Under the supervision of the instructor, the trainee should be able to:

1. Identify the typical lubrication system components of a gas compressor.
2. Describe the preventive maintenance requirements for a gas compressor.
3. Demonstrate common troubleshooting techniques for a gas compressor.
4. Demonstrate the common steps to perform each of the following:

   - Prepare a gas compressor for shutdown and repair.
   - Isolate a gas compressor from a pipeline system.
   - Repair a rotary and reciprocating gas compressor.
   - Prepare the gas compressor for start-up and operational check after maintenance or repair has been completed.

NCCER STANDARDIZED CRAFT TRAINING PROGRAM

The National Center for Construction Education and Research (NCCER) provides a standardized national program of accredited craft training. Key features of the program include instructor certification, competency-based training, and performance testing. The program provides trainees, instructors, and companies with a standard form of recognition through a National Craft Training Registry. The program is described in full in the *Guidelines for Accreditation*, published by the NCCER. For more information on standardized craft training, contact the NCCER by writing us at P.O. Box 141104, Gainesville, FL 32614-1104; calling 352-334-0911; or e-mailing info@nccer.org. More information may be found at our Web site, www.nccer.org.

HOW TO USE THIS ANNOTATED INSTRUCTOR'S GUIDE

Each page presents two sections of information. The larger section displays each page exactly as it appears in the Trainee Module. The narrow column ties suggested trainee and instructor actions to each page and provides icons to call your attention to material, safety, audiovisual, or testing requirements. The bottom of each page includes space for your notes.

If you see the Teaching Tip icon, that means there is a teaching tip associated with this section. Also refer to any suggested teaching tips at the end of the module.

SAFETY CONSIDERATIONS

Ensure that the trainees are equipped with appropriate personal protective equipment. Emphasize the importance of following all safety precautions and procedures when working with power tools.

PREPARATION

Before teaching this module, you should review the Module Outline, Objectives, Performance Tasks, and the Materials and Equipment List. Be sure to allow ample time to prepare your own training or lesson plan and gather all required equipment and materials.

MATERIALS AND EQUIPMENT LIST

Transparencies

Markers/chalk

Blank acetate sheets

Transparency pens

Pencils and scratch paper

Module Examinations*

Performance Profile Sheets*

Overhead projector and screen

Whiteboard/chalkboard

Appropriate personal protective equipment

Copies of *49 CFR Part 192* (Gas) and/or *49 CFR Part 195* (Liquid)

Copies of your company policy and procedures manual

Copies of *Figures 1, 2,* and *3* with the callouts covered

Copies of manufacturers' specifications for gas compressors

Copies of manufacturers' maintenance manuals, including recommended startup checklists

Gas compressors on which to practice maintenance and repair techniques

Tools and materials to repair common rotary compressor components, including:
- Mechanical seals
- O-rings
- Thread caulking
- Silicone caulking
- Oil pumps
- Couplings
- Lubricant
- Fasteners
- Oil
- Straightedges

Tools and materials to repair common reciprocating compressor components, including:
- Crossheads
- Drive chains
- Screws
- Nuts
- Dust plugs
- Sprockets
- Drift punches
- Washers
- Valve caps
- Bolts
- O-rings
- Machinist's rules
- Lube oil pump chain sprockets
- Sprocket setscrews
- Woodruff keys
- Gaskets
- Metal files
- Lubricator
- Pry bars
- Screwdrivers
- Valve tools
- Anti-seize lubricant
- Valve retainer
- Lubricant

Sample documentation for startup and operational checks

* Located in the Test Booklet.

ADDITIONAL RESOURCES

This module is intended to present thorough resources for task training. The following reference works are suggested for both instructors and motivated trainees interested in further study. These are optional materials for continued education rather than for task training.

Heavy Duty Oil Flooded Twin Screw Rotary Compressors Technical Manual For Models: AR166, AR208, AR260K & AR260, March 2002. Mount Vernon, Ohio: Ariel Corporation.

Heavy Duty Balanced Opposed Compressors Technical Manual For Models: JGM, JGN, JGP and JGQ; and JGI (Vertical Non-Balanced), January 1999. Mount Vernon, Ohio: Ariel Corporation.

Heavy Duty Balanced Opposed Compressors Technical Manual For Models: JGW, JGR and JGJ, January 2001. Mount Vernon, Ohio: Ariel Corporation.

TEACHING TIME FOR THIS MODULE

An outline for use in developing your lesson plan is presented below. Note that each Roman numeral in the outline equates to one session of instruction. Each session has a suggested time period of 2½ hours. This includes 10 minutes at the beginning of each session for administrative tasks and one 10-minute break during the session. Approximately 15 hours are suggested to cover *Maintain, Troubleshoot, and Repair Gas Compressors.* You will need to adjust the time required for hands-on activity and testing based on your class size and resources.

Topic	Planned Time
Session I. Lubrication Systems	
A. Oil-Flooded Design	___________
B. Force-Feed Design	___________
1. Performance Testing	___________
Session II. Performing Preventive Maintenance on Gas Compressors	
A. Preventive Maintenance for Rotary Compressors	___________
B. Preventive Maintenance for Reciprocating Compressors	___________
1. Performance Testing	___________
Session III. Troubleshooting, Preparing, and Isolating Gas Compressors	
A. Troubleshooting Gas Compressors	___________
1. Performance Testing	___________
B. Preparing for Shutdown and Repair	___________
C. Isolating Gas Compressors for Repair	___________
Session IV. Repairing Gas Compressor Components, Part One	
A. Repairing Common Rotary Compressor Components	___________
B. Practicing Procedures	___________
1. Laboratories	___________
Session V. Repairing Gas Compressor Components, Part Two	
A. Repairing Common Reciprocating Compressor Components	___________
B. Practicing Procedures	___________
1. Laboratories	___________
Session VI. Preparing for Startup and Operational Check	
A. Startup and Operational Check for Rotary Compressors	___________
B. Startup and Operational Check for Reciprocating Compressors	___________
C. Summary	
1. Summarize module	___________
2. Answer questions	___________

D. Module Examination ______________

1. Trainees must score 70% or higher to receive recognition from the NCCER.
2. Record the testing results on Craft Training Report Form 200 and submit the results to the Training Program Sponsor.

E. Performance Testing ______________

1. Trainees must perform each task to the satisfaction of the instructor to receive recognition from the NCCER.
2. Record the testing results on Craft Training Report Form 200 and submit the results to the Training Program Sponsor.

Module 63305-02

Maintain, Troubleshoot, and Repair Gas Compressors

Instructor's Notes:

Instructor's Notes:

Introduction to the Pipeline Operator Qualification Rule, API, and NCCER

It could be said that petroleum products keep America running. Petroleum products like oil and natural gas are the principal sources of energy in the United States and are used in many manufacturing and chemical processes. Petroleum products are primarily transported by pipeline. There are more than a million miles of liquid and natural gas pipeline in this country, ranging in diameter from two inches to four feet. These pipelines bring crude oil and natural gas from production fields to processing plants and refineries—and ultimately to the consumer.

Pipeline Operator Qualification

Petroleum products are considered hazardous, so pipeline personnel must take extra precautions and follow specific procedures when working with them. To help ensure the safe handling and transportation of petroleum products, the United States Department of Transportation (DOT) mandates safe practices as federal regulations.

One such regulation, the Operator Qualification (OQ) rule, requires pipeline operators to develop a written qualification plan to evaluate their personnel and their contract personnel (see *49 CFR Part 192 Subpart N—Gas,* and *49 CFR Part 195 Subpart G—Liquid*). The evaluation hinges on the operator's ability to perform certain covered tasks and to be able to recognize and respond to abnormal operating conditions that may be encountered while performing these covered tasks.

Covered Tasks

A pipeline performance task is deemed *covered* if the activity passes the following four-part test:

1. Is it performed on a pipeline facility?
2. Is it an operations or maintenance task?
3. Is it performed as a requirement of *49 CFR Part 195 (Liquid)* or *49 CFR Part 192 (Gas)*?
4. Does it affect the operation or integrity of the pipeline facility?

If the answer to all four questions is *yes*, then the activity is considered *covered* and must figure into the evaluation process of pipeline personnel.

American Petroleum Institute

The OQ rule and its related covered tasks provided the catalyst to develop a unified pipeline training and assessment program—so that pipeline operators and owners have ready access to the assessment and training tools they need. This effort was the direct result of the vision and leadership provided by The American Petroleum Institute (API) and API members.

Efforts to unify the industry are nothing new to API. Indeed, the impetus for forming the Institute in 1919 was the need to standardize engineering specifications for drilling and production equipment. Today, the most pressing issues revolve around public perceptions and government policies toward the industry. Speaking with one voice on these issues has become as essential as having interchangeable parts in the field. API's members determine the petroleum industry's positions on public policy as well as the standards that govern its day-to-day operations worldwide.

Consortium on Operator Qualification

API assembled a focus group whose primary purpose was to identify covered tasks and provide compliance recommendations to pipeline operators.

The Consortium on Operator Qualification (COOQ) became the guiding force for the identification and development of covered tasks for the liquid pipeline industry. Their work product, API publication 1161, *Guidance Document for the Qualification of Liquid Pipeline Personnel,* formed the basis for the resulting Pipeliner Training and Assessment Program.

The covered task list created in API publication 1161 has now evolved to include gas covered tasks and is maintained by NCCER. The most current covered task list can be downloaded at www.nccer.org.

National Center for Construction Education and Research

With the guidance and support of API, its members, and the Consortium on Operator Qualification, the National Center for Construction Education and Research (NCCER) was able to design and implement the Pipeliner Training and Assessment Program.

NCCER is a not-for-profit 501(c)(3) education foundation established in 1995 by the world's largest and most progressive construction and maintenance contractors and national contractor associations. NCCER was founded to address the

severe workforce shortage facing the construction and maintenance industries and to develop a standardized training process and curricula. Today, NCCER is supported by hundreds of leading construction and maintenance companies, manufacturers, and national associations and developed more than 35 construction and maintenance titles.

NCCER also maintains a National Registry, which contains the records of trainees and instructors who have received NCCER training. All completion information for every trainee is sent to the NCCER and filed within the Registry. The Registry can then confirm training and skills for workers as they move from company to company, state to state, or even within their own company.

Standardized Curriculum (NCCER's Contren™ Learning Series)

NCCER curricula are developed using the expertise of industry subject matter experts (SMEs). The use of SMEs as an integral part of every new development or revision ensures that the curriculum is current, industry-driven, and indicative of industry practices. The development and/or revision process is facilitated from start to finish by NCCER project managers, who work closely with the SMEs and a select team of technical writer(s) on the development and presentation of the curriculum.

Pipeline: Five Training Paths

The pipeline curriculum focuses on essential skills and competencies within five skill areas:

- Corrosion Control
- Electrical and Instrumentation
- Maintenance
- Mechanical
- Operations (Field and Control Center)

SMEs from the pipeline industry have determined the skills and competencies needed based on API's covered tasks and the consensus of the group/industry. Specific company information is generally not incorporated into the curriculum unless the SMEs determine that the information is the industry standard.

Training Modules and Covered Tasks

The smallest unit of instruction in the curriculum is a *module*. Each curriculum is composed of modules (the equivalent of stand-alone chapters) that are logically sequenced into complete *levels* of instruction. Every module is task-oriented and requires the successful completion of a written (knowledge-based) examination and a performance (hands-on) test. Modules may be individually selected and/or combined for efficient task training. Modules in the pipeline curricula, to the extent possible, will align with API-developed covered tasks. While not all modules will relate to covered tasks, all covered tasks will relate to at least one module.

Benefits of Training to the Pipeline Marketplace

Applying NCCER's standardized training process allows pipeline owners and contractors in the United States to establish the benchmark for acceptable skill and competency levels within the pipeline industry. This process will assist training organizations in developing and/or improving standardized pipeline training programs that effectively match their resources and operating conditions while responding to regulatory demands and to the needs of the industry. The process will also recognize training organizations that meet the minimum standards of competence, warranting regulatory, public, and professional confidence. This in turn will foster national unity within the pipeline industry for the mutual benefit of individual technicians, owners, contractors, and ultimately pipeline-users and society.

Additional Resources

To review the Pipeliner Training and Assessment Program Implementation and Resource Guide and the latest version of pipeline covered tasks, visit NCCER's Web site at www.nccer.org, or contact:

National Center for Construction Education and Research
PO Box 141104
Gainesville, FL 32614-1104
(888) 622-3720
Web site: www.nccer.org

For more information on API, please contact:

American Petroleum Institute
1220 L Street, Northwest
Washington, D.C. 20005
(202) 682-8000
Web site: www.api.org

To order *API Pub 1161*, contact Global Engineering Documents by calling 1-800-854-7149 or visiting their Web site at www.global.his.com.

Additional sources of information on the pipeline operator qualification regulation include:

DOT – Office of Pipeline Safety
www.ops.dot.gov

Transportation Safety Institute
www.tsi.dot.gov

Instructor's Notes:

COURSE MAP

This course map shows all of the modules in the third level of the Pipeline Mechanical curriculum. The suggested training order begins at the bottom and proceeds up. Skill levels increase as you advance on the course map. The local Training Program Sponsor may adjust the training order.

PIPELINE MECHANICAL LEVEL THREE

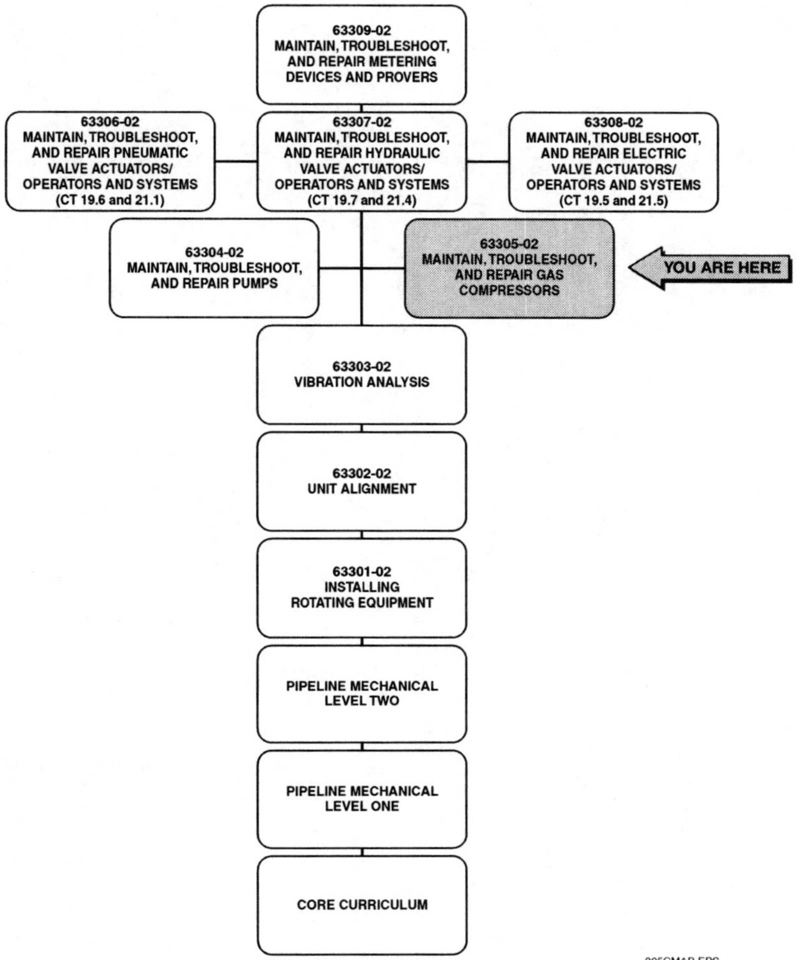

305CMAP.EPS

Homework

Assign reading of Module 63305-02.

Instructor's Notes:

MODULE 63305-02 CONTENTS

MODULE 63305-02 CONTENTS (Continued)

Figures

Tables

Instructor's Notes:

MODULE 63305-02

Maintain, Troubleshoot, and Repair Gas Compressors

Objectives

When you have completed this module, you will be able to do the following:

1. Identify the typical lubrication system components of a gas compressor.
2. Describe the preventive maintenance requirements for a gas compressor.
3. Identify the common troubleshooting techniques for a gas compressor.
4. Identify the common steps required to prepare for shutdown and repair of a gas compressor.
5. Identify the common steps required to isolate a gas compressor from a pipeline system.
6. Identify the common steps required to repair a rotary and reciprocating gas compressor.
7. Identify the common steps required to prepare the gas compressor for start-up and operational check after maintenance has been completed.

Prerequisites

Before you begin this module, it is recommended that you successfully complete the following: Core Curriculum; Pipeline Core, Modules 66101-02 and 66102-02; Pipeline Mechanical Level One; Pipeline Mechanical Level Two; Pipeline Mechanical Level Three, Modules 63301-02 through 63303-02.

Required Trainee Materials

1. Pencil and paper
2. Appropriate personal protective equipment

1.0.0 ◆ INTRODUCTION

Gas must be compressed in order to achieve gas flow through a pipeline. Gas compressors are used for this purpose. Compression requirements must be carefully calculated to size the compressors and determine the number and spacing of compressors along the pipeline system to ensure pipeline safety and efficiency.

There are many different types and designs of gas compressors. The most common to gas transportation are the rotary and reciprocating compressors. This module will introduce you to gas compressor troubleshooting and maintenance procedures.

These repair procedures for a rotary gas compressor are generalized from a technical manual from Ariel Corporation for their heavy duty oil flooded twin screw rotary compressors. The repair procedures for a reciprocating gas compressor are generalized from a technical manual from Ariel Corporation for their heavy duty balanced opposed compressors. Permission has been granted from Ariel Corporation to use this material. The actual procedures have been generalized to give an overview of the steps required to repair a typical rotary or reciprocating compressor.

Always remember to follow the compressor manufacturer's maintenance manual and specifications as well as company policy and procedures when performing any work on a gas compressor. Ensure that the compressor is not under any type of manufacturer's warranty before repairing or disassembling the compressor.

Materials

Ensure that you have everything required to teach the course. Check the Materials and Equipment List at the front of this module.

Classroom

Explain that terms shown in bold are defined in the Glossary at the back of this module.

Audiovisual

Show Transparency 1, Course Objectives.

Show Transparency 2, Performance Tasks.

Teaching Tip

See the general Teaching Tip at the end of this module.

Classroom

Review the common types and designs of gas compressors.

Stress the importance of following manufacturers' maintenance manuals when working with compressors.

Classroom

Review the functions of lubrication in a compressor. Explain the concepts of viscosity index, lubricity, and wetability.

Compare oil-flooded and force-feed designed compressors. Answer any questions trainees have.

Audiovisual

Show Transparency 3 (Figure 1). Review the components in a typical rotary screw compressor system.

Teaching Tip

See the Teaching Tip for Section 2.0.0 at the end of this module.

Performance Testing

Have each trainee perform Performance Task 1 to your satisfaction. Fill out a Performance Profile Sheet for each trainee.

2.0.0 ◆ LUBRICATION SYSTEMS

Lubrication is vital for the successful operation of a compressor and deserves special attention for long operational life. The lubrication performs at least six functions in a compressor:

- *Reduces friction* – Decreasing friction decreases energy requirement and heat buildup.
- *Reduces wear* – Decreasing wear increases equipment life expectancy and decreases maintenance costs.
- *Cools rubbing surfaces* – Cooling of rubbing parts maintains working tolerances, extends oil life, and removes heat from the system.
- *Prevents corrosion* – Minimization of surface corrosion decreases friction, heat, and component wear. Corrosion protection is generally provided by additives rather than the base lubricant.
- *Seals and reduces contaminant buildup* – Lubrication improves gas seal on piston rings and packing rings, as well as flushes away contaminants from moving parts.
- *Dampens shock* – Shock loads are cushioned, thereby reducing vibration and noise, increasing component life.

Liquid lubricants commonly used in compressors include petroleum-based lubricants and synthetic fluids. Lubricant additives improve the **viscosity index** and **lubricity**, inhibit oxidation and rust formation, improve detergency, provide antiwear protection, increase **wetability**, and resist washing of the lubricant.

If a compressor is exposed to cold temperatures, the oil system should be designed so that the unit may be safely started with adequate oil flow to the main bearings. Temperature-controlled cooler bypass valves, oil heaters, cooler louvers, and even buildings may be required to ensure successful operation. Cold weather installation may require the use of multiviscosity oils in the compressor frame as long as the oil is certified as shear-stable. The viscosity of shear-stable oil does not degrade through use. Multiviscosity oils are subject to a shorter oil life than single grade oils by 30% to 50%.

Electric-motor-driven compressors must have a prelube pump to ensure oil flow prior to startup. A **start permissive** should disable the startup sequence if oil pressure is below 10 pounds per square inch gauge (psig). An automatic prelube cycle is recommended for all electric drive systems and is required with unattended start sequences or frequent starts.

You should always refer to the gas compressor manufacturer's maintenance manual and your lubricant supplier to identify the specific lubricant that is best suited to your application and environment. Compressors are typically lubricated using the oil-flooded design and the force-feed design.

2.1.0 Oil-Flooded Design

Rotary screw compressors are oil flooded and normally use a pressurized, filtered, recirculation oil system. Oil is supplied to the compressor through rotor injection and various oil connections. *Figure 1* shows a typical rotary screw compressor lube oil system.

Oil flooding of the compression chamber is supplied at rotor injection. The oil manifold supplies lubrication to the bearings, oil to the mechanical seal, and hydraulic power to position the slide valve.

2.2.0 Force-Feed Design

Most reciprocating compressors use a force-feed lubrication system. The force-feed lubrication system provides oil to the compressor cylinders and the piston rod packing. *Figure 2* shows a typical reciprocating compressor force-feed lubrication system.

The force-feed lubricator needs to be adjusted before startup and after the compressor has been in operation. An indicator pin on the distributor block shows the rate at which the block is cycling. To adjust the lubricator, screw down the feed regulator until the indicator strokes at the proper rate. *Figure 3* shows typical force-feed lubricators.

Most compressors will have a no-oil or low-oil shut-down switch and a high-oil temperature cut-off switch installed in the oil system. These are installed to protect the compressor from the lack of lubrication or excessive oil temperature. The settings and adjustments will vary depending on the type of compressor used and the environment where the compressor is installed.

3.0.0 ◆ PERFORMING PREVENTIVE MAINTENANCE ON GAS COMPRESSORS

Preventive maintenance inspections are critical to efficient and safe operation of gas compressors on pipelines. These inspections are typically broken down into two actions: inspections and corrective actions.

Homework

Have trainees review Sections 3.0.0–3.2.7 for the next session.

Materials

Ensure that you have everything required for testing during this session.

Instructor's Notes:

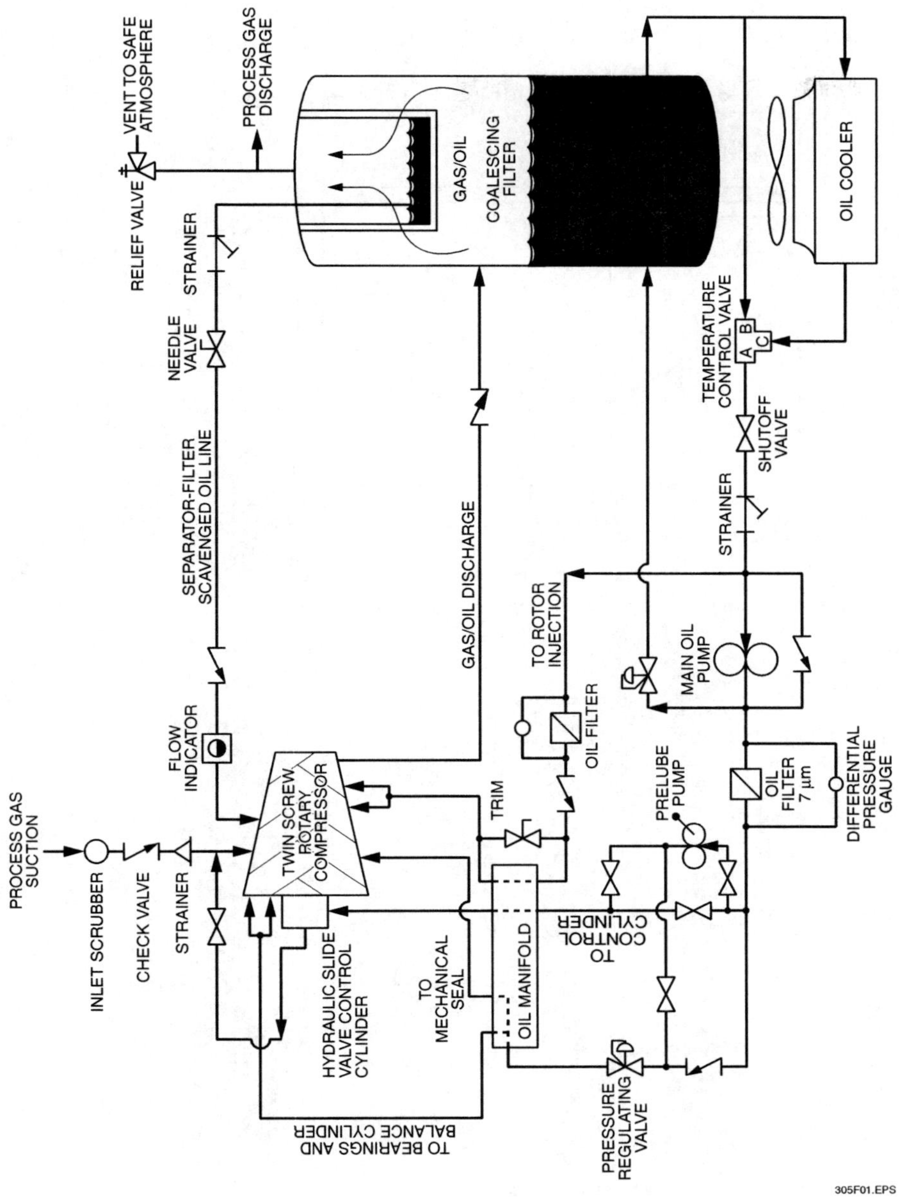

Figure 1 ◆ Rotary screw compressor oil lube system.

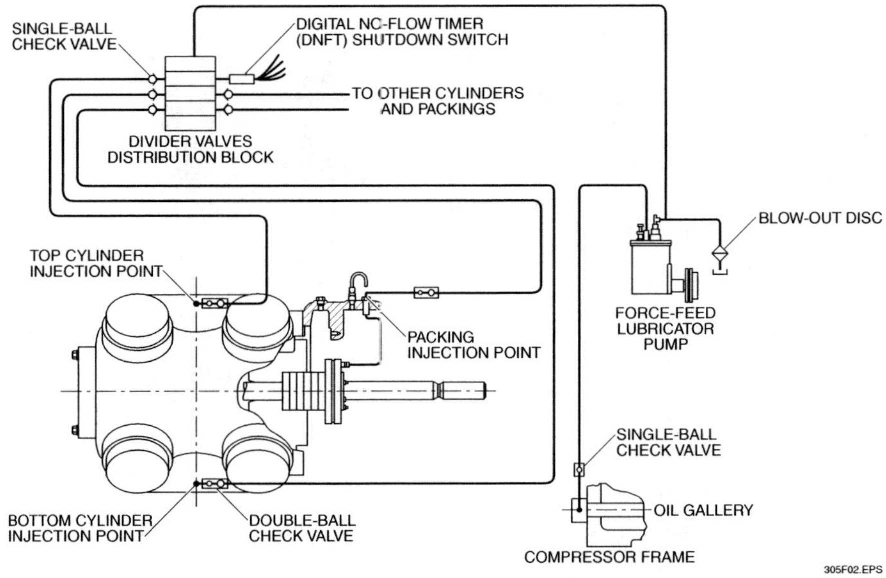

Figure 2 ◈ Reciprocating compressor force-feed system.

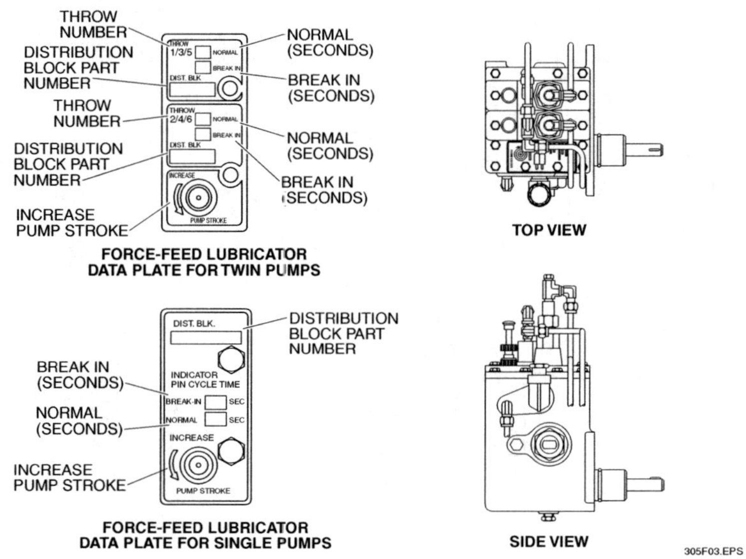

Figure 3 ◈ Force-feed lubricators.

Instructor's Notes:

Inspections are checks to evaluate the condition of the equipment. When such inspections determine that the equipment is in good condition, no corrective action is needed. When the inspections identify a problem, action must be taken to resolve the problem immediately. In some situations, the problem may be solved by the inspector. In other situations, the inspector may have to call someone else authorized to correct the problem or initiate a work order to correct the problem.

> **NOTE**
> Supervisory Control and Data Acquisition (SCADA) monitors compressor operations and detects faulty operations indicating that a compressor may need maintenance.

3.1.0 Preventive Maintenance for Rotary Compressors

The following are some common preventive maintenance tasks and intervals that are recommended for a rotary compressor. Always refer to the manufacturer's maintenance manual for specific preventive maintenance tasks for your compressor. Document the findings of the inspection per company policy and procedures as well.

Preventive maintenance for rotary compressors includes:

- Daily inspection
- Monthly inspection
- Semiannual inspection
- Annual inspection

3.1.1 Daily Inspection

Follow these steps to perform a daily inspection of any type of gas compressor:

Step 1 Check oil pressure and temperatures.

> **NOTE**
> If there is a significant change in the recorded oil pressure and temperature readings, an inspection of the compressor should be completed as soon as possible.

Step 2 Check the gas/oil separator oil level when the compressor is in operation. It should be in middle range of the sight glass. If it is not, determine and correct the cause.

Step 3 Check the seal vent for blowing or excessive leakage. If either exists, determine the cause and, if necessary, replace the mechanical seal parts.

Step 4 Check and repair any gas leaks.

Step 5 Check and repair any oil leaks.

Step 6 Check and record the operating pressure and temperatures.

> **NOTE**
> If there is a significant change in the recorded operating pressure and temperature readings, an inspection of the compressor should be completed as soon as possible.

Step 7 Check for unusual noises or vibrations.

Step 8 Check the gas/oil separator scavenged oil flow indicator for oil and entrained gas flow. Adjust as required.

Step 9 Check the oil filter differential pressures.

Step 10 Check the slide valve position. Adjust as required.

Step 11 Check the counter thrust gas balance pressure. Adjust as required.

Step 12 Check the gas suction strainer for differential pressure.

3.1.2 Monthly Inspection

Follow these steps to perform a monthly inspection of a gas compressor:

Step 1 Complete the daily inspection requirements.

Step 2 Check and confirm the safety shut-down functions per any applicable company procedures.

Step 3 Adjust the counterthrust gas balance pressure for current operating conditions.

Step 4 Review the oil analysis report to determine if the oil should be changed.

Step 5 Inspect the anchor bolts and shims for looseness. Tighten as required.

Classroom

Explain how to respond to problems identified during inspections.

Review the daily and monthly preventive maintenance tasks for rotary compressors.

Classroom

Discuss the elements of a semiannual and annual inspection for rotary compressors. Go over any questions trainees have.

Refer to company policies. Ask trainees to review policies for documenting inspection findings. Have trainees practice documenting information according to company policy.

Compare the daily, monthly, and semiannual preventive maintenance tasks for reciprocating compressors with those for rotary compressors.

3.1.3 Semiannual Inspection

Follow these steps to perform a semiannual inspection of a gas compressor twice a year or after the compressor has operated for 4,000 hours:

Step 1 Complete the daily and monthly inspection requirements.

Step 2 Check the oil viscosity requirements for current gas analysis and operating conditions.

Step 3 Verify that the oil is suitable for continued use. A more frequent oil change interval may be required if operating in an extremely dirty environment, if the oil supplier recommends it, or if an oil analysis dictates it. A less frequent oil change interval may be required if the oil is replenished on a regular basis due to usage.

Step 4 Clean the oil and gas strainer.

Step 5 Retighten the hold-down studnuts to proper torque values, and perform a soft foot check.

> **NOTE**
> More than 0.002 inch pull-down requires reshimming.

Step 6 Inspect the coupling discs for cracking and/or axial distortion.

> **NOTE**
> Realign if necessary to hold coupling alignment to within 0.002 inch.

3.1.4 Annual Inspection

Follow these steps to perform a yearly inspection of a gas compressor once a year or after the compressor has operated for 8,000 hours:

Step 1 Complete the daily, monthly, and semiannual inspections.

Step 2 Inspect the compressor mounting for twist or bending by checking shimming of the compressor feet.

Step 3 Check and recalibrate all temperature and pressure gauges.

Step 4 Change the gas/oil separator filter element now, before the differential pressure exceeds collapse pressure, or as specified by the manufacturer or oil consultant.

3.2.0 Preventive Maintenance for Reciprocating Compressors

The following are some common preventive maintenance tasks and intervals that are recommended for a reciprocating compressor. Always refer to the manufacturer's maintenance manual for specific preventive maintenance tasks for your compressor. Document the findings of the inspection per company policy and procedures.

Preventive maintenance for reciprocating compressors includes:

- Daily inspection
- Monthly inspection
- Semiannual inspection
- Annual inspection
- Two-year inspection
- Four-year inspection
- Six-year inspection

3.2.1 Daily Inspection

Follow these steps to perform a daily inspection of a gas reciprocating compressor:

Step 1 Check the compressor oil pressure.

> **NOTE**
> For most compressors the oil pressure should be 50 to 60 psig when at operating temperature. Compressor inlet oil temperature is 190°F maximum.

Step 2 Check the compressor oil level.

> **NOTE**
> The oil level should be near the middle of the sight glass when the compressor is running. If not, determine and correct the problem. Do not overfill.

Step 3 Check the oil makeup tank for sufficient oil supply.

Step 4 Check the lubricator block cycle movement indicator.

> **NOTE**
> Refer to the information plate on top of the lubricator box for the correct cycle time. Very dirty or wet gas may require a more frequent cycle time than normal.

Step 5 Check and correct any gas leaks.

Instructor's Notes:

Step 6 Check the primary and secondary packing vents for blowing.

> **NOTE**
> If blowing exists, determine the cause and, if necessary, replace the internal parts of the packing.

Step 7 Check and correct any oil leak.

Step 8 Check the operating pressure and temperatures, and compare them with previously recorded pressures and temperatures.

> **NOTE**
> If not normal, determine the cause of the abnormality.

Step 9 Check the shutdown setpoints.

Step 10 Verify that the low-oil pressure shutdown is a minimum of 35 psi.

Step 11 Verify that the high-temperature shutdown is set as specified by the manufacturer's specifications.

Step 12 Verify that the high-low pressure shutdowns are set as close as practical.

> **NOTE**
> Consideration should be given to the rod load capacity of the compressor.

Step 13 Check the lubricator box oil level.

Step 14 Check for any unusual noises or vibrations.

Step 15 If balance valves are used in the force-feed lubrication system, readjust as required to keep the pump system pressures balanced.

3.2.2 Monthly Inspection

Follow these steps to perform a monthly inspection of a gas reciprocating compressor:

Step 1 Complete the daily inspection.

Step 2 Check and confirm the safety shutdown functions.

Step 3 For cylinders rated greater than 3,500 psi, remove the cylinder heads, and check the cylinder for oil presence to verify lubrication of the cylinder.

3.2.3 Semiannual Inspection

Follow these steps to perform a semiannual inspection of a gas reciprocating compressor twice a year or after the compressor has operated for 4,000 hours:

Step 1 Complete the daily and monthly inspections.

Step 2 Drain and replace the lubricator box oil.

Step 3 Change the oil filter now, when the differential pressure exceeds 10 psi, or as specified by the manufacturer's specifications.

Step 4 Change the compressor oil.

> **NOTE**
> A more frequent oil change interval may be required if operating in an extremely dirty environment, if the oil supplier recommends it, or if an oil analysis dictates it. A less frequent oil change interval may be allowed if the oil is replenished on a regular basis due to the force-feed lubricator usage.

Step 5 Clean the element in the small oil filter supplied on the force-feed lubrication system.

> **NOTE**
> This filter should be cleaned every time the main oil filter is changed.

Step 6 Clean the strainer when the oil is changed.

Step 7 Open the frame when oil is changed and visually inspect for foreign material.

> **NOTE**
> Disassembly of the frame is not recommended unless a problem has been identified.

Step 8 Check the fluid level in the damper, if applicable.

Step 9 Retighten the hold-down studnuts to proper torque valves, and perform a soft foot check as specified by the manufacturer's specifications.

Classroom

Review the procedures for performing annual, two-year, four-year, and six-year inspections of gas reciprocating compressors.

Distribute copies of manufacturers' specifications for gas compressors. Ask trainees to locate and review the inspection requirements. Discuss the preventive maintenance activities outlined in the specifications.

NOTE

More that 0.002 inch pull-down requires reshimming. If reshimming is required, realign if necessary to hold the coupling alignment to within 0.005 inch total indicated runout (TIR).

Step 10 For cylinders rated greater than 3,500 psi, inspect the piston ring end gap as specified by the manufacturer's specifications.

NOTE

Replace the rings that are outside the maximum limits as specified in the manufacturer's maintenance manual.

3.2.4 Annual Inspection

Follow these steps to perform an annual inspection of a gas reciprocating compressor once a year or after the compressor has operated for 8,000 hours:

Step 1 Complete the daily, monthly, and semi-annual inspections.

Step 2 Check the main bearing clearance, connecting rod bearing clearance, and crank thrust clearance with a bar and an indicator as specified by the manufacturer's specifications.

NOTE

If the clearance is outside the limits as specified in the manufacturer's maintenance manual, replace the affected bearings.

Step 3 Check the **crosshead** guide clearance with a feeler gauge, and if it is outside the limits, replace the affected parts.

Step 4 Inspect the valves for broken plates and loose center bolts and replace any broken parts.

Step 5 Tighten the center bolts to the manufacturer's specified torque value.

Step 6 Inspect the cylinder bores for damage and/or wear.

NOTE

If gouged so that the total cross sectional area of the gouges is greater than 0.001 square inch per inch of cylinder circumference, the cylinder should be replaced or rebored to a maximum of 0.020 inch oversize. The cylinder should also be replaced or rebored if the bore is more than 0.001 inch per inch of the cylinder diameter out of round or if it is tapered. Reboring removes the nitride surface of the cylinder bore, so the nitride surface must be reapplied after the boring is completed and then inspected.

Step 7 Inspect the piston ring end gap.

NOTE

Replace the rings if they are outside the maximum limit as specified by the manufacturer's specifications.

Step 8 Inspect the piston rods for damage and excessive wear.

NOTE

If gouged or scratched, replace the rod. If the rod is more than 0.005 inch under size, out of round by more than 0.001 inch, or tapered by more than 0.002 inch, replace the rod.

Step 9 Rebuild the cylinder packing cases.

Step 10 Inspect the frame for twisting or bending by checking the shimming of the compressor feet.

Step 11 Realign the compressor if necessary to hold the coupling alignment to within 0.005 inch TIR.

Step 12 Check and recalibrate all the temperature and pressure gauges.

Step 13 Check and record the compressor rod run-out as specified by the manufacturer's specifications.

Step 14 Grease the stem threads at the grease fittings with two to three pumps of multi-purpose grease using a standard hand-pump grease gun.

Instructor's Notes:

Step 15 Clean the crankcase breather filter.

Step 16 Adjust the drive chains as specified by the manufacturer's specifications.

Step 17 Pressure test the distribution block(s).

3.2.5 Two-Year Inspection

Follow these steps to perform a two-year inspection of a gas reciprocating compressor once every two years or following 16,000 hours of operation:

Step 1 Complete the daily, monthly, semiannual, and annual inspections.

Step 2 Check the auxiliary end chain drive for sprocket teeth undercutting and the chain for excessive stretching as specified by the manufacturer's specifications.

Step 3 Rebuild the oil wiper cases.

3.2.6 Four-Year Inspection

Follow these steps to perform a four-year inspection of a gas reciprocating compressor once every four years or following 32,000 hours of operation:

Step 1 Complete the daily, monthly, semiannual, annual, and two-year inspections.

Step 2 Check the main and connecting rod bearing clearances as specified by the manufacturer's specifications.

> **CAUTION**
> Disassembly to check the clearances is not recommended. Disassembly should only be performed if excessive clearance is indicated.

Step 3 Check the crosshead guide clearance with a feeler gauge.

Step 4 Check the crosshead pins to the crosshead pin bore and connecting rod bushing bore by removing the crosshead pins.

Step 5 Check for excessive wear in the auxiliary end drive chain tightener.

Step 6 Check for excessive ring groove wear in the pistons as specified by the manufacturer's specifications.

3.2.7 Six-Year Inspection

Follow these steps to perform a six-year inspection of a gas reciprocating compressor once every six years or following 48,000 hours of operation:

Step 1 Complete the daily, monthly, semiannual, annual, two-year, and four-year inspections.

Step 2 Replace the main and connecting rod bearing shells and bushings.

Step 3 Replace the lubricator distribution blocks.

Step 4 Replace the crosshead bushings.

4.0.0 ◆ TROUBLESHOOTING A GAS COMPRESSOR

Troubleshooting is the key to solving problems with the compressor operation quickly. You should have a good understanding of the gas pipeline system and the compressors that are installed. The following are common causes of compressor failures:

- Improper lubrication
- Improper alignment and adjustments
- High temperature
- Weak or broken valves

Look for these possible failures first, before you go to other more complex items. If you eliminate these possibilities, start checking other items in the gas compressor system.

Troubleshooting is a matter of identifying a problem and then determining its possible causes. The best method for troubleshooting is the process of elimination. After the problem and its possible causes are identified, the causes should be eliminated one by one, starting with the one that requires the least amount of downtime.

Minor problems for a rotary compressor are often traced to liquid, dirt, improper alignment, and improper adjustments. These problems can usually be corrected by cleaning, adjustment, elimination of an adverse condition, or replacement of relatively minor parts.

Major problems for a rotary compressor can usually be traced to long periods of operation with unsuitable lubrication, improper counterthrust gas balance pressure, or the lack of routine maintenance. It is very important to record the pressures and temperatures at a given load point. This will show any variation in the operation and will indicate a problem. It is not possible to list all the problems that may occur with a rotary compressor; however, *Table 1* shows some of the common problems that may occur.

Safety

Emphasize that disassembly to check the clearances is recommended only if excessive clearance is indicated.

Teaching Tip

See the Teaching Tip for Section 3.0.0 at the end of this module.

Performance Testing

Have each trainee perform Performance Task 2 to your satisfaction. Fill out a Performance Profile Sheet for each trainee.

Homework

Have trainees review Sections 4.0.0–6.0.0 for the next session.

Materials

Ensure that you have everything required for testing during this session.

Classroom

Review the common causes of compressor failure.

Classroom

Refer to Tables 1 and 2. Discuss problems that may occur with rotary and reciprocating compressors and the possible causes.

On the whiteboard/ chalkboard, write a list of symptoms associated with rotary and reciprocating compressors. Have trainees identify the potential problems. Ask whether trainees have any questions.

Table 1 Troubleshooting Chart for a Rotary Compressor (1 of 2)

Problem	Possible Cause
Low oil pressure	• Driven coupling-half loose on pump shaft (coupling disengaged) • Defective pressure gauge • Oil regulating valves improperly adjusted for conditions • Dirty or plugged oil strainer • Dirty oil filter • Improper low oil pressure switch setting • Low oil viscosity • Excessive leakage at seals or bearings • Leaking oil pump check valve • Oil pump wear • Improper end clearance in oil pump
High oil pressure	• Cold oil • Oil regulating valves improperly adjusted for conditions • High oil viscosity • Oil temperature too low
Low oil temperature	• Thermostatic valve element defective • Oil heater or thermostat defective
High oil temperature	• Thermostatic valve element defective • Cooler fan not working • Oil cooler dirty • Low oil viscosity
High gas discharge temperature	• Inadequate rotor injection oil rate • Plugged oil strainer • High oil temperature • High suction temperature • High suction or discharge pressure
Excessive vibration and/or noise in compressor	• Loose compressor hold down bolting • Coupling not properly aligned • Improper lube oil and/or insufficient lube rate • Lubrication failure • Liquids in gas • Excessive rotor end play • Loose or worn bearings • Erratic cylinder wear • Rotor contacting cylinder or heads • Swelled or warped rotor blades
High oil consumption	• Oil system overfilled • Separator-scavenged oil line plugged or throttling valve closed • Excessive oil dilution from process gas • Collapsed coalescing filter element • Oil/gas separator coalescing filter gaskets not sealed or defective • Improper coalescing filter element installed
Oil leakage at slide valve axial position indicator case mounting	• Position indicator drive rod seal damaged
Excessive leakage at mechanical seal weep tube	• Mechanical seal not effective and should be replaced
First-stage discharge air temperature excessive	• Fouled intercooler, causing high first-stage discharge pressure • Unloading valve not fully opened or clogged • Clogged intake filter • Insufficient or high temperature inlet water • Scale or residue built up in water jacket • High inlet air temperature • Insufficient or wrong lubrication oil • Swelled or warped rotor blades

304T01A.EPS

Instructor's Notes:

Table 1 Troubleshooting Chart for a Rotary Compressor (2 of 2)

Problem	Possible Cause
Second-stage discharge air temperature excessive	• First-stage malfunction • Fouled intercooler, causing high inlet air temperature • Fouled aftercooler, causing high second-stage discharge pressure • Scale or residue buildup in water jacket • Insufficient or wrong lubrication oil • Swelled or warped rotor blades
Compressor will not unload properly	• Faulty pressure switch in three-way solenoid pilot valve • Clogged control air line or ports in three-way solenoid pilot valve • Unloading valve dirty or internal parts worn

304T01B.EPS

A reciprocating compressor may develop similar problems during normal operations. A key to long service life for reciprocating compressors is the lubrication systems. Always follow the manufacturer's recommendations. The oil should be analyzed on a regular basis, and the pressures and temperatures should be recorded as well. It is not possible to list all the problems that may occur with a reciprocating compressor either; however, *Table 2* shows some of the common problems that may occur.

5.0.0 ◆ PREPARING FOR SHUTDOWN AND REPAIR

Before repairing or replacing any compressor, you must take steps to prepare for this maintenance activity. It is crucial to relieve all pressure from the system and ensure the working environment is safe. This procedure provides general instructions for preparing a gas compressor for shutdown and repair. Always follow company policy and procedures and the compressor manufacturer's service manual for specific instructions for the compressor that you are repairing.

Follow these steps to prepare the compressor to be shut down and repaired:

Step 1 Obtain and review the past service and inspection reports for this compressor.

> **NOTE**
> During this review, look for any unusual conditions that might need to be addressed during the repair process and how often service has been completed for the compressor. Review the pressure and temperature readings and the lubrication service record for the compressor.

Step 2 Obtain the manufacturer's service manual for the compressor that is to be serviced. Review the repair procedure described in the compressor service manual with particular attention to the components' specifications and tolerances.

Step 3 Obtain all the hand tools, special tools, and instruments required for the repair process.

Step 4 Inspect the tools and instruments for serviceability, and replace those tools or instruments that are not serviceable.

Step 5 Obtain any personal protective equipment (PPE) that is required for the repair process as required by your company's policies.

Step 6 Obtain all the general equipment and supplies required for the repair process, such as a fire extinguisher, waste containers, cleaning solvents, rags, lubricants, and seals.

Step 7 Verify the location of the compressor to be repaired.

Step 8 Survey the area of the gas compressor to be repaired for any unusual conditions, such as a spill, that may present a fire hazard or a toxic leak. Ensure that the area around the compressor is safe to enter and work in.

> **NOTE**
> If the compressor area is not safe to work in, contact the control center and take safe and appropriate action for the condition as directed by the control center and the company's policies and procedures.

Performance Testing

Have each trainee perform Performance Task 3 to your satisfaction. Fill out a Performance Profile Sheet for each trainee.

Classroom

Go over the procedures to prepare a compressor for shutdown and repair.

Emphasize the importance of reviewing past service and inspection reports.

Classroom

Discuss the procedures to isolate a gas compressor for repair. Answer any questions trainees may have.

Review the proper procedures for notifying the operations control center.

Table 2 Troubleshooting Chart for a Reciprocating Compressor

Problem	Possible Cause
Low intercooler pressure (Full load operation)	• Intercooler drain valve open • Plugged filter • Low-pressure suction and/or discharge valves leaking. (Hottest valve is the probable cause.) • Low-pressure unloader plunger spring broken, allowing plunger to rest on valve disc, retarding valve action • Leaking diaphragm in high-pressure suction valve, allowing high-pressure gas to backflow through unloading lines and partly unloading low-pressure valve • Leaking intercooler relief valve • Badly worn low-pressure piston rings. (Check for excessive blow-by through crankcase breather.)
High intercooler pressure (Full operation pressure)	• Leaking high-pressure suction and/or discharge valve. (Hottest valve is the probable cause) • High-pressure unloader plunger spring broken, allowing plunger to rest on valve disc, restricting valve action • Plugged intercooler core
Unit will not unload (Receiver relief valve blows)	• Control stuck; check for dirt or varnish • Diaphragm ruptured in high-pressure unloading valves • Air to control shut off • Filter in control line or control plugged • Unloader line leaking
Low gas pressure	• Compressor too small for demand • Gas leak in system • Plugged or restricted air filters • Suction unloading valves partly open • Badly worn piston rings • Unit under speed
Short valve life	• Dirt entering unit • Lift of valve too great • Corrosive vapors entering unit • Maladjustment of suction unloading plunger • Uneven or worn valve seat • Weak or collapsed valve spring
Noisy valves	• Loose assembly • Unloader spring broken • Broken or weak valve spring • Maladjustment of unloader plunger travel

304T02.EPS

Step 9 Verify the accessibility to the gas compressor by removing any obstruction.

Step 10 Verify the gas compressor equipment number and nameplate data to ensure that you are working on the correct gas compressor.

> **NOTE**
> If the nameplate is missing, contact the control center to verify the compressor and replace the nameplate according to company policy and procedure.

6.0.0 ◆ ISOLATING GAS COMPRESSORS FOR REPAIR

Gas compressors need to be isolated pneumatically, electrically, and mechanically from the pipeline system before any repairs can be made. To isolate a gas compressor for repairs:

Step 1 Verify that all the steps for proper shutdown and preparation for the gas compressor repair have been completed.

Step 2 Notify the operations control center and/or affected personnel that you are beginning the repair in order to maintain awareness of the activities and to ensure safety of the pipeline and the personnel.

Instructor's Notes:

> **NOTE**
> Follow company policy and procedure to complete this task. The purpose of these requirements is to ensure the safety of the pipeline and the personnel who maintain the pipeline system.

Step 3 Isolate all the energy sources.

> **WARNING!**
> All energy sources must be isolated to prevent them from causing equipment damage or personal injury when the gas pressure is relieved.

Step 4 Shut down the power to the compressor prime mover, if applicable.

Step 5 Relieve the pressure throughout the pipeline system, or isolate the compressor from the pipeline system.

Step 6 Lock out and tag out the gas compressor per your company policy.

Step 7 Obtain a line-opening permit, and purge the compressor and lines as required by your company's policies.

> **NOTE**
> The line-opening permit will vary by company policy. Always follow your company policy for opening a gas pipeline system.

Step 8 Ensure that all of the pressure from the gas and oil system has been relieved from the compressor before the repair process begins.

> **WARNING!**
> Pressure left in the system or compressor when a connection is broken could cause personal injury.

Step 9 Clean the compressor with an approved cleaning solvent and lint-free wiping cloths before the repair process begins to ensure that no dirt or contaminant enters the compressor.

> **NOTE**
> The repair process should be completed in a clean work area to prevent dirt or contaminants from coming in contact with the compressor components. Any dirt or contaminant that is not removed from the compressor component will have an adverse effect on the compressor repair and operation. Discard the cleaning solvent according to company policies.

7.0.0 ◆ REPAIRING GAS COMPRESSOR COMPONENTS

The following section describes some common repair procedures for rotary and reciprocating compressors. Always follow the gas compressor manufacturer's service manual for specific guidance and specifications for the model of compressor that you are repairing. Additionally, follow your company's polices and procedures.

7.1.0 Repairing Common Rotary Compressor Components

The major components of a rotary compressor that wear out and require maintenance include the rotor housing, discharge housing, rotors, bearings, mechanical seal, slide valve, thrust balance piston, and oil pump. The mechanical seal, slide valve, slide valve position indicator, and oil pump and coupling may be replaced in the field. If other components need to be replaced, the compressor should be removed, replaced, and returned to the manufacturer or a company-approved vendor for repairs.

Absolute cleanliness, including the use of lint-free wiping cloths, is a necessity during any maintenance activity on the compressor. When access covers have been removed, keep the compressor covered to protect the internal surfaces from debris except when you are actually working within it. Any components that have been removed should be protected from falling objects that might mar or chip machined surfaces.

Whenever a compressor is disassembled, O-rings at nonpressure positions must be carefully inspected before reuse. If the O-ring shows signs of wear or damage, they must be replaced. O-rings at pressure positions should also be replaced. O-ring lubricants must be compatible with O-ring materials, lube oil, and process gas.

Classroom

On the whiteboard/chalkboard, write the first step for preparing a compressor for shutdown. Call on trainees to explain subsequent steps until all of the steps are listed in order. Repeat this same task for the steps to isolate a gas compressor.

Safety

Explain why all energy sources must be isolated.

Emphasize that pressure left in the system can cause serious injuries.

Homework

Have trainees review Sections 7.0.0–7.1.3 for the next session.

Materials

Ensure that you have everything required for the laboratories during this session.

Classroom

Review the major components of a rotary compressor. Stress the importance of cleanliness when performing maintenance activities.

Classroom

Explain the role of the mechanical seal in a rotary compressor.

Audiovisual

Show Transparency 4 (Figure 4). Review the procedure for replacing a mechanical seal.

7.1.1 Replacing the Mechanical Seal

The mechanical seal prevents loss of internal process gas pressure at the male rotor drive around the shaft to the compressor exterior. The seal consists of a rotating inner ring that turns with the rotor and a spring-loaded carbon seal assembly held stationary in the seal housing cover. Sealing is generated by the spring force and the hydraulic force of the lubricating oil pressure acting on the seal faces.

The mechanical seal's critical surfaces, such as seal faces and O-ring seats, must be handled with extreme care. The replacement seal components must be transported and stored in the original unopened packaging. If the replacement seals have been subjected to impact, such as being dropped, they must not be used. The storage of the seal must be dry, dust free, and at a reasonably constant temperature. If the seal is over three years old, it should be inspected by the seal manufacturer before being used or replaced. Refer to *Figure 4* when replacing the mechanical seal.

Follow these steps to replace a mechanical seal:

Step 1 Remove the thread protection caulking compound located between the shaft seal locknut and coupling hub, being careful not to damage the threads.

Step 2 Hold the rotor from turning with the drive train assembled coupling in place, and loosen the shaft seal locknut using a spanner wrench.

> **NOTE**
> Heating the locknut to 200°F maximum may be required to break the seal. Follow the manufacturer's specifications.

Step 3 Disassemble the coupling center section and remove the coupling hub half and retaining bolt from the male rotor.

Step 4 Remove the shaft seal locknut.

Step 5 Remove the screws that secure the seal housing cover to the discharge cover.

Step 6 Using two jack bolts, jack the seal housing cover loose.

Step 7 Slide the seal housing cover and the spacer off the rotor.

> **NOTE**
> Be careful to protect the rotor threads from damage when removing and installing the mechanical seal.

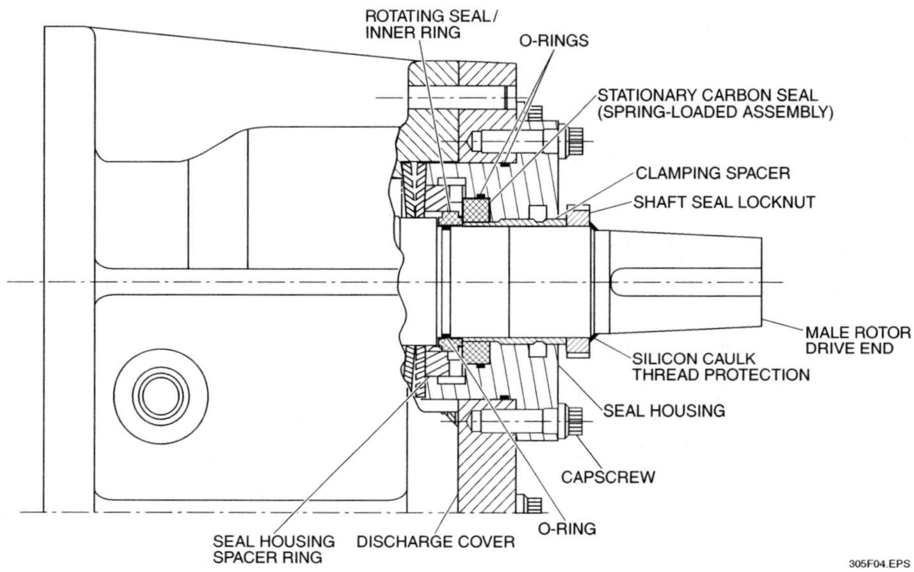

Figure 4 ◆ Top view of a mechanical seal.

Instructor's Notes:

Step 8 Remove the inner seal ring and O-ring from the rotor.

Step 9 Disassemble the seal housing cover internal parts by removing the spacer ring, seal-spring loaded assembly, the O-rings, and the jack bolts.

Step 10 Discard both the mechanical seal elements and all the O-rings and locknuts.

Step 11 Clean the parts while being careful to remove the caulking compound residue from the rotor threads, external and internal, and the coupling retaining bolt threads.

Step 12 Reassemble using a new mechanical seal, new O-rings, and a new locknut.

> **NOTE**
> Lubricate the O-ring with a high-viscosity silicone fluid.

Step 13 Place the O-ring and the inner seal ring in position on the rotor.

> **NOTE**
> On some models, a dowel may be used to help hold it stationary in the seal housing as specified by the manufacturer specifications.

Step 14 Replace the O-rings on the seal housing.

Step 15 Press the seal-spring loaded assembly into the seal housing.

> **NOTE**
> Do not push against the seal face or tap it into place.

Step 16 Reassemble the spacer ring and the clamping spacer into seal housing.

Step 17 Slide the seal housing on to the rotor, being careful to position the leakage weep hole downward while aligning the bolting holes.

Step 18 Install the screws and tighten to the recommended torque value.

Step 19 Apply thread protection caulking compound to the first two external rotor shaft threads, and install the nut on the rotor, but only hand-tight.

Step 20 Temporarily reassemble the drive coupling hub and the key, hand-tight only.

Step 21 Hold the coupling hub to keep the rotor from turning, and tighten the shaft seal locknut to the proper torque value as specified by the manufacturer's specifications.

Step 22 Reassemble the complete coupling to manufacturer's specifications.

Step 23 Apply a bead of silicone caulking compound at the shaft to seal the locknut face, covering all of the exposed shaft threads to help protect it from impact damage and thread corrosion.

> **NOTE**
> The coupling retaining bolt torque is based upon the coupling design and the power to be transmitted. Consult the manufacturer for specific torque value.

7.1.2 Replacing the Oil Pump

Follow these steps to replace the oil pump:

Step 1 Mark the pump suction and the discharge end connections.

Step 2 Remove the oil piping from the pump.

Step 3 Remove the screws or studnuts and slide the pump out, separating the coupling halves at the sleeve.

> **NOTE**
> The coupling-driven half should come out attached to the pump shaft, and the coupling sleeve should be retained by the drive half of the coupling. The drive coupling should be bolted to the female rotor.

Step 4 Loosen the key setscrew and remove the gear coupling half from the pump shaft with a gear puller.

Step 5 Inspect the coupling's geared teeth for excessive wear and damage.

> **NOTE**
> If there is excessive wear or damage, replace the entire coupling with a new assembly.

Laboratory

Set up several stations with the tools and materials necessary to replace a mechanical seal. Ask trainees to practice the procedure. Go over any questions trainees have.

Classroom

Explain how to replace an oil pump in a rotary compressor.

Audiovisual

Show Transparency 5 (Figure 5). Review the procedures for replacing the oil pump coupling.

Step 6 Replace the pump with a new pump that is recommended for the compressor application.

> **NOTE**
> Always refer to the manufacturer's recommendations for the correct replacement pump.

Step 7 Install the coupling-driven half to the new pump shaft with the key installed and firmly seated.

> **NOTE**
> There should be a slight clearance of 0.005 inch between the top of the key and the mating coupling or as specified by the manufacturer specifications.

Step 8 Position the coupling half on the pump shaft so the axial spacing between the coupling halves is 3⁄32 inch or as specified by the manufacturer specifications.

Step 9 Tighten the setscrew.

Step 10 Replace the O-rings.

Step 11 Lubricate the adapter cover plate bore and the coupling sleeve teeth with clean oil.

Step 12 Slide the pump into place with the shaft position rotated to properly engage the gear teeth.

> **NOTE**
> Be sure that the pump flow is in the correct direction.

Step 13 Replace the fasteners and tighten to the proper torque.

Step 14 Pour a small quantity of oil into the pump intake and reattach the oil piping.

7.1.3 Replacing the Oil Pump Coupling

Follow these steps to replace the oil pump coupling. Refer to *Figure 5* when performing this procedure:

Step 1 Mark the pump suction and the discharge end connections.

Step 2 Remove the oil piping from the pump.

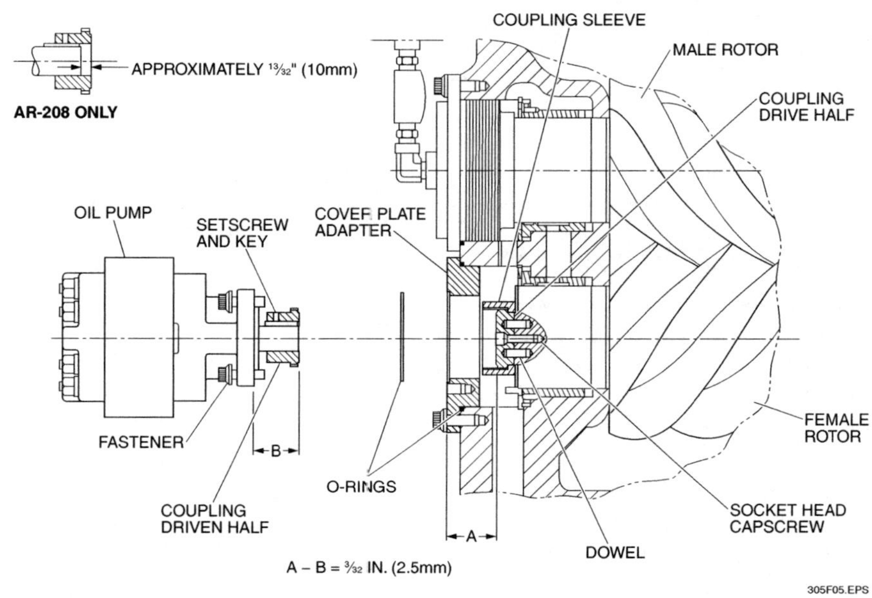

Figure 5 ◆ Oil pump and coupling.

Instructor's Notes:

Step 3 Remove the screws or studnuts and slide the pump out, separating the coupling halves at the sleeve.

> **NOTE**
> The coupling driven half should come out attached to the pump shaft, and the coupling sleeve should be retained by the drive half of the coupling. The drive coupling should be bolted to the female rotor.

Step 4 Loosen the key set screw and remove the gear coupling half from the pump shaft with a gear puller.

Step 5 Inspect the coupling's geared teeth for excessive wear or damage.

> **NOTE**
> If there is excessive wear or damage, replace the entire coupling with a new coupling assembly.

Step 6 Remove the pump.

Step 7 Remove the adapter cover plate fasteners, the cover plate, and the O-ring.

Step 8 Discard the O-ring.

Step 9 Remove the drive coupling half socket head screw.

Step 10 Remove the drive coupling half and sleeve assembly, being careful not to cock the coupling half and break the hardened dowel pins.

Step 11 Replace the coupling drive half and sleeve with new parts.

Step 12 Position the coupling drive half with the sleeve installed over the dowel pins and draw up by tightening the screw to the proper torque as specified by the manufacturer's specifications.

Step 13 Verify that the sleeve is free to pivot on the drive coupling half gear teeth.

Step 14 Lubricate the rotor housing bore with clean oil and install the adapter cover plate with a new O-ring.

Step 15 Replace the fasteners and tighten to the proper torque as specified by the manufacturer specifications.

Step 16 Position the new coupling driven half on the end of the pump shaft.

Step 17 Slide the pump into place with the shaft position rotated to properly engage the gear teeth.

> **NOTE**
> Be sure that the pump flow is in the correct direction.

Step 18 Replace the fasteners and tighten to the proper torque as specified by the manufacturer's specifications.

Step 19 Check the oil level in the pump and fill to the proper level.

7.2.0 Repairing Common Reciprocating Compressor Components

The major components of a reciprocating compressor frame assembly are the crankcase or frame with integral crosshead guides, crankshaft and bearings, connecting rods, chain drive system, and crossheads.

Absolute cleanliness, including the use of lint-free wiping cloths is a necessity during any maintenance on the compressor. When access covers have been removed, keep the frame covered to protect the interior from dust except when you are actually working on the compressor. Any component that has been removed should be protected from falling objects that might mar or chip the surfaces.

Whenever the machine is disassembled, gaskets at nonpressure positions must be carefully inspected before reuse. Damaged gaskets must be replaced. Gaskets at pressure locations should be replaced. Always apply an anti-seize lubricant to both sides of the gaskets for easy removal at a later date. During a major overhaul, drain and flush the crankcase.

7.2.1 Removing and Installing the Crosshead

Follow these steps to remove and install a crosshead. Refer to *Figure 6* when performing this procedure:

Step 1 Remove the crosshead guide side covers and cylinder head.

> **CAUTION**
> Before removing the cylinder head, back off all screws to ⅛ inch. Ensure that the head is loose and that the cylinder is vented.

Laboratory

Ask a trainee to demonstrate the steps for replacing an oil pump and the oil pump coupling. Have trainees explain what was done correctly and/or incorrectly.

Homework

Have trainees review Sections 7.2.0–7.2.6 for the next session.

Materials

Ensure that you have everything required for the demonstrations and laboratories during this session.

Classroom

Review the major components of a reciprocating compressor.

Safety

Explain how to back off all screws before removing the cylinder head.

Audiovisual

Show Transparency 6 (Figure 6). Explain how to remove and install a crosshead.

Classroom

Refer to Figure 7. Explain how to replace the bushing in a crosshead.

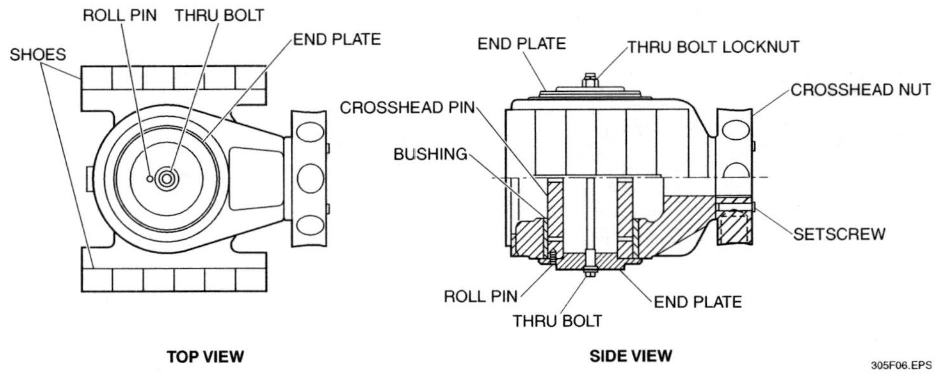

Figure 6 ◆ Crosshead.

Step 2 Move the crosshead to its inner dead center position and back off, but do not remove the crosshead set screws.

Step 3 Loosen the crosshead nut.

Step 4 Screw the piston rod out of the crosshead.

Step 5 Turn the crosshead nut off the piston rod.

Step 6 Push the rod end forward to the edge of the packing to provide a clearance for the crosshead removal.

Step 7 With the crosshead in its outer dead center position, remove the crosshead pin thru bolt, the locknut, the end-plates, and the pin.

Step 8 Turn the crankshaft to its inner dead center position.

Step 9 Move the crosshead to its outer dead center position to be free of the connecting rod.

NOTE
Ensure the connecting rod does not drop and damage the crosshead guide surface.

Step 10 Roll the crosshead 90 degrees and remove it through the guide opening.

Step 11 Check the crosshead pin to crosshead/bushing clearance per the manufacturer's specifications.

NOTE
Wear on the pin can be determined by visual inspection. Replace the pin and bushing as required. If the bushing is replaced (see *Figure 7*), the side of the crosshead receiving the new bushing should be directly supported to prevent possible crushing of the crosshead by the press.

Step 12 Visually inspect the shoe surfaces for scoring.

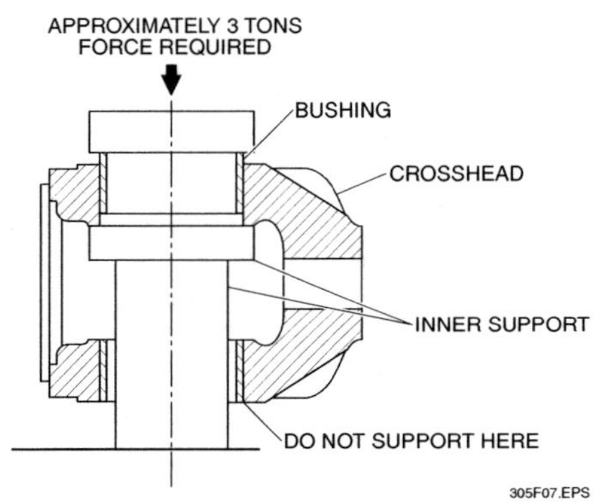

Figure 7 ◆ Crosshead bushing replacement.

Instructor's Notes:

> **NOTE**
> Since they are constantly lubricated under pressure during operation, there should be no signs of wear.

Step 13 Ensure that the crossheads are returned to their original **throw** location for installation.

Step 14 With the crosshead on its side, slip the crosshead into the crosshead guide. Once it is inside, the whole assembly can be rolled to the upright position.

> **NOTE**
> Ensure that it does not become cocked. Should the crosshead become wedged, do not force it. Ease it off and start the installation process again. Be careful not to damage the crosshead shoe surface during the installation process.

Step 15 Turn the crankshaft to its outer dead center position to locate the connecting rod in position and insert the crosshead pin.

Step 16 Install the end plates' thru bolt and the locknut.

Step 17 Tighten the locknut to the torque recommended by the manufacturer.

Step 18 Reinstall the crosshead nut on the piston rod.

> **NOTE**
> Ensure that the setscrew cup points to the crosshead side of the nut.

Step 19 Screw the piston/rod assembly into the crosshead.

> **NOTE**
> Ensure that all the threads are well lubricated with clean, fresh oil to ensure a smooth installation.

Step 20 Set the piston end clearance to the manufacturer's specifications.

> **CAUTION**
> If the piston end clearance is not properly set, serious damage could occur to the gas compressor.

Step 21 Tighten the crosshead nut to the manufacturer's specifications.

Step 22 Apply anti-seize lubricant to the gaskets.

Step 23 Replace the crosshead guide side covers and tighten all screws.

7.2.2 Adjusting the Drive Chain

The chain drive system (see *Figure 8*) is crankshaft-driven at the auxiliary end of the frame. The chain runs the lube oil pump and force-feed lubricator. Chain tightness is controlled by an idler sprocket attached to the eccentric adjustment cap. The chain dips into the crankcase oil and, as a result, is constantly lubricated. Follow these steps to adjust the drive chain:

Step 1 Roll the machine to the tightest position on the chain.

> **NOTE**
> This prevents snugging up on the chain at a slack position and breaking rollers or ruining the pump and the lubricator bearings when the chain goes through its tightest position.

Step 2 Remove the screws and the plugs from the eccentric adjustment cap.

Step 3 Rotate the cap clockwise, facing the gap from the outside, to line up two new screw holes.

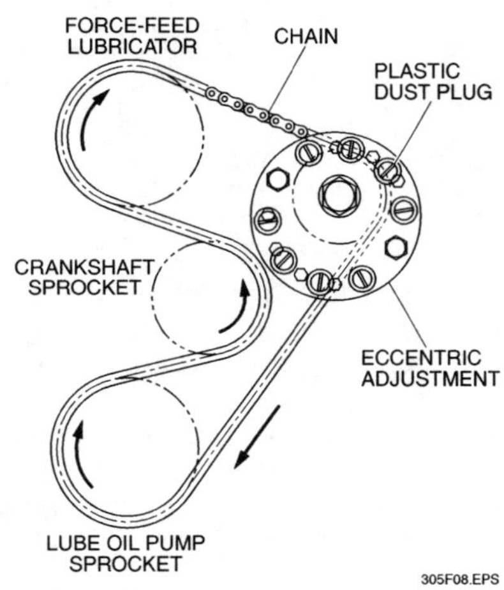

Figure 8 ◆ Chain drive system.

Classroom

Ensure that trainees understand how to return the crossheads to their original throw location. Ask whether trainees have any questions.

Safety

Discuss the consequences of failing to properly set the piston end clearance.

Audiovisual

Show Transparency 7 (Figure 8). Explain how to adjust the drive chain.

Classroom

Review the procedures for replacing a drive and a sprocket.

Refer to Figure 9. Explain how to remove the chain from the idler sprocket.

> **NOTE**
> If this makes the chain too tight, try turning the cap around in a counterclockwise direction for a different hole lineup.

Step 4 Tighten the screws to the recommended torque.

Step 5 Roll the machine to check the tightness in a number of positions.

> **NOTE**
> As its tightest position, the chain should be adjusted so that the deflection of the most accessible span is within 0.036 to 0.060 inch. This deflection is measured from a straightedge held on the chain wrapped over the two sprockets. The allowable deflection is measured at the center of the span while a force of 2 to 3 pound finger pressure is applied to the chain.

Step 6 Replace the dust plugs in the eccentric adjustment cap to keep dirt out of the unused holes in the end cover.

7.2.3 Replacing the Drive Chain and Sprocket

The drive chain should be replaced if the elongation exceeds 0.084 inch over a 10 pitch length. The section of chain to be measured should be stretched tight, in place, in the compressor and measured with a vernier caliper. A reading made outside of the rollers at 10 pitches should be added to a reading between the inside of the same rollers and then divided by two. If this calculation exceeds 3.834 inches for 3/8 pitch, the chain should be replaced to the manufacturer's specifications. Any sprocket showing any undercutting should be replaced. Follow these steps to replace a drive chain and a sprocket:

Step 1 Remove the frame top cover.

Step 2 Remove the dust plugs and the screws that hold the eccentric adjustment cap to the end cover.

Step 3 Rotate the eccentric cap to loosen the chain for removal.

Step 4 Remove the chain from the idler sprocket and remove the entire assembly from the end cover (see *Figure 9*).

Step 5 Measure the exact distance from the outer face of the chain sprocket to the closest face of the eccentric cap with a machinist's rule.

> **NOTE**
> Note the measurement for the proper positioning of the new sprocket.

Step 6 Use a drift punch to remove the roll pin from the idler nut.

Step 7 Remove the nut, screw, sprocket, and washer.

Step 8 Discard all four of these items.

Step 9 Remove and discard the cap O-ring.

Step 10 Reassemble all parts using a new screw, washer, sprocket, and nut.

Step 11 Compare the sprocket to the cap measurement previously taken.

> **NOTE**
> If necessary, shim between the cap and the idler spacer to make the measurements equal.

Step 12 Install the assembly on the frame, but do not install the chain.

Step 13 Check the alignment to the crankshaft drive sprocket with the crankshaft centered in end play using a straightedge.

Step 14 If it is not in alignment to within 1/32 inch, shim as necessary.

Step 15 Remove the entire assembly when the alignment is within 1/32 inch.

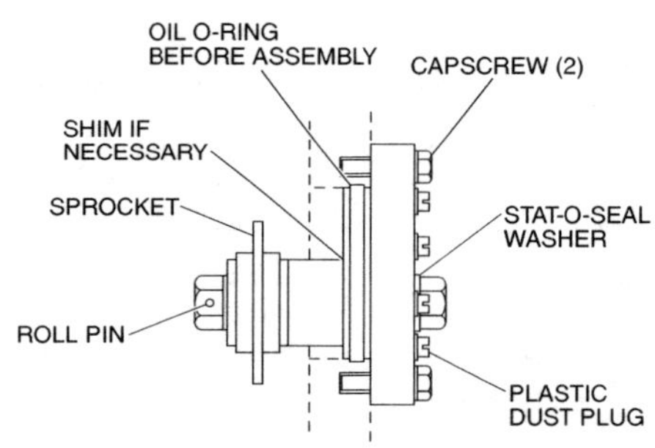

Figure 9 ◆ Chain idler sprocket.

Instructor's Notes:

Step 16 Tighten the idler nut to the recommended torque, and drill a 1/8-inch hole through the nut.

Step 17 Tap in a new roll pin.

Step 18 Apply oil and install a new O-ring.

Step 19 Install the assembly and adjust the chain to the proper tension.

7.2.4 Replacing the Lube Oil Pump Chain Sprocket

Refer to *Figure 10* as you follow these steps to replace a lube oil pump chain sprocket:

Step 1 Remove all piping from the pump.

Step 2 Remove the fasteners from the pump.

> **NOTE**
> The pump with the sprocket should pass through the hole in the end cover after the removal of the chain.

Step 3 Measure the exact distance from the sprocket drive face to the pump mounting adapter face with a machinist's rule.

> **NOTE**
> Record this measurement for future reference.

Step 4 Remove the sprocket setscrew from the sprocket and then remove the sprocket from the oil pump.

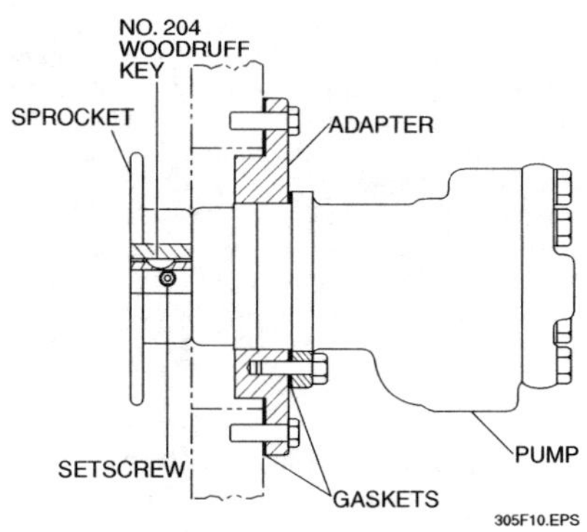

Figure 10 ◆ Lube oil pump and sprocket.

Step 5 Remove the Woodruff key.

Step 6 Install a new Woodruff key after checking to make sure that the key will fit into the new sprocket.

> **NOTE**
> If the key is too thick, it can be polished with an emery cloth on a flat surface until it can easily slide into the notch. It may also be a little high and require filing of the top edge.

Step 7 Install a new sprocket to the original measurement between the sprocket drive face and the pump mounting flange face.

Step 8 Tighten the setscrew when it is in position.

Step 9 Install new gaskets.

> **NOTE**
> Apply an anti-seize lubricant to the gasket surface before installing the pump.

Step 10 Install the pump onto the frame.

Step 11 Check the alignment to the crankshaft drive sprocket to within 1/32 inch or to the manufacturer's specifications using a straightedge.

> **NOTE**
> If it is not in alignment, adjust the sprocket position as necessary.

Step 12 Reinstall and adjust the drive chain.

Step 13 Reinstall all the tubing to the oil pump.

7.2.5 Replacing the Force-Feed Lubricator Chain Sprocket

Refer to *Figure 11* as you follow these steps to replace a force-feed lubricator chain sprocket:

Step 1 Remove the sprocket setscrew and sprocket.

Step 2 Remove the tubing from the lubricator.

Step 3 Remove the mounting bracket screws and lubricator.

Step 4 Remove the Woodruff key from the shaft and file the shaft to remove any burrs raised by the cup point of the setscrew.

Demonstration

Demonstrate how to adjust a drive chain to the proper tension.

Audiovisual

Show Transparency 8 (Figure 10). Explain how to replace a lube oil pump chain sprocket.

Laboratory

Set up stations with machinist's rules and lube oil pump chain sprockets. Ask trainees to practice measuring the distance from the sprocket drive face to the pump mounting adapter face.

Classroom

Refer to Figure 11. Review the steps to replace a force-feed lubricator chain sprocket.

Demonstration

Demonstrate how to properly remove the Woodruff key and to file the shaft. Answer any questions trainees have.

Audiovisual

Show Transparency 9 (Figure 12). Review the components in a valve assembly.

Classroom

Discuss the procedure to remove, repair, and reassemble a valve.

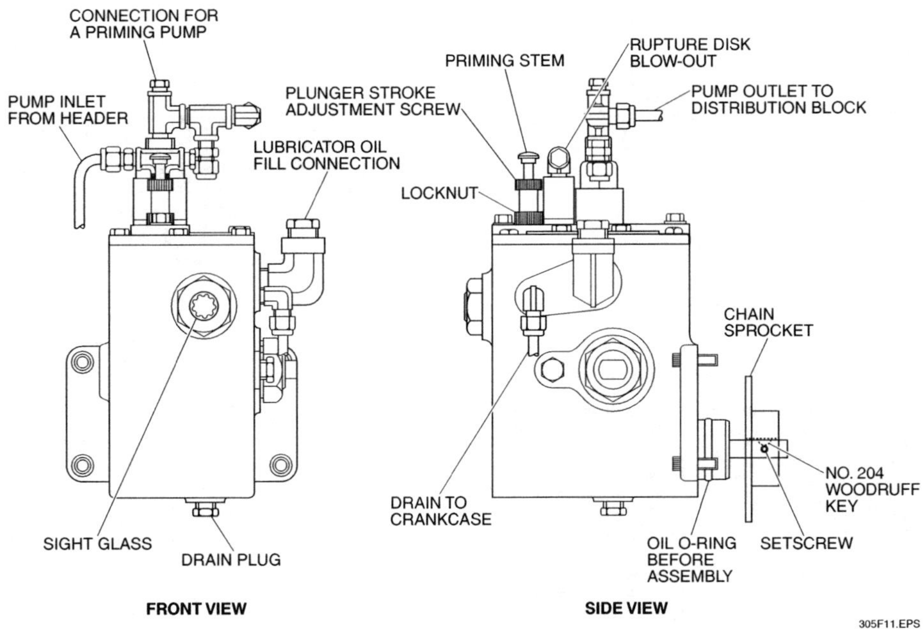

Figure 11 ◆ Force-feed lubricator and sprocket.

Step 5 Install a new O-ring.

Step 6 Install a new Woodruff key after checking to make sure that the key will fit into the new sprocket.

> **NOTE**
> If it is too thick, it can be polished with an emery cloth on a flat surface until it can easily slide into the notch. It may also be a little high and require filing of the top edge.

Step 7 Oil the new O-ring and reinstall the lubricator on the frame after the new key has been installed and the new sprocket will fit properly.

Step 8 Place the new sprocket onto the shaft.

Step 9 Check the alignment to the crankshaft drive sprocket using a straightedge to within $\frac{1}{32}$ inch or to the manufacturer's specifications.

Step 10 Tighten the setscrew.

Step 11 Install the drive chain and adjust to the proper tension.

Step 12 Reattach all tubing to the lubricator.

7.2.6 Removing, Repairing, and Reassembling a Valve

The valves (see *Figure 12*) will use springs with different ratings for different pressure levels. If different operating pressures are used, different springs may be required. The suction valve is selected on the basis of operating suction pressure and the discharge valve for operating **discharge pressure**. Proper valve spring selection is also based upon the operating speed, gas specific gravity, and suction temperature of the gas. Always refer to the manufacturer's maintenance manual for the correct selection of valves and springs.

Follow these steps to remove, repair, and reassemble a valve:

Step 1 Slightly loosen all of the bolts on each valve cap.

Instructor's Notes:

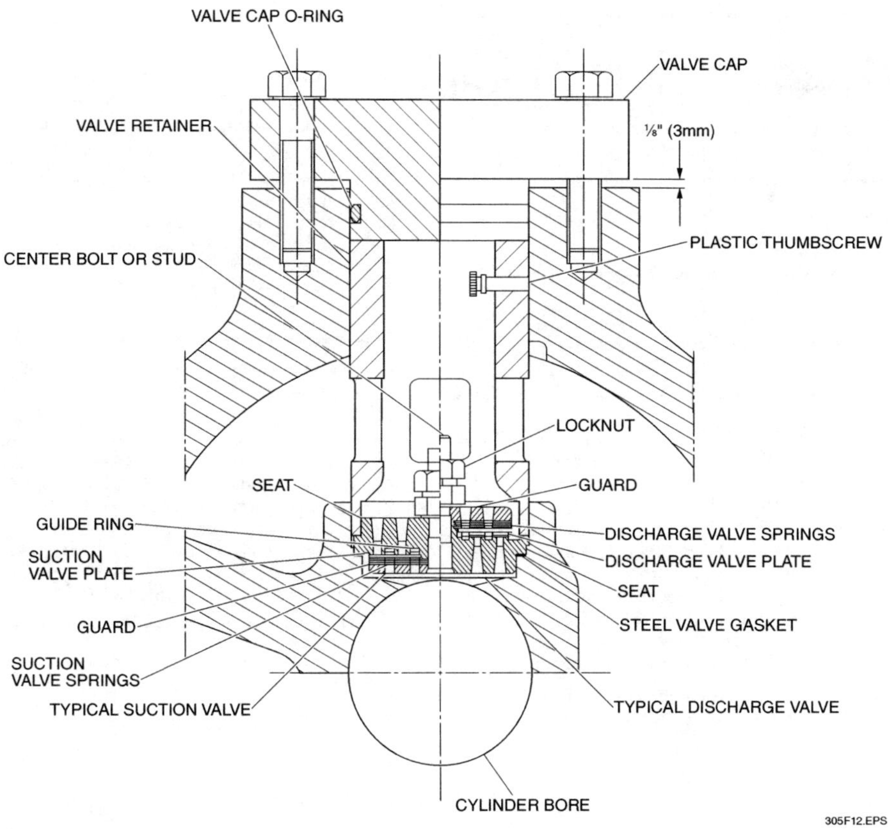

Figure 12 ◆ Valve assembly.

> **NOTE**
> With all the bolts loose, the cap should stay in its original position. If the cap pushes out on its own, immediately stop and take steps necessary to relieve the pressure.

> **WARNING!**
> Before attempting to remove any valve cap, be sure that all the pressure has been vented from the compressor cylinder. The pressure must be completely vented from both the suction and discharge passages of the cylinder.

Step 2 Remove the valve cap bolts. A pry bar or screwdriver may be used to pry them loose.

Step 3 Screw a valve tool over the valve center bolt.

Step 4 Remove the valve and retainer from the compressor. On some models, the suction and discharge piping and cylinder head must be removed to gain access to the concentric valve. A concentric valve combines the suction and discharge valves into one assembly.

Step 5 Remove the flat metal gasket.

Safety

Explain problems that can result if the pressure is not properly vented from the suction and discharge passages of the cylinder.

Safety

Ask a trainee to explain why the flat metal gasket must be removed from the compressor.

Emphasize the importance of following proper techniques for fastener tightening.

Classroom

Refer to Figure 13. Review the components in a high-pressure valve cap assembly.

Laboratory

Set up several stations with valve caps, bolts, and O-rings. Ask trainees to practice tightening the valve caps to the recommended torque.

Teaching Tip

See the Teaching Tip for Section 7.0.0 at the end of this module.

Homework

Have trainees review Sections 8.0.0–8.2.0 and answer the Review Questions for the next session.

> **CAUTION**
> The flat metal gasket must be removed to prevent damage to the compressor. The gasket may remain in the pocket or may fall into the cylinder gas passage. It may be difficult to see. This gasket should be replaced after removal.

Step 6 Coat the metallic flat gasket with an anti-seize lubricant.

> **NOTE**
> The gasket can be inserted into the valve pocket or stuck onto the valve. You must take care to keep this gasket from falling into the gas passage.

Step 7 Screw in the retainer keeper far enough to provide friction so that the bottom retainers will not fall out while the cap is being installed.

Step 8 Insert the valve and retainer into the pocket together. When installed properly, the distance from the outer retainer face to the surface of the valve boss on the cylinder should be ⅛ inch shorter than the length of the nose on the valve cap or to the manufacturer's specifications.

Step 9 Inspect the valve cap O-ring for cuts or gashes, and replace it if necessary.

Step 10 Lubricate the O-ring and the nose of the valve cap.

Step 11 Insert the cap and tighten the bolts evenly to the recommended torque.

> **NOTE**
> If the assembly is put together correctly, the distance from the underside of the cap to the valve boss surface on the cylinder will be ⅛ inch. Be certain all parts, gasket faces, and mating surfaces are absolutely clean, and always use clean, fresh oil on the treads before reinstalling the bolts.

> **CAUTION**
> Proper fastener tightening technique is essential for sealing valve caps with the soft metallic wire gaskets used in some high-pressure cylinders. It is important to draw up bolting to full torque in even and gradual steps. Do not allow the valve cap to be tightened with bias on one bolt or to become cocked in the bore. Such bias or cocking can cause uneven crush of the gasket, which could cause a leak and possible bolt failure.

Figure 13 shows a high-pressure valve cap assembly.

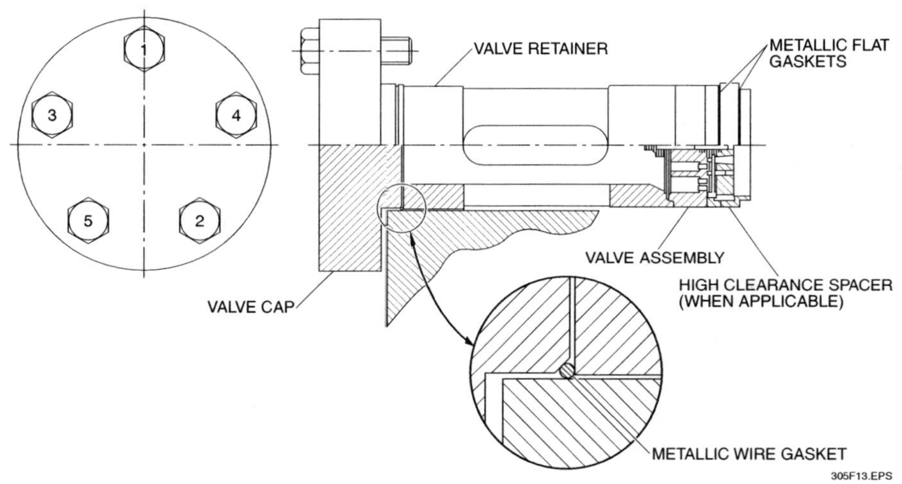

Figure 13 ◆ High-pressure valve cap assembly.

Instructor's Notes:

8.0.0 ◆ PREPARING FOR STARTUP AND OPERATIONAL CHECK

After a major repair of a compressor, be sure to complete the manufacturer's recommended startup checklist and comply with company policy and procedures. These will vary by company operational requirements. The following are general startup and operational checks for rotary and reciprocating compressors.

8.1.0 Startup and Operational Check for Rotary Compressors

Follow these general steps to prepare a rotary compressor for startup and operational check:

Step 1 Check that the compressor soft foot is not more than 0.002 inch or more than the manufacturer's specifications.

Step 2 Check the torque on the compressor's anchor bolts.

Step 3 Check the piping and supports to verify that they do not bend or stress the compressor.

Step 4 Check the coupling bolt's torque value.

Step 5 Verify that the compressor to driver alignment is within 0.002 inch TIR or to the manufacturer's specifications.

Step 6 Check that the compressor rotor turns freely.

Step 7 Check that the compressor oil system has been filled with oil to the proper level without overfilling.

Step 8 Check that the compressor oil supply isolation valve is in the OPEN position.

Step 9 Verify that the oil filter and oil system have been primed.

Step 10 Verify that the suction and discharge pressure shutdown is set.

Step 11 Ensure that the safety relief valves are installed and working properly.

Step 12 Verify that the gas suction lines have been blown out to remove water, slag, and dirt.

Step 13 Verify that a strainer screen has been installed upstream of the compressor suction.

Step 14 Purge the piping and compressor to remove all the air.

Step 15 Ensure that the area is clear of tools, debris, and personnel.

Step 16 Notify the control center of the operational test and obtain permission for the operational test. Continue with the operational test at the time given by the control center.

Step 17 Start the compressor to the idle speed and immediately load the compressor to 25% minimum slide value axial position. Increase the minimum driver speed if necessary to stabilize the load.

Step 18 Verify that the bearing oil pressure came up immediately.

Step 19 Verify that all the pressure gauges are working.

Step 20 Check for any strange noises or shaking in the compressor or piping.

Step 21 Verify that the bearing and mechanical seal oil pressure is above minimum pressure required for current operating condition.

Step 22 Check the compressor for oil leaks.

Step 23 Check the compressor piping for gas leaks.

Step 24 Verify that the mechanical seal is properly sealing gas at the shaft.

Step 25 Notify the control center after the tests are complete.

Step 26 Clean the work area and return all tools and equipment to the appropriate locations.

NOTE
Dispose of waste products per company policy.

Step 27 Complete all required company documentation.

8.2.0 Startup and Operational Check for Reciprocating Compressors

Follow these general steps to prepare a reciprocating compressor for startup and operational check:

Step 1 Check that the compressor's soft foot has been shimmed to the manufacturer's specifications.

Materials

Ensure that you have everything required for testing during this session.

Classroom

Review the procedures for preparing a rotary compressor for startup and operational check.

Distribute sample documentation for startup and operational checks. Ensure that trainees understand how to properly complete the documentation.

Explain how to prepare a reciprocating compressor for startup and operational check. Ask whether trainees have any questions.

Teaching Tip

See the Teaching Tip for Section 8.0.0 at the end of this module.

Step 2 Check the bottom crosshead clearance on all corners.

Step 3 Verify that the piping and supports do not bend or stress the compressor.

Step 4 Check the coupling bolt's torque value.

Step 5 Verify that the compressor to driver alignment is within 0.002 inch TIR or to the manufacturer's specifications.

Step 6 Verify that the compressor crankcase has been filled with oil to the proper level without overfilling.

Step 7 Verify that the compressor oil supply isolation valve is in the OPEN position.

Step 8 Verify that the oil filter and all lube oil piping has been primed with oil.

Step 9 Verify that the crankcase breather element is clean.

Step 10 Verify that the force-feed lubricator box is filled with oil and primed.

Step 11 Verify that the proper lube feed rate has been established.

Step 12 Verify that the primary and secondary packing vents and distance piece vents are open.

Step 13 Verify that the gas suction lines have been blown out to remove water, slag, and dirt.

Step 14 Purge the piping and compressor of air.

Step 15 Verify that the area is clear of tools, debris, and personnel.

Step 16 Notify the control center of the operational test and obtain permission for the test. Continue the operational test at the time given by the control center.

Step 17 Start the compressor.

Step 18 Verify that the oil pressure came up immediately.

Step 19 Check for any strange noises or shaking in the compressor or piping.

Step 20 Verify that the lubricator is properly set and the distribution block indicator pin is moving.

Step 21 Check for oil leaks.

Step 22 Verify that the rod packings are sealing properly.

Step 23 Check all gauges and verify that the readings are within normal specifications of the compressor.

Step 24 Check for leaks at all the connections.

Step 25 Notify the control center when the operational check is complete.

Step 26 Clean the work area and return all tools and equipment to their appropriate location.

NOTE
Dispose of waste products per company policy.

Step 27 Complete all required company documentations.

Summary

Common problems for a gas compressor include improper lubrication, improper alignment and adjustments, high temperature, and weak or broken valves. Look for the simple problems first and then proceed to the more time-consuming components. Use the manufacturer's checklist to help locate and solve the compressor malfunctions in a timely manner.

The compressor must be isolated from the pipeline system before any work can begin. In some situations, a backup compressor may be used to maintain the pipeline function, but the compressor requiring repairs must be shut down and isolated. Always follow your company's policy for lockout and tagout of the energy sources and the gas. The pressure must be relieved at the compressor and its components before work can begin. The repair work on a compressor must be done in a clean work area to prevent contamination of the compressor.

During any repair and installation process, always follow the manufacturer's recommendations and specifications. During the repair process, careful handling of the components is required to prevent damage. If nicks or scratches occur to the mating surfaces, they need to be repaired or replaced prior to installation.

Instructor's Notes:

Review Questions

1. Gas must be _____ in order to achieve gas flow through a pipeline.
   a. liquefied
   b. heated
   c. compressed
   d. cooled

2. Lubrication _____ friction, _____ energy requirements.
   a. Decreases; increasing
   b. Decreases; decreasing
   c. Increases; decreaing
   d. Increases; increasing

3. Lubricant additives are used to improve the _____ and _____.
   a. shock load; friction
   b. efficiency; output
   c. washing; working tolerance
   d. viscosity index; lubricity

4. Electric-motor-driven compressors must have a(n) _____ to ensure oil flow prior to startup.
   a. prelube pump
   b. start permissive
   c. warmup cycle
   d. priming pump

5. Oil flooding of the compression chamber is supplied at _____.
   a. the rotary screw
   b. rotor injection
   c. the slide valve
   d. the feed regulator

6. Most reciprocating compressors use a(n) _____ lubrication system.
   a. oil-flooded
   b. continuous
   c. force-feed
   d. impulse feed

7. Preventive maintenance inspections are typically divided into two actions: _____.
   a. inspections and corrective actions
   b. remove and rebuild
   c. repair and replace
   d. call outs and shut down

8. The gas/oil separator oil level should be in the _____ range of the sight glass when the compressor is in operation.
   a. lower
   b. zero
   c. upper
   d. middle

9. Perform a semiannual inspection of a gas compressor twice a year or after the compressor has operated for _____ hours.
   a. 3,000
   b. 4,000
   c. 5,000
   d. 6,000

10. For most reciprocating compressors, the oil pressure should be _____ psig at operating temperature.
    a. 30 to 40
    b. 40 to 50
    c. 50 to 60
    d. 60 to 70

11. A reciprocating compressor's low oil pressure shutdown is a minimum of _____ psi.
    a. 35
    b. 40
    c. 45
    d. 50

12. A more frequent oil change interval may be required if operating in a(n) _____ environment.
    a. very hot
    b. extremely dirty
    c. cold
    d. humid

13. Check the crosshead guide clearance of a reciprocating compressor with a _____.
    a. caliper
    b. scale
    c. micrometer
    d. feeler gauge

14. _____ is the key to solving problems with the compressor operation quickly.
    a. Troubleshooting
    b. Corrective action
    c. Inspection
    d. Replacement

Classroom

Go over the Review Questions for this module, and answer any questions trainees have.

Examination

Administer the Module Examination. Be sure to record the results of the Exam on Craft Training Report Form 200 and submit the results to the Training Program Sponsor.

Performance Testing

Have each trainee perform Performance Task 4 to your satisfaction. Fill out a Performance Profile Sheet for each trainee.

Ensure that all Performance Tests have been completed and Performance Profile Sheets for each trainee are filled out. Be sure to record the results of the testing on Craft Training Report Form 200 and submit the results to the Training Program Sponsor.

15. The best method for troubleshooting is the process of _____.

a. trial and error
b. replacement
c. elimination
d. adjustment

16. A key to long service life for a reciprocating compressor is the _____.

a. operating temperature
b. lubrication system
c. inspection frequency
d. adjustment tolerance

17. When preparing a gas compressor for shutdown and repair, it is critical to _____ from the system.

a. remove it
b. expel gas
c. purge all liquid
d. relieve all pressure

18. Obtain a(n) _____ permit to purge the compressor and lines.

a. line-opening
b. work order
c. repair
d. purging

19. The _____ on a rotary compressor may be replaced in the field.

a. rotor housing
b. mechanical seal
c. thrust balance piston
d. bearings

20. The _____ prevents loss of internal process gas pressure at the male rotor drive.

a. coupling
b. O-rings
c. mechanical seal
d. slide valve

21. Always apply a(n) _____ to both sides of a gasket for easy removal at a later date.

a. anti-seize lubrication
b. gasketol
c. grease
d. silicone

22. Ensure that the crossheads are returned to their original _____ location for installation.

a. outer dead center
b. inner dead center
c. cocked
d. throw

23. Drive chain tightness is controlled by a(n) _____ attached to the eccentric adjustment cap.

a. roll pin
b. idler sprocket
c. set screw
d. locknut

24. The drive chain should be replaced if the elongation exceeds _____ inches over a 10 pitch length.

a. 3.834
b. 1.834
c. 0.84
d. 0.084

25. The valves will use springs with different rating for different _____.

a. temperature levels
b. speeds
c. pressure levels
d. load limits

Instructor's Notes:

GLOSSARY

Trade Terms Introduced in This Module

Crosshead: A machined device, usually of cast iron or other heavy metal, used to attach the connecting rod to the piston in a compressor.

Discharge pressure: The total gas pressure at the discharge flange of the compressor.

Lubricity: The slipperiness or ability of a lubricant to decrease friction.

Start permissive: A device that will prevent start-up of a piece of equipment if certain conditions are not met.

Throw: The radius of a circle described by a crankshaft or similar part. The maximum displacement of a machine part moved by a crank or cam. In compressors, the word is sometimes used to refer to the offset portion of the crank to which the piston is attached.

Viscosity index: Measure of the ability of an oil to resist the thinning effect caused by increasing the oil temperature.

Wetability: A measure of the ability of the lubricant to adhere to a metal surface. An increase in wetability increases the lubricant's resistance to washing effects.

Instructor's Notes:

Additional Resources and Acknowledgments

This module is intended to present thorough resources for task training. The following reference works are suggested for further study. These are optional materials for continued education rather than for task training. NCCER would also like to thank the following company for permitting the use of their technical manuals.

Heavy Duty Oil Flooded Twin Screw Rotary Compressors Technical Manual For Models: AR166, AR208, AR260K & AR260, March 2002. Mount Vernon, Ohio: Ariel Corporation.

Heavy Duty Balanced Opposed Compressors Technical Manual For Models: JGM, JGN, JGP and JGQ; and JGI (Vertical Non-Balanced), January 1999. Mount Vernon, Ohio: Ariel Corporation.

Heavy Duty Balanced Opposed Compressors Technical Manual For Models: JGW, JGR and JGJ, January 2001. Mount Vernon, Ohio: Ariel Corporation.

Figure Credits

Ariel Corporation 305F01, 305F02, 305F03, 305F04, 305F05, 305F06, 305F07, 305F08, 305F09, 305F10, 305F11, 305F12, 305F13

MODULE 63305-02 — TEACHING TIPS

The following are suggested activities or instructional methods to help you teach the material in this AIG.

General

When you call on someone to answer a question, the rest of the class relaxes or even tunes out because they expect that the question and answer will take place only between you and the trainee you called on. Instead, use this technique to involve more trainees in answering questions and to keep them on their toes.

1. Ask trainees to define a term or explain a concept.
2. After one trainee has answered, ask a trainee seated nearby if the answer is right. Then ask whether a trainee in the back of the room agrees.
3. Ask trainees to explain why they think an answer is right or wrong.
4. Use the session to clear up incorrect ideas and encourage trainees to learn from their mistakes.

Section 2.0.0 *Lubrication Systems*

This exercise will familiarize trainees with the typical lubrication system components of a gas compressor. For this exercise, you will need copies of *Figures 1, 2,* and *3* with the callouts covered. Trainees will need their Trainee Guides and pencils and paper to take notes for discussion. Allow 20 minutes for the exercise.

1. Distribute copies of *Figures 1, 2,* and *3* with the callouts covered.
2. Ask trainees to identify the lubrication system components in each figure.
3. Have trainees explain the role of each component and review the functions performed by lubrication.
4. Answer any questions trainees have about lubrication systems.

Section 3.0.0 *Performing Preventive Maintenance on Gas Compressors*

This exercise will give trainees the opportunity to review preventive maintenance requirements for gas compressors. For this exercise, trainees will need their Trainee Guides, pencils, and paper to take notes for discussion. Allow 30 minutes for this exercise.

1. Create a variety of hypothetical scenarios for which preventive maintenance will be required on a gas compressor.
2. Divide the class into small groups according to the number of scenarios.
3. Assign a scenario to each group who will then describe the preventive maintenance tasks required for their gas compressor.
4. Review and discuss trainees' answers. Ask trainees to correct and/or add any necessary tasks.
5. Answer any questions trainees have about performing preventive maintenance on gas compressors.

Section 7.0.0 *Repairing Gas Compressor Components*

Trainees have reviewed a variety of procedures for repairing rotary and reciprocating gas compressor components. For this exercise, trainees will need their Trainee Guides, pencils, and paper to take notes for discussion. Allow 20 to 30 minutes for this exercise.

1. On the whiteboard/chalkboard, write a list of repair procedures for rotary and reciprocating compressors.
2. Ask trainees to select and locate one of the procedures in their company policies.
3. Have trainees compare company policies with the recommended procedures. Discuss differences and note any additional safety precautions.
4. Emphasize the importance of following company procedures. Go over any questions trainees have about repairing gas compressor components.

Section 8.0.0 *Preparing for Startup and Operational Check*

This exercise will give trainees the opportunity to review manufacturers' recommended startup checklists. For this exercise, you will need several manufacturers' maintenance manuals. Trainees will need their Trainee Guides, company policies, pencils, and paper to take notes for discussion. Allow 20 to 30 minutes for this exercise.

1. Divide the class into small groups according to the number of manufacturers' maintenance manuals.
2. Ask trainees to locate and review the recommended startup checklists.
3. Have trainees compare the manufacturer's checklist with their company policies. Discuss any differences and ensure that trainees understand the procedures.
4. Explain which set of procedures takes precedence. Answer any questions trainees have about preparing for startup and operational check.

MODULE 63305-02 — ANSWERS TO REVIEW QUESTIONS

Answer	Section
1. c	1.0.0
2. b	2.0.0
3. d	2.0.0
4. a	2.0.0
5. b	2.1.0
6. c	2.2.0
7. a	3.0.0
8. d	3.1.1
9. b	3.1.3
10. c	3.2.1
11. a	3.2.1
12. b	3.2.3
13. d	3.2.4
14. a	4.0.0
15. c	4.0.0
16. b	4.0.0
17. d	5.0.0
18. a	6.0.0
19. b	7.1.0
20. c	7.1.1
21. a	7.2.0
22. d	7.2.1
23. b	7.2.2
24. d	7.2.3
25. c	7.2.6

The NCCER makes every effort to keep these textbooks up-to-date and free of technical errors. We appreciate your help in this process. If you have an idea for improving this textbook, or if you find an error, a typographical mistake, or an inaccuracy in NCCER's Contren™ textbooks, please write us, using this form or a photocopy. Be sure to include the exact module number, page number, a detailed description, and the correction, if applicable. Your input will be brought to the attention of the Technical Review Committee. Thank you for your assistance.

Instructors – If you found that additional materials were necessary in order to teach this module effectively, please let us know so that we may include them in the Equipment/Materials list in the Instructor's Guide.

Write: Curriculum Revision and Development Department
National Center for Construction Education and Research
P.O. Box 141104, Gainesville, FL 32614-1104

Fax: 352-334-0932

E-mail: curriculum@nccer.org

Craft ______ Module Name ______

Copyright Date ______ Module Number ______ Page Number(s) ______

Description ______

(Optional) Correction ______

(Optional) Your Name and Address ______

Module 63306-02

Maintain, Troubleshoot, and Repair Pneumatic Valve Actuators/Operators and Systems

(CT 19.6 and 21.1)

Maintain, Troubleshoot, and Repair Pneumatic Valve Actuators/Operators and Systems Annotated Instructor's Guide

Module 63306-02

MODULE OVERVIEW

This module discusses pneumatic system maintenance procedures, system components, and schematic diagrams. Trainees will learn how to troubleshoot, repair, and adjust pneumatic systems and their components.

PREREQUISITES

Please refer to the Course Map in the Trainee Module. Prior to training with this module, it is recommended that the trainee shall have successfully completed the following:

Core Curriculum; Pipeline Core, Modules 66101-02 and 66102-02; Pipeline Mechanical Levels One and Two; Pipeline Mechanical Level Three, Modules 63301-02 through 63305-02

OBJECTIVES

Upon completion of this module, the trainee will be able to:

1. Perform pneumatic system preventive maintenance procedures.
2. Inspect pneumatic system components.
3. Read pneumatic system schematic diagrams.
4. Troubleshoot pneumatic systems.
5. Repair pneumatic system components.
6. Adjust pneumatic valve actuators/operators (CT 19.6).
7. Repair pneumatic valve actuators/operators (CT 21.1).

PERFORMANCE TASKS

Under the supervision of the instructor, the trainee should be able to:

1. Perform preventive maintenance procedures on a pneumatic system.
2. Inspect pneumatic system components.
3. Troubleshoot pneumatic systems.
4. Repair pneumatic system components.
5. Adjust pneumatic valve actuators/operators (CT 19.6).
6. Repair pneumatic valve actuators/operators (CT 21.1).

NCCER STANDARDIZED CRAFT TRAINING PROGRAM

The National Center for Construction Education and Research (NCCER) provides a standardized national program of accredited craft training. Key features of the program include instructor certification, competency-based training, and performance testing. The program provides trainees, instructors, and companies with a standard form of recognition through a National Craft Training Registry. The program is described in full in the *Guidelines for Accreditation*, published by the NCCER. For more information on standardized craft training, contact the NCCER by writing us at P.O. Box 141104, Gainesville, FL 32614-1104; calling 352-334-0911; or e-mailing info@nccer.org. More information may be found at our Web site, www.nccer.org.

HOW TO USE THIS ANNOTATED INSTRUCTOR'S GUIDE

Each page presents two sections of information. The larger section displays each page exactly as it appears in the Trainee Module. The narrow column ties suggested trainee and instructor actions to each page and provides icons to call your attention to material, safety, audiovisual, or testing requirements. The bottom of each page includes space for your notes.

Teaching Tip

If you see the Teaching Tip icon, that means there is a teaching tip associated with this section. Also refer to any suggested teaching tips at the end of the module.

SAFETY CONSIDERATIONS

Ensure that the trainees are equipped with appropriate personal protective equipment. Emphasize the importance of following all safety precautions and procedures when working with power tools.

PREPARATION

Before teaching this module, you should review the Module Outline, Objectives, Performance Tasks, and the Materials and Equipment List. Be sure to allow ample time to prepare your own training or lesson plan and gather all required equipment and materials.

MATERIALS AND EQUIPMENT LIST

Transparencies

Markers/chalk

Blank acetate sheets

Transparency pens

Pencils and scratch paper

Module Examinations*

Performance Profile Sheets*

Copies of Quick Quiz**

Copies of Performance Verifications for Covered Tasks 19.6 and 21.1†

Overhead projector and screen

Whiteboard/chalkboard

Appropriate personal protective equipment

Copies of *49 CFR Part 192* (Gas) and/or *49 CFR Part 195* (Liquid)

Copies of your company policy and procedures manual

Sample manufacturer's warranties

Sample preventive maintenance records and schedules

Sample equipment records

Pneumatic schematic diagrams

Sample actuators/operators and the necessary tools for disassembly and reassembly

Manufacturers' instruction manuals for the sample actuators/operators

A variety of new and damaged gauges

Sample valve maintenance logs

* Located in the Test Booklet.
**Located at the end of this module.
† Available for download from www.nccer.org.

TEACHING TIME FOR THIS MODULE

An outline for use in developing your lesson plan is presented below. Note that each Roman numeral in the outline equates to one session of instruction. Each session has a suggested time period of 2½ hours. This includes 10 minutes at the beginning of each session for administrative tasks and one 10-minute break during the session. Approximately 15 hours are suggested to cover *Maintain, Troubleshoot, and Repair Pneumatic Valve Actuators/Operators and Systems (CT 19.6 and 21.1)*. You will need to adjust the time required for hands-on activity and testing based on your class size and resources.

Topic	Planned Time
Session I. Preventive Maintenance	
A. PM Methods	____________
1. Laboratory	____________
B. Scheduling	____________
C. Manufacturer's Instructions and Warranties	____________
1. Laboratory	____________
D. Training	____________
E. Documentation	____________
1. Performance Testing	____________
Session II. Inspecting Pneumatic System Components and Reading Diagrams	
A. Inspecting Intake Filters	____________
B. Inspecting Intercoolers	____________
C. Inspecting Aftercoolers	____________
D. Inspecting Separators	____________
E. Inspecting Receivers	____________
F. Inspecting In-Line Filters	____________
G. Inspecting Pressure Regulators	____________
H. Inspecting Lubricators	____________
1. Performance Testing	____________
I. Reading Diagrams	____________
1. Laboratory	____________
Session III. Troubleshooting Pneumatic Actuators/Operators	
A. Examining the Air Supply	____________
B. Inspecting the Actuator/Operator	____________
C. Inspecting the Control Valve	____________
D. Inspecting Sequence Circuits	____________
E. Inspecting Interlocks	____________
F. Troubleshooting Charts	____________
1. Laboratory	____________
2. Performance Testing	____________
Session IV. Repairing Pneumatic System Components	
A. Preparing for Shutdown and Repair	____________
B. Repairing and Overhauling	____________
1. Laboratory	____________
C. Replacing Gauges	____________
1. Laboratory	____________
2. Performance Testing	____________

Session V. Pneumatic Valve Actuators/Operators, Part One

A. Rack and Pinion Actuators/Operators ____________

B. Fail-Safe Actuators/Operators ____________

C. Adjusting Valve Actuators/Operators ____________

D. Repairing Valve Actuators/Operators ____________

   1. Laboratory ____________

E. Abnormal Operating Conditions ____________

Session VI. Pneumatic Valve Actuators/Operators, Part Two

A. Guest Speaker ____________

B. Summary

   1. Summarize module ____________
   2. Answer questions ____________

C. Module Examination ____________

   1. Trainees must score 70% or higher to receive recognition from the NCCER.
   2. Record the testing results on Craft Training Report Form 200 and submit the results to the Training Program Sponsor.

D. Performance Testing ____________

   1. Trainees must perform each task to the satisfaction of the instructor to receive recognition from the NCCER.
   2. Record the testing results on Craft Training Report Form 200 and submit the results to the Training Program Sponsor.

Module 63306-02

Maintain, Troubleshoot, and Repair Pneumatic Valve Actuators/Operators and Systems

(CT 19.6 and 21.1)

Instructor's Notes:

Instructor's Notes:

Introduction to the Pipeline Operator Qualification Rule, API, and NCCER

It could be said that petroleum products keep America running. Petroleum products like oil and natural gas are the principal sources of energy in the United States and are used in many manufacturing and chemical processes. Petroleum products are primarily transported by pipeline. There are more than a million miles of liquid and natural gas pipeline in this country, ranging in diameter from two inches to four feet. These pipelines bring crude oil and natural gas from production fields to processing plants and refineries—and ultimately to the consumer.

Pipeline Operator Qualification

Petroleum products are considered hazardous, so pipeline personnel must take extra precautions and follow specific procedures when working with them. To help ensure the safe handling and transportation of petroleum products, the United States Department of Transportation (DOT) mandates safe practices as federal regulations.

One such regulation, the Operator Qualification (OQ) rule, requires pipeline operators to develop a written qualification plan to evaluate their personnel and their contract personnel (see *49 CFR Part 192 Subpart N—Gas,* and *49 CFR Part 195 Subpart G—Liquid*). The evaluation hinges on the operator's ability to perform certain covered tasks and to be able to recognize and respond to abnormal operating conditions that may be encountered while performing these covered tasks.

Covered Tasks

A pipeline performance task is deemed *covered* if the activity passes the following four-part test:

1. Is it performed on a pipeline facility?
2. Is it an operations or maintenance task?
3. Is it performed as a requirement of *49 CFR Part 195 (Liquid)* or *49 CFR Part 192 (Gas)*?
4. Does it affect the operation or integrity of the pipeline facility?

If the answer to all four questions is *yes*, then the activity is considered *covered* and must figure into the evaluation process of pipeline personnel.

American Petroleum Institute

The OQ rule and its related covered tasks provided the catalyst to develop a unified pipeline training and assessment program—so that pipeline operators and owners have ready access to the assessment and training tools they need. This effort was the direct result of the vision and leadership provided by The American Petroleum Institute (API) and API members.

Efforts to unify the industry are nothing new to API. Indeed, the impetus for forming the Institute in 1919 was the need to standardize engineering specifications for drilling and production equipment. Today, the most pressing issues revolve around public perceptions and government policies toward the industry. Speaking with one voice on these issues has become as essential as having interchangeable parts in the field. API's members determine the petroleum industry's positions on public policy as well as the standards that govern its day-to-day operations worldwide.

Consortium on Operator Qualification

API assembled a focus group whose primary purpose was to identify covered tasks and provide compliance recommendations to pipeline operators.

The Consortium on Operator Qualification (COOQ) became the guiding force for the identification and development of covered tasks for the liquid pipeline industry. Their work product, API publication 1161, *Guidance Document for the Qualification of Liquid Pipeline Personnel,* formed the basis for the resulting Pipeliner Training and Assessment Program.

The covered task list created in API publication 1161 has now evolved to include gas covered tasks and is maintained by NCCER. The most current covered task list can be downloaded at www.nccer.org.

National Center for Construction Education and Research

With the guidance and support of API, its members, and the Consortium on Operator Qualification, the National Center for Construction Education and Research (NCCER) was able to design and implement the Pipeliner Training and Assessment Program.

NCCER is a not-for-profit 501(c)(3) education foundation established in 1995 by the world's largest and most progressive construction and maintenance contractors and national contractor associations. NCCER was founded to address the

severe workforce shortage facing the construction and maintenance industries and to develop a standardized training process and curricula. Today, NCCER is supported by hundreds of leading construction and maintenance companies, manufacturers, and national associations and developed more than 35 construction and maintenance titles.

NCCER also maintains a National Registry, which contains the records of trainees and instructors who have received NCCER training. All completion information for every trainee is sent to the NCCER and filed within the Registry. The Registry can then confirm training and skills for workers as they move from company to company, state to state, or even within their own company.

Standardized Curriculum (NCCER's Contren™ Learning Series)

NCCER curricula are developed using the expertise of industry subject matter experts (SMEs). The use of SMEs as an integral part of every new development or revision ensures that the curriculum is current, industry-driven, and indicative of industry practices. The development and/or revision process is facilitated from start to finish by NCCER project managers, who work closely with the SMEs and a select team of technical writer(s) on the development and presentation of the curriculum.

Pipeline: Five Training Paths

The pipeline curriculum focuses on essential skills and competencies within five skill areas:

- Corrosion Control
- Electrical and Instrumentation
- Maintenance
- Mechanical
- Operations (Field and Control Center)

SMEs from the pipeline industry have determined the skills and competencies needed based on API's covered tasks and the consensus of the group/industry. Specific company information is generally not incorporated into the curriculum unless the SMEs determine that the information is the industry standard.

Training Modules and Covered Tasks

The smallest unit of instruction in the curriculum is a *module*. Each curriculum is composed of modules (the equivalent of stand-alone chapters) that are logically sequenced into complete *levels* of instruction. Every module is task-oriented and requires the successful completion of a written (knowledge-based) examination and a performance (hands-on) test. Modules may be individually selected and/or combined for efficient task training. Modules in the pipeline curricula, to the extent possible, will align with API-developed covered tasks. While not all modules will relate to covered tasks, all covered tasks will relate to at least one module.

Benefits of Training to the Pipeline Marketplace

Applying NCCER's standardized training process allows pipeline owners and contractors in the United States to establish the benchmark for acceptable skill and competency levels within the pipeline industry. This process will assist training organizations in developing and/or improving standardized pipeline training programs that effectively match their resources and operating conditions while responding to regulatory demands and to the needs of the industry. The process will also recognize training organizations that meet the minimum standards of competence, warranting regulatory, public, and professional confidence. This in turn will foster national unity within the pipeline industry for the mutual benefit of individual technicians, owners, contractors, and ultimately pipeline-users and society.

Additional Resources

To review the Pipeliner Training and Assessment Program Implementation and Resource Guide and the latest version of pipeline covered tasks, visit NCCER's Web site at www.nccer.org, or contact:

National Center for Construction Education and Research
PO Box 141104
Gainesville, FL 32614-1104
(888) 622-3720
Web site: www.nccer.org

For more information on API, please contact:

American Petroleum Institute
1220 L Street, Northwest
Washington, D.C. 20005
(202) 682-8000
Web site: www.api.org

To order *API Pub 1161*, contact Global Engineering Documents by calling 1-800-854-7149 or visiting their Web site at www.global.his.com.

Additional sources of information on the pipeline operator qualification regulation include:

DOT – Office of Pipeline Safety
www.ops.dot.gov

Transportation Safety Institute
www.tsi.dot.gov

Instructor's Notes:

COURSE MAP

This course map shows all of the modules in the third level of the Pipeline Mechanical curriculum. The suggested training order begins at the bottom and proceeds up. Skill levels increase as you advance on the course map. The local Training Program Sponsor may adjust the training order.

PIPELINE MECHANICAL LEVEL THREE

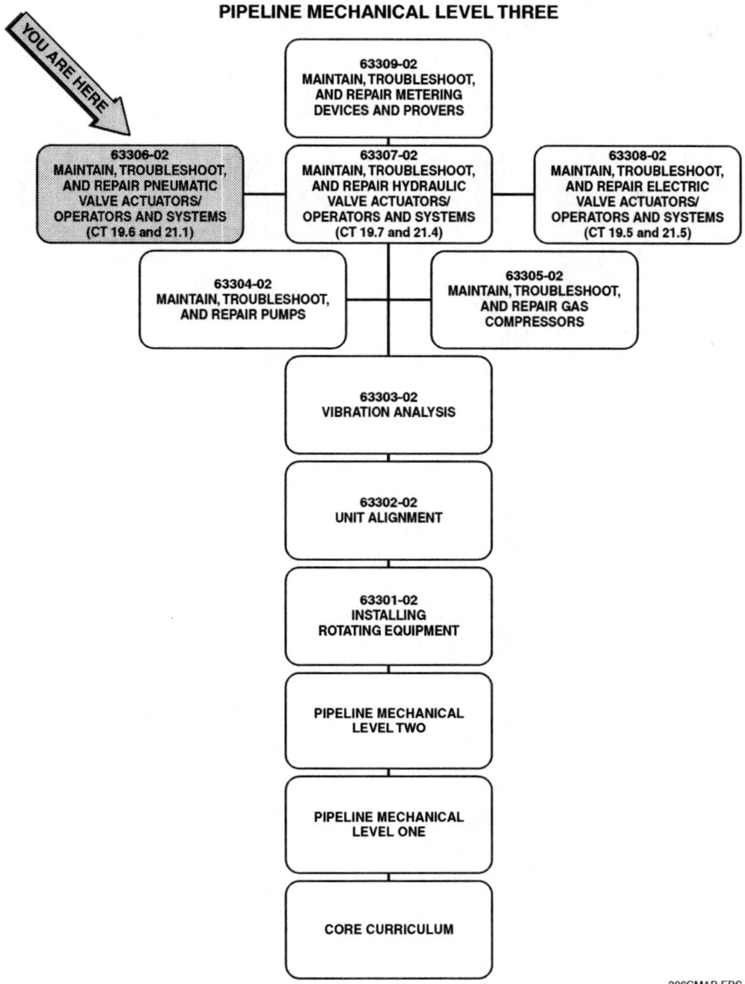

306CMAP.EPS

Homework

Assign reading of Module 63306-02.

Instructor's Notes:

MODULE 63306-02 CONTENTS

MODULE 63306-02 CONTENTS (Continued)

Figures

Table

Instructor's Notes:

MODULE 63306-02

Maintain, Troubleshoot, and Repair Pneumatic Valve Actuators/Operators and Systems (CT 19.6 and 21.1)

Objectives

When you have completed this module, you will be able to do the following:

1. Perform pneumatic system preventive maintenance procedures.
2. Inspect pneumatic system components.
3. Read pneumatic system schematic diagrams.
4. Troubleshoot pneumatic systems.
5. Repair pneumatic system components.
6. Adjust pneumatic valve actuators/operators (CT 19.6).
7. Repair pneumatic valve actuators/operators (CT 21.1).

Prerequisites

Before you begin this module, it is recommended that you successfully complete the following: Core Curriculum; Pipeline Core, Modules 66101-02 and 66102-02; Pipeline Mechanical Level One; Pipeline Mechanical Level Two; Pipeline Mechanical Level Three, Modules 63301-02 through 63305-02.

Required Trainee Materials

1. Pencil and paper
2. Appropriate personal protective equipment

1.0.0 ◆ INTRODUCTION

Pneumatic systems utilize compressed air to perform work throughout the pipeline industry. Pneumatic systems have many uses on the pipeline including the powering of actuators/operators, driving pneumatic hand tools and lifts, and providing the energy for many types of operations.

Pneumatic systems use compressors to capture air from the atmosphere and compress it, raising the pressure of the air. The air is treated by other components of the pneumatic systems and is then supplied to the actuator/operators or other equipment. Depending on the application, compressed air may be required to be clean and free of oil or to contain some oil for lubrication, to be supplied at high or low pressure, or to be supplied at high or low volume.

A valve actuator/operator is a mechanical device that is used to open, close, or position a valve manually or automatically. Manual actuators/operators come in the form of handwheels, levers, and wrenches that are attached to the valve stem, either directly or by means of a gearbox. Automatic actuators/operators are used when the valve is remote from the main work area, when the frequency of operation would require unreasonable human effort, or if rapid opening and closing of the valve is required. In some installations, automatic remote control actuator/operators are installed as a matter of convenience, reducing the number of employees required for facility operation. Large valves that require great effort to turn must be equipped with some type of powered mechanism, such as electric, pneumatic, or hydraulic motors.

Note that in some locations the gearbox is called an actuator. In other locations, it is called an actuator/operator. Technically speaking, the operator is the gearbox unit or hand device that actually moves the valve. The actuator is the power unit that is added to turn the operator

Materials

Ensure that you have everything required to teach the course. Check the Materials and Equipment List at the front of this module.

Classroom

Explain that terms shown in bold are defined in the Glossary at the back of this module.

Audiovisual

Show Transparency 1, Course Objectives.

Show Transparency 2, Performance Tasks.

Teaching Tip

See the general Teaching Tip at the end of this module.

Classroom

Review the uses of pneumatic systems. Explain how pneumatic systems use compressors.

Discuss the uses for automatic actuators/operators.

Classroom

Discuss the duties involved in preventive maintenance for pneumatic systems.

Laboratory

Refer to *49 CFR 192.605.* Emphasize the importance of following a manual of written procedures. Ensure that trainees understand and read these procedures.

Audiovisual

Show Transparency 3 (Figure 1). Review the priorities of a good preventive maintenance program.

Classroom

Review the basic methods for implementing preventive maintenance procedures. Answer any questions trainees have.

automatically. In this module, the term actuator/operator will be used collectively to refer to the entire unit.

You must also understand that many actuators/operators are devices that experience heavy loads although they may be idle for many days at a time. Some valves, if turned manually, would require so much time and effort that an opening or closing operation would be self-defeating if it were not an emergency. For example, some large valves require more than 10,000 handwheel revolutions from open to close. The time that would be required for a power unit to open or close such a valve would be extensive. For that reason, some mechanical technicians and crews carry gasoline engine, air, or electric powered portable valve actuators/operators to assist them with cycling valves.

You must have a good understanding of the principles of pneumatic systems along with strong troubleshooting skills. Troubleshooting pneumatic actuators/operators and systems effectively is a necessity for maintaining the safe and efficient operation of the pipeline.

The components of pneumatic systems must be properly maintained and repaired to ensure that the system provides properly treated air at the required pressure and volume for the application. The maintaining of pneumatic systems consists of performing regularly scheduled preventive maintenance procedures, inspecting the system to detect problems, troubleshooting, and repairing the system as needed.

2.0.0 ◆ PREVENTIVE MAINTENANCE

Preventive maintenance (PM) is the periodic inspection of facility assets and equipment to uncover conditions that may lead to failure or shorten anticipated life. The long-term goal for a good PM program is to increase reliability, decrease unplanned equipment downtime, reduce overall maintenance costs, and support improvements in quality and production. The duties involved in PM are generally grouped into the following activities:

- Inspecting components
- Lubricating components
- Replacing components
- Cleaning components
- Adjusting components
- Documenting work accomplished

The start of any PM program involves perform ing a detailed analysis of past equipment failures including their costs, their causes, any remedia actions, and the changing demands on the equip ment. A detailed maintenance history of all of th equipment within the system must be developed After this information is gathered, it is divided and prioritized according to the maintenanc item's effect, ranging from maximum to mini mum. The criteria used to determine effec includes safety, product costs, overhead costs, and equipment replacement costs.

A good PM program targets the items tha cause economic damage as first priority. The sec ond priorities are the maintenance items that need to be done but can wait until the first prioritie have been dealt with. The lowest priority mainte nance items are those that have little monetar and absolutely no safety impact; these can b attended to after they have failed. *Figure 1* show a PM record.

In a general application of *49 CFR 192.605*, eac pipeline operator is required by federal law t prepare and follow a manual of written proce dures for regular maintenance of valve actua tor/operators. This manual must be updated a least once every fifteen months. Applicable por tions of it must be kept at locations where work i performed. Such manuals must provide for safet of operations during these maintenance proce dures. You must follow company policies and procedures during maintenance activities.

The law only applies to maintenance activitie on the pipeline that could result in danger to per sonnel, the public, facilities, or the environmen Inspections, diagnostics, repair of units located i areas isolated from the pipeline, bench work in shop, and how information gathered about suc units is managed is not included.

However, sound management practice require that all equipment be maintained in good work ing condition. It is not only a matter of safe and efficient operation but also economic operatior Valve actuators/operators represent a sizeabl investment that must be protected.

2.1.0 PM Methods

PM methods are systematic maintenance proce dures that reduce emergencies, downtime, and costly production delays. PM procedures var from one piece of pneumatic equipment t another, but the basic principles are the same. Th

Instructor's Notes:

PNEUMATIC SYSTEM
PREVENTIVE MAINTENANCE PROCEDURE

Equipment No. ____________________ Equipment Name ____________________

Inspection Frequency ____________________

1. Check coolant and oil levels.
2. Change filters.
3. Check all bearings for noise or excessive temperature.
4. Check belt tension and condition.
5. Check operation of valves and air pressure.
6. Check for leaks in unit around pipes, and correct.
7. Clean area.

Inspection Frequency ____________________

1. Shut down and open up for inspection and cleaning.
2. Remove guard on belt.
3. Take tension off belts.
4. Check bearing condition.
5. Replace belts and take up proper tension.
6. Check sheave alignment with yardstick, and correct.
7. Replace guard.
8. If new belts are required, return in 24 hours to take up stretch.

Inspection Frequency ____________________

1. Grease fan at two fittings.
2. Wipe up excess grease.
3. Remove plugs from electric motor; insert special fitting on fill.
4. With motor running, force grease into bearing until grease appears at drain hole.
5. Remove special fitting and replace fill plug.
6. Run 20 minutes and replace drain plug.
7. Wipe up excess grease.

Repair Record

Date	Work Accomplished	Cost

306F01.EPS

Figure 1 ◆ PM record.

basic methods for implementing PM procedures can be remembered using the word CARE:

- C = *Cleaning* the system and equipment
- A = *Adjusting* the components
- R = *Replacing* disposable components
- E = *Examining* or inspecting equipment

2.2.0 Scheduling

Scheduling PM is an important element in the PM program. PM performed too frequently increases costs, as does PM that is performed too infrequently. Companies use computerized systems, card files, and maintenance logs to assign individuals to manage the responsibilities of maintenance.

Classroom

Discuss problems that can result if preventive maintenance is performed too frequently or too infrequently.

Classroom

Refer to Figure 2. Discuss the elements of a typical preventive maintenance schedule.

Laboratory

Distribute sample manufacturer's warranties. Ask trainees to review the warranties and identify the consequences of failing to perform preventive maintenance activities in accordance with the warranty requirements.

Classroom

Discuss the importance of conducting training and of maintaining proper training records.

Some company computers generate work orders at regular intervals for the equipment to make certain that legally required schedules are met and that equipment in the most isolated locations is included. Schedules must be reviewed on a continual basis to check their accuracy and effectiveness. The schedule is always changing and depends on consistent documentation.

The maintenance schedule, formed within the company, specifies all the recommended maintenance required to keep all equipment in good operating condition. *Figure 2* shows a typical PM schedule for pneumatic systems.

2.3.0 Manufacturer's Instructions and Warranties

Preventive maintenance must be performed according to manufacturer's instructions and warranty requirements. Many actuators/operators are supplied with warranties, and in order to retain that warranty, they must be maintained according to the manufacturer's instructions. If it is still under warranty, the work will need to be carried out only as authorized by the manufacturer. In some cases, the manufacturer may require that the actuator/operator to be returned to the factory for warranty service. The record for the particular actuator/operator should be reviewed to see if the warranty is still in effect before any work is attempted. The specified maintenance procedures should be followed to ensure proper operation and to maintain warranty requirements.

2.4.0 Training

You should become completely familiar with the types of actuators/operators being used and the types of maintenance products required for their job assignments. An ongoing training program with regular updates and reviews is considered essential by some companies. Training should also include the use of proper methods of documentation.

ACTION	PART OR ITEM	RUNNING HOURS	TIME INTERVALS						
			DAILY	WEEKLY	MONTHLY	3 MO.	6 MO.	YEARLY	2 YRS.
CHECK	OIL LEVEL	8	X						
INSPECT	COOLANT LEVEL	8	X						
INSPECT	DISCHARGE TEMPERATURE (AIR)	8	X						
INSPECT	SEPARATOR ELEMENT DIFFERENTIAL	8	X						
INSPECT	AIR FILTER	8	X						
INSPECT	OIL FILTER	8	X						
REPLACE	COOLANT FILTER	150		X (INITIAL CHANGE ONLY)					
CHECK	TEMPERATURE SENSOR	1,000			X				
REPLACE	COOLANT FILTER	2,000					X		
CLEAN	SEPARATOR SCAVENGE SCREEN AND ORIFICE	4,000							
CLEAN	COOLER CORES	4,000					X		
REPLACE	AIR FILTER	4,000						X	
REPLACE	SEPARATOR ELEMENT	4,000						X	
REPLACE	COOLANT	8,000							X
INSPECT	STARTER CONTACTS	8,000						X	

306F02.EPS

Figure 2 ◆ Typical PM schedule for a pneumatic system.

Instructor's Notes:

2.5.0 Documentation

Records are also a very important element of the PM system. Records are maintained to build a detailed and accurate historical file of the equipment in relation to hours of operation, parts usage, and cost of maintenance (see *Figure 3*). The records system must store accurate information in the following categories:

- Listing of each piece of equipment requiring PM
- Location of each piece of equipment
- Running history of maintenance actions on the equipment, including the parts replaced
- Time expended on PM
- Cost of replacement parts

3.0.0 ◆ INSPECTING PNEUMATIC SYSTEM COMPONENTS

Along with scheduled PM, regular inspection of pneumatic systems is important. Pneumatic system PM inspections should be performed regularly to ensure that the components are in proper working order and to maintain a long service life for components. In some cases, scheduled PM will miss a problem that a follow-up inspection will find. Company procedures usually specify how often PM inspections should be performed.

> **WARNING!**
> Before inspecting or maintaining any pneumatic equipment, ensure that the equipment is properly shut down and locked out and tagged. All pressure must be relieved to prevent personal injury.

Pneumatic system PM inspections include inspecting the following components:

- Intake filters
- **Intercoolers**
- Aftercoolers
- Separators
- **Receivers**
- In-line filters
- Pressure regulators
- Lubricators

EQUIPMENT RECORD

TAG NO. **K-G4002**		OLD PL NO. **For Juice Box**	
TAG LOCATION **On Compressor**		EQUIP LOCATION **Gr. F1 Refrig. Rm. (by Laundry)**	
MFG. **Tecumseh**		VENDOR	
MFG. SERIAL NO. **87811**		DATE **11/18/69**	COST
JOB NO.		FLOW DIA. BOOK NO.	CAT. NO.

DESCRIPTION OF EQUIPMENT	SPARE PARTS IN STOCK	
	PART NO.	PART NAME
Compressor - Condenser (Belt Driven - 2 Belts)		
Belt Size: 56 500 Pulley Shaft		
Alco Evaporation Valve 435		
Sporlan Catch-All Filter (Type C-164)		
Ranco Pressure Control P.B. Reset 012-1546		
Penn Water Valve 1/2" Pipe		
Type 246F04AR Model 2300 All-Range 70-260 psig		
Commercial Service Frame U-W110S Model 3089-54B		
Serial # Part# AB F-12 (Change 1 lb.)		
Unit Test 235-SA652		
ANNUAL: 4.0		
Change oil. Check belts/pulleys for alignment.		
Clean water side of condenser.		

	TITLE

K-G4002

306F03.EPS

Figure 3 ◆ Equipment record.

Classroom

Refer to Figure 3. Review the information that must be recorded for preventive maintenance documentation. Go over any questions trainees have.

Teaching Tip

See the Teaching Tip for Sections 2.0.0–2.5.0 at the end of this module.

Performance Testing

Have each trainee perform Performance Task 1 to your satisfaction. Fill out a Performance Profile Sheet for each trainee.

Homework

Have trainees review Sections 3.0.0–4.0.0 for the next session.

Materials

Ensure that you have everything required for testing and the laboratory during this session.

Safety

Emphasize the importance of locking out and tagging pneumatic equipment before it is inspected.

Classroom

Explain when it is necessary to replace intake filters.

Audiovisual

Show Transparency 4 (Figure 4). Review the components of an air intercooler.

Classroom

Refer to Figures 5 and 6. Compare the types of aftercoolers.

Audiovisual

Show Transparency 5 (Figure 7). Explain how to ensure that the filter element is not clogging.

3.1.0 Inspecting Intake Filters

Intake filters filter the intake air to prevent dirt and moisture from entering the compressor. The filters should be inspected at each startup and periodically to ensure that they are clean, dry, and undamaged. Replace the filter as needed and on a regular basis as part of a good PM program.

3.2.0 Inspecting Intercoolers

The air in a multistage air compressor is heated each time it is compressed. The hotter the air becomes, the more it expands. Compressing the hot air heats the air even more, reducing the efficiency of the compressor. To overcome this problem, the air must be cooled between compression stages. Intercoolers are devices designed to cool the air. The two types of intercoolers used in multistage air compressors are air and liquid.

An air intercooler (see *Figure 4*) consists of finned tubing that runs between the stages of the compressor. The air around the tubing and fins dissipates the heat into the atmosphere. The air intercooler should be inspected to ensure that the fins are clean and not bent or damaged, interfering with heat transfer and causing the compressor to overheat.

A liquid intercooler uses water to cool the air as it passes through the intercooler. The water is circulated through tubing inside the tank-like intercooler structure containing the compressed air. Cool water is brought in through tubing from one end of the intercooler. Hot compressed air enters the intercooler structure from the end of the other to allow maximum heat transfer. Because the cooling of the air causes some condensation, a drain is installed on the intercooler to remove the condensation. A liquid intercooler should be inspected to ensure that it is clean and that the condensate is draining properly.

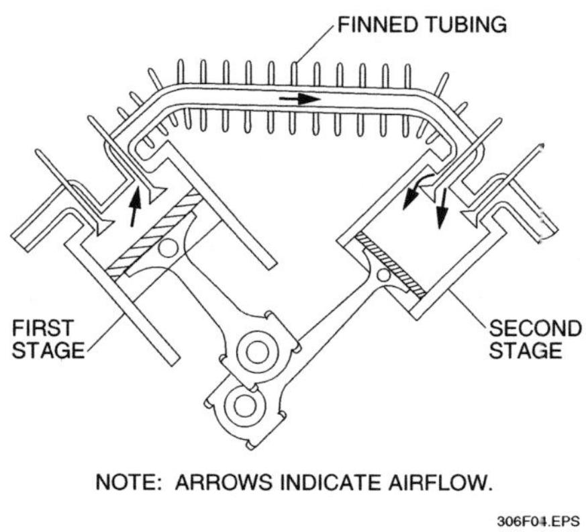

Figure 4 ◆ Air intercooler.

3.3.0 Inspecting Aftercoolers

Aftercoolers reduce the discharge temperature of the compressed air and enable water vapor and oil vapor to be removed. The primary types of aftercoolers are the air-cooled aftercooler, the water-cooled aftercooler, and the refrigeration dryer.

An air-cooled aftercooler consists of finned tubing through which compressed air passes on its way to the air receiver. The aftercooler is situated adjacent to the compressor flywheel. The compressor flywheel draws the cooling air over the tubes to cool the air and condense the moisture. Air-cooled aftercoolers are used as belt guards on some air compressor models. Air-cooled aftercoolers should be inspected to ensure that they are clean and the fins are not bent or damaged.

A water-cooled aftercooler (see *Figure 5*) consists of copper tubing with multiple fins located within a steel shell. Headers on each end of the aftercooler provide connections for water, air, and condensate piping. The water-cooled aftercooler should be inspected to ensure that it is clean and the condensate is draining properly.

A refrigeration dryer (see *Figure 6*) is a self-contained, cooling unit that cools the compressed air and lowers the water vapor content of the air to the level required by the system. When inspecting a refrigeration dryer, you should:

- Ensure that the filter is clean and dry. Replace as needed.
- Ensure that the refrigerant coil is clean and not bent or damaged.
- Ensure that the condensate drain is draining properly.
- Ensure that the overall unit is clean and not damaged.

3.4.0 Inspecting Separators

Separators remove water vapor or oil droplets carried along from the aftercooler. Some separators have automatic drains that prevent the need for routine draining of the collected water and oil. The separator should be inspected to ensure that it is clean and there are no air leaks. Inspect the separator drain to ensure that it is draining properly, and inspect the filter element for clogging. *Figure 7* shows a separator.

Instructor's Notes:

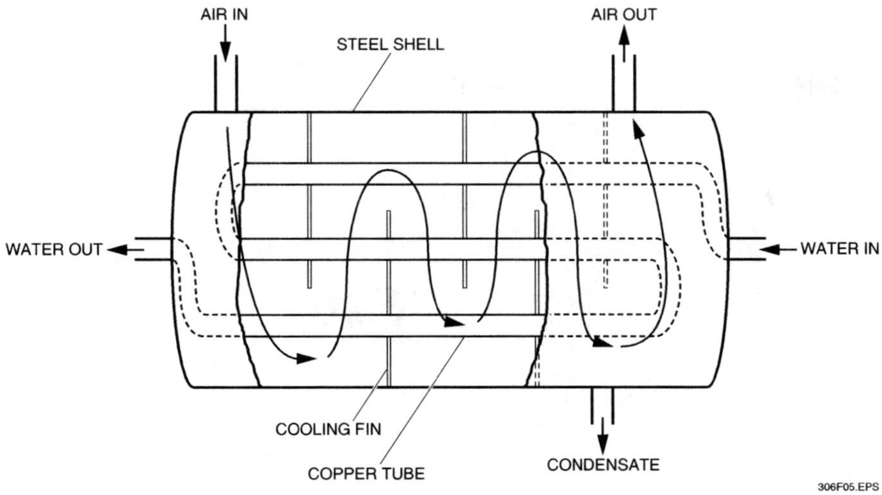

Figure 5 ◆ Water-cooled aftercooler.

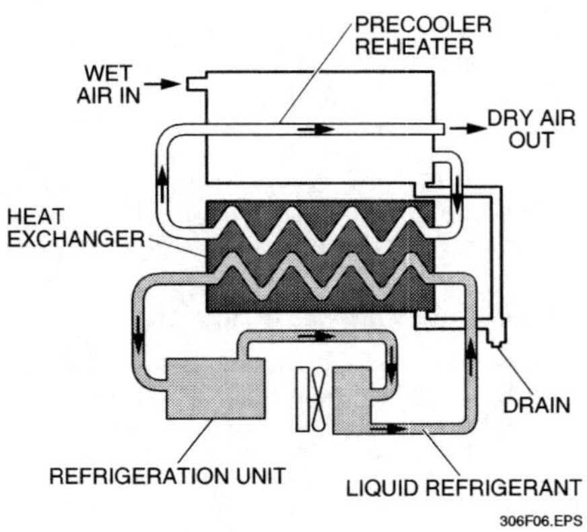

Figure 6 ◆ Refrigeration dryer.

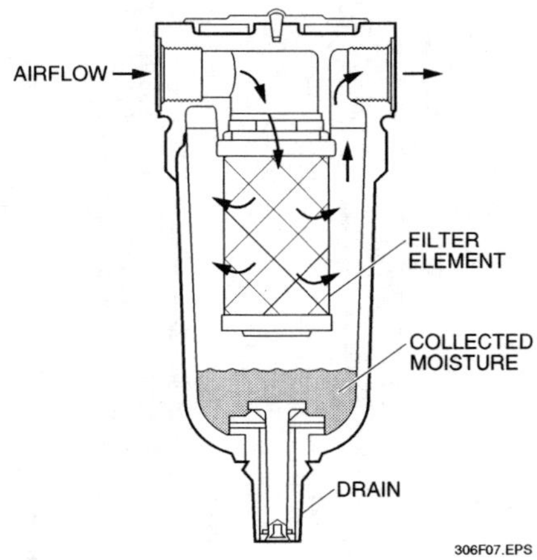

Figure 7 ◆ Separator.

3.5.0 Inspecting Receivers

A receiver (see *Figure 8*) is a tank-like structure that stores compressed air. The receiver dissipates the heat from the compressed air, collects moisture from the air, and helps provide smooth airflow from the system. A receiver should be inspected to ensure that it is clean and there are no air or water leaks. The drain should be inspected to verify that it is draining properly.

3.6.0 Inspecting In-Line Filters

In-line filters (see *Figure 9*) are used in compressed air systems to remove most solid and liquid particles from the compressed air. In-line filters should be inspected to ensure that they are clean and free of air leaks. The filter drain should be inspected to verify that it is draining properly.

Classroom

Explain how a clean receiver provides smooth airflow from the system. Ask whether trainees have any questions.

Audiovisual

Show Transparency 6 (Figure 9). Review the procedure to ensure that an in-line filter is draining properly.

Classroom

Discuss the symptoms of internal regulator malfunctions.

Audiovisual

Show Transparency 7 (Figure 11). Explain how to ensure that the lubricator's oil reservoir is properly filled.

Teaching Tip

See the Teaching Tip for Sections 3.0.0–3.8.0 at the end of this module.

Performance Testing

Have each trainee perform Performance Task 2 to your satisfaction. Fill out a Performance Profile Sheet for each trainee.

Classroom

Explain the importance of being able to read pneumatic schematic diagrams.

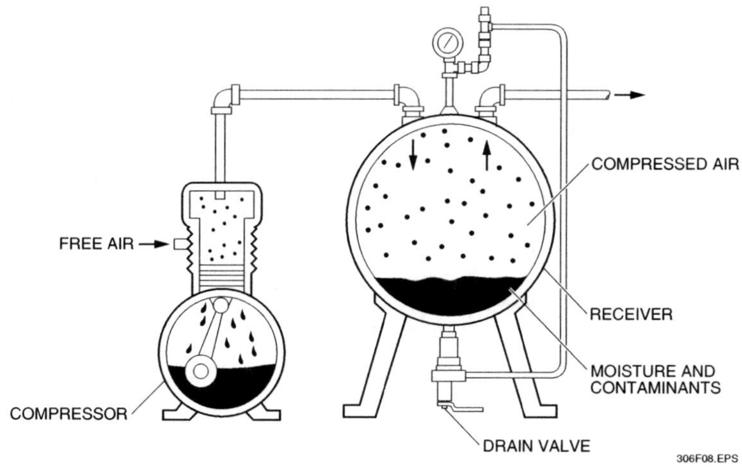

Figure 8 ◆ Receiver.

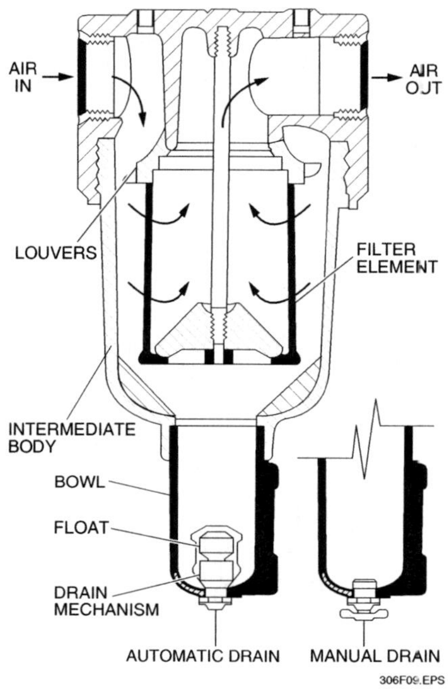

Figure 9 ◆ In-line filter.

3.7.0 Inspecting Pressure Regulators

A pressure regulator (see *Figure 10*) is a normally open valve that is positioned after the receiver in a pneumatic system. It regulates the pressure of the air to the user. During routine inspection, regulators should be examined to ensure that they are clean and free of air leaks. Internal regulator malfunctions are evidenced by erratic pressures downstream of the regulator.

3.8.0 Inspecting Lubricators

Lubricators (see *Figure 11*) inject a fine mist of oil into the air immediately before the user to provide seal lubrication, prevent sticking of moving parts, and control wear of the user components. During routine inspection, lubricators should be inspected to ensure that the oil reservoir is at the full mark and that the lubricators are clean and free of oil and air leaks.

4.0.0 ◆ READING PNEUMATIC SCHEMATIC DIAGRAMS

Any problems found when inspecting a pneumatic system and any problems that occur during normal operation must be corrected. Troubleshooting consists of locating and eliminating the cause(s) of problems or malfunctions in a machine or system. To effectively troubleshoot a pneumatic system, a good understanding of how the system operates is required. You must be able to read and

Instructor's Notes:

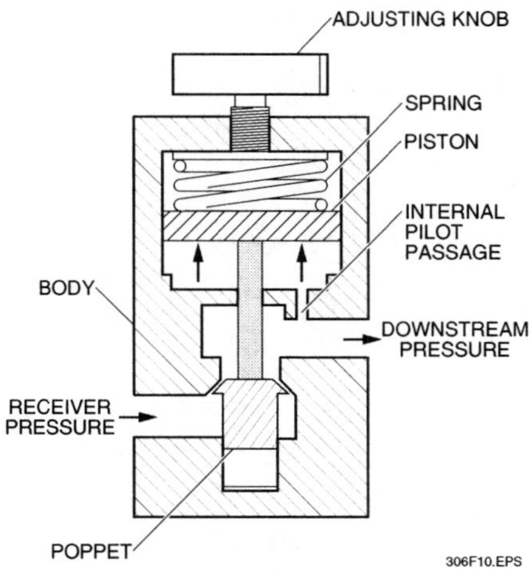

Figure 10 ◆ Pressure regulator.

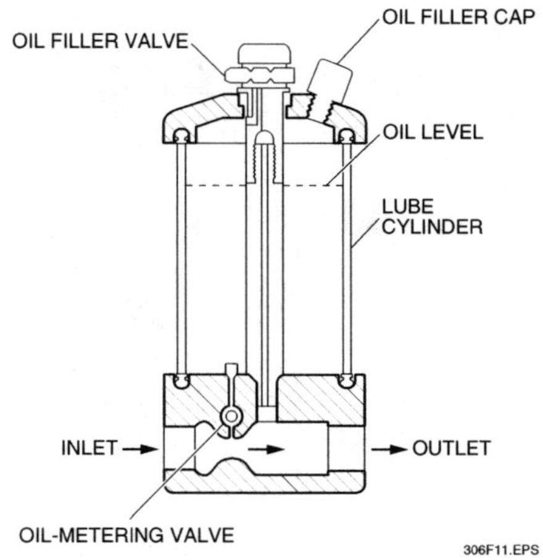

Figure 11 ◆ Lubricator.

understand a schematic diagram of the pneumatic system. By using a schematic diagram, you can determine how the system is designed to operate and can locate the various components to determine where the problem is occurring.

Reading a schematic diagram consists of recognizing the standard symbols used and being able to trace the system lines from one component to another. The symbols (see *Figure 12*) represent the components and air lines of the system.

When tracing a system, begin at the intake filter and trace the line through the system until the entire system has been traced and all of the components identified. Being able to trace the system and identify the components and their functions helps pinpoint problems when troubleshooting pneumatic systems.

5.0.0 ◆ TROUBLESHOOTING PNEUMATIC ACTUATORS/OPERATORS AND SYSTEMS

Good troubleshooting is based on following a sequential, logical method for locating the problem in a failing system. A good troubleshooter establishes and follows a set procedure for pinpointing problems instead of using a random, hit-or-miss approach, which takes much longer and may accomplish very little. For example, the following list for troubleshooting an actuator/operator highlights the importance of being methodical and gathering lots of information. To diagnose a problem with an actuator/operator, you should:

WARNING!
The control center must be notified before any work is begun. Company safety policy must be followed when troubleshooting a valve actuator/operator.

Step 1 Understand how the valve and its actuator/operator function within the system.

Step 2 Learn about the problem by investigating the symptoms. Include input from local operations personnel, the unit's historical information, and personal assessment.

Step 3 If it is a repeat problem, contact the manufacturer's technical assistance department for information and suggestions about the particular actuator/operator.

Step 4 Take note of all attached equipment influencing the operation of the unit. The operation of limiting devices and other related units are especially important in diagnosing problems.

Step 5 Notify the control center, upstream and downstream pumping stations, and any other affected personnel on the pipeline.

Laboratory

Distribute sample schematic diagrams. Ask trainees to interpret the symbols in the diagrams using the legend in Figure 12. Answer any questions trainees have.

Homework

Have trainees review Sections 5.0.0–6.0.0 for the next session.

Materials

Ensure that you have everything required for testing during this session.

Safety

Ask a trainee to explain why the control center must be notified before work begins on pneumatic systems.

Classroom

Review the procedures for diagnosing a problem with an actuator/operator.

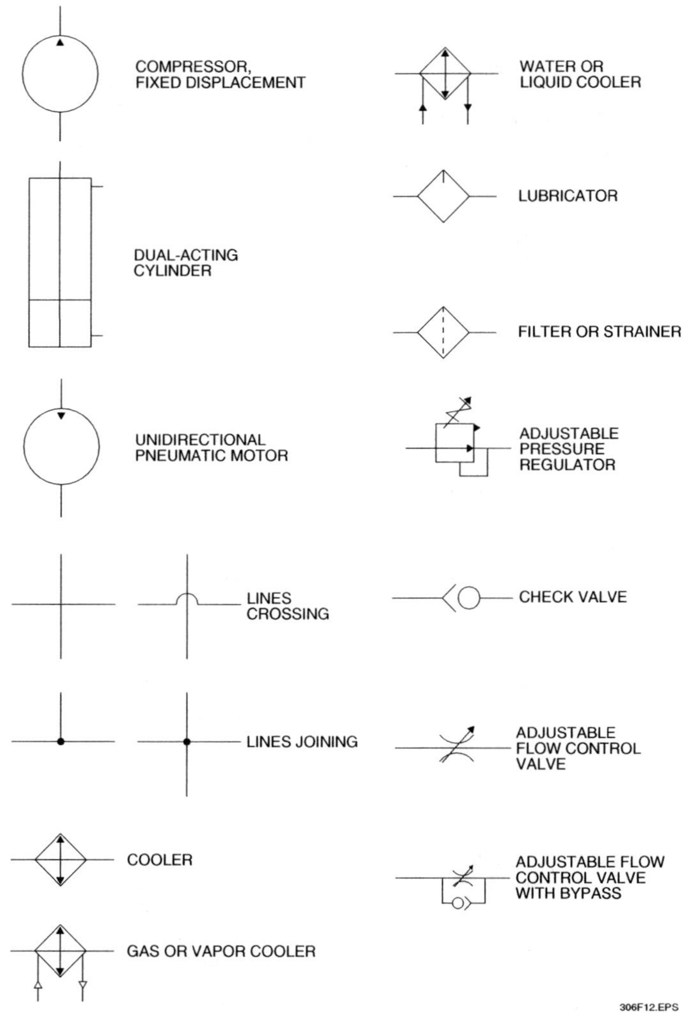

Figure 12 ◆ Common pneumatic symbols.

Step 6 Switch the actuator/operator to manual control, and run the unit through its cycle completely several times to observe the actual problem.

Step 7 Form a personal theory, supported by facts, about the cause of the problem.

Step 8 Form a plan of action for the most efficient solution to the problem and the return of the actuator/operator to normal operation.

Step 9 Initiate all applicable lockout/tagout procedures and follow company policy for PPE and safety procedures.

Instructor's Notes:

Step 10 Perform repairs/adjustments according to company policy and manufacturer's recommendations.

Step 11 Inspect the unit for worn or damaged parts. They should be repaired or replaced while the unit is disassembled.

CAUTION

When disassembling an actuator/operator for inspection and repairs, clean the area, keep it clean, and keep the actuator/operator parts clean.

Step 12 Test the valve for proper operation.

Step 13 Return the unit to normal use.

Step 14 Notify the control center and other affected personnel, and complete all required documentation per company policy.

Troubleshooting the pneumatic system serving a machine that fails to operate can be accomplished by starting at the actuator/operator of the machine and working toward the air supply or by starting at the air supply and working toward the actuator. The method used is a matter of preference. The important thing is to inspect each component in sequence until the problem is pinpointed. This section explains how to troubleshoot a system from the actuator/operator to the air supply. Troubleshooting a common pneumatic system includes inspecting the following components:

- Air supply
- Actuator/operator
- Control valve
- Sequence circuits
- Interlocks

5.1.0 Examining the Air Supply

One of the first things to examine in a failing system is to verify that there is a supply of air at a sufficient rate of flow and pressure. It is possible to have an air supply at proper pressure but low flow. This could be caused by a clogged filter, which can allow enough air through it to indicate an adequate reading on a pressure gauge but does not allow enough flow when the system is operating. A leaking air supply line can produce the same results. Verify that electric power to the machine and the system controls is turned on. Examine the air supply to ensure that there is no water in the system. If the air supply is saturated with water, it can cause severe damage to a pneumatic system.

5.2.0 Inspecting the Actuator/Operator

The next step in troubleshooting the system is to inspect the actuator/operator. Disconnect the actuator/operator from the load, and verify that it operates properly. Disconnecting and manually operating it indicates whether the problem is in the pneumatic system or the mechanical components. If the actuator/operator cycles properly when disconnected, the problem is probably in the machine. If it does not operate properly, the cause must be determined and corrected. The actuator/operator may not be cycling properly because of a lack of air caused by a crimped or broken airline. Another possible cause may be that the cylinder is sticking due to dried-out lubricant, corrosion, or the deterioration of rubber parts.

5.3.0 Inspecting the Control Valve

If the actuator/operator is working properly, loosen or disconnect the air lines to it, and verify that there is airflow at that point. If there is no airflow, the problem is either in or upstream of the control valve. A sluggish control valve may be caused by inadequate lubrication, swollen seals, heat, swollen or deteriorated packing, or binding caused by excessive operating pressure. Additionally, worn packing, valve seats, and valve bodies can cause internal leaking in the valve and can affect actuator movement.

5.4.0 Inspecting Sequence Circuits

Sequence and logic circuits used in a pneumatic actuator/operator system sense time, sequence, and interlock and direct signals to power valves. Valves types used in logic circuits include check, shuttle, flow-control, quick-release, cam-limit, and three- and four-way sequence. Malfunctions in these circuits can result in false stroking or the malfunction of the actuator/operator. These circuits can be complicated to troubleshoot, but using the process of elimination and logical troubleshooting techniques, the problem can be pinpointed and corrected.

5.5.0 Inspecting Interlocks

Interlocks in a pneumatic system protect the actuator/operator from being operated at the wrong time, thereby protecting the actuator/operator from being damaged. They may also be used to protect personnel from injury. Interlocks may be two-, three-, or four-way valves that either block or permit airflow to the logic circuits. They may be operated by a button, lever, cam, or pedal. Interlocks may be pilot-operated by hydraulic or pneumatic

Safety

Explain problems that can result if actuator/operator parts are not kept clean during inspection and repairs.

Classroom

Review the components to inspect when troubleshooting pneumatic systems.

Explain how to identify common malfunctions when troubleshooting pneumatic system components. Go over any questions trainees have.

Classroom

Refer to Table 1. Present hypothetical problems that may occur with a pneumatic system. Ask trainees to identify the possible cause and a solution for each.

Performance Testing

Have each trainee perform Performance Task 3 to your satisfaction. Fill out a Performance Profile Sheet for each trainee.

Homework

Have trainees review Sections 7.0.0–7.3.0 for the next session.

Materials

Ensure that you have everything required for the demonstration, laboratories, and testing during this session.

Safety

Emphasize the importance of carefully monitoring all valve and actuator/operator activities.

Discuss the consequences of failing to shut off the air supply when performing maintenance on a pneumatic system.

power or an electric solenoid. To troubleshoot an interlock valve, verify that the actuator/operator moves freely and that there are no obstructions.

> **WARNING!**
> Never allow an actuator/operator to be operated without the interlocks in service. This could result in damage to the valve or actuator/operator or injury to personnel.

Continue to troubleshoot the system in a logical, sequential manner until the problem is identified. After the problem and possible causes are identified, the causes should be eliminated one by one, starting with the one that requires the least amount of downtime or most minor disassembly. Inspections that involve extensive downtime or a major disassembly should be saved until last.

6.0.0 ◆ PNEUMATIC TROUBLESHOOTING CHARTS

In troubleshooting a pneumatic system, a thorough understanding of pneumatic principles and how the system works is a necessity. However, a good troubleshooting chart is also a useful tool. Troubleshooting charts are sometimes provided by the manufacturer. There are also generic charts that may be used with common systems.

Troubleshooting charts describe some of the problems, probable causes, and possible solutions for a common system. Once a problem has been identified, the troubleshooting chart can be used to help determine the cause and the solution that should be taken. *Table 1* shows a pneumatic system troubleshooting chart.

7.0.0 ◆ REPAIRING PNEUMATIC SYSTEM COMPONENTS

When a pneumatic system component becomes worn past the point of efficiently performing the function for which it was intended, it must be repaired or overhauled. There are numerous types and makes of pneumatic components; it is essential to refer to the specific manufacturer's maintenance and repair manual when repairing or overhauling. This module provides general procedures that must be performed in conjunction with the manual. Repairing pneumatic system components includes performing the following procedures:

- Preparing the system for shutdown and repair
- Repairing and overhauling components
- Replacing gauges

7.1.0 Preparing for Shutdown and Repair

Every company has safety policies and procedures that must be followed when performing maintenance and repairs. Working on pneumatic actuators/operators and their systems includes certain procedures that are required because maintenance on a pneumatic actuator/operator necessitates the operation of a valve and therefore affects the pipeline system. These procedures must be followed before every inspection and repair procedure on pneumatic actuators/operators discussed in this module:

Step 1 Secure all required work orders, permits, and PPE.

Step 2 Follow all company safety policies and procedures.

Step 3 Verify the location of the pneumatic valve actuator/operator to be inspected.

Step 4 Verify its accessibility.

Step 5 Verify its number and nameplate data.

> **NOTE**
> If the nameplate is missing or illegible, it must be replaced according to company policy and procedures.

Step 6 Notify the control center that work is beginning.

> **WARNING!**
> You must maintain an ongoing awareness of the valve and actuator/operator activities being performed to ensure safety during the performance of the task.

Before repairing or replacing any pneumatic actuator/operator system component, certain steps must be taken.

> **WARNING!**
> Before performing maintenance on any component in a pneumatic actuator/operator system, the air supply must be shut off and all pressure relieved from the system to avoid damage and/or serious injury to personnel.

Instructor's Notes:

Table 1 Pneumatic System Troubleshooting Chart

Problem	Possible Cause	Possible Solution
The actuator cylinder is drifting.	The piston seal is leaking.	Replace the piston seal.
	The operating valve is leaking.	Repair or replace the operating valve.
	The connecting lines are leaking.	Tighten or repair the connecting lines.
The cylinder is sticking or binding.	The packing on the cylinder rod is over tightened.	Loosen or replace the packing.
	The cylinder is misaligned.	Align the cylinder.
	The seals are defective.	Replace the seals.
	The pilot pressure is low.	Adjust the pilot pressure.
The cylinder body seal is leaking.	The tie rods are loose.	Retorque the tie rods.
	The pressure at the cylinder is excessive.	Adjust the pressure.
	The cylinder body seal is pinched or extruded.	Replace the cylinder body seal.
The rod gland seal is leaking.	The seal is deteriorating.	Replace the seal.
A pneumatic valve is blowing to exhaust when not actuated.	The inlet poppet is not sealing.	Repair or replace the poppet.
A pneumatic valve is blowing to exhaust when actuated.	The valves seals are faulty.	Replace the valve seals.
	The valve-to-base gasket is faulty.	Replace the valve-to-base gasket.
	The spool is damaged.	Repair or replace the spool.
A solenoid is failing to activate its valve, but manual override activates the valve.	The solenoid is faulty.	Replace the solenoid.
A solenoid is failing to activate its valve, and manual override fails to activate the valve.	The voltage at the solenoid is low.	Restore the voltage to the proper level.
	There are varnish deposits on the valve.	Clean the varnish deposits from the valve.
A solenoid is buzzing.	The pilot pressure is too low.	Adjust the pilot pressure.
	The valve seals are defective.	Replace the valve seals.
	The solenoid is faulty.	Replace the solenoid.
A solenoid is burning out prematurely.	The voltage to the solenoid is low.	Restore the voltage to the proper level.
	The voltage to the solenoid is incorrect.	Restore the voltage to the proper level.
	There is varnish in a direct-operated spool valve.	Remove the varnish from the valve.
The pilot section is blowing to exhaust.	The pilot cover is loose.	Tighten the pilot cover.
	The poppet is not sealing.	Repair or replace the poppet.
A poppet is chattering.	The air supply pressure is too low.	Adjust the air supply pressure.
	The pilot pressure or signal pressure is too low.	Adjust the pilot or signal pressure.
	The muffler is faulty.	Replace the muffler.

306T01.EPS

You should follow these general steps to prepare the pneumatic system for repairs:

Step 1 Gather all the necessary tools, including hoisting equipment, as required.

Step 2 Obtain the manufacturer's service manual for the exact unit to be serviced, and review the repair and installation procedures.

Step 3 Lower all vertical cylinders.

Step 4 Discharge all accumulators.

Step 5 Isolate all energy sources, electric, hydraulic, and pneumatic.

WARNING!
All energy sources must be isolated to prevent them from causing equipment damage and/or serious injury when pressure is relieved.

Classroom

Review the procedure for preparing a pneumatic system for shutdown and repair.

Safety

Ask a trainee to explain why all energy sources must be isolated.

Safety

Review problems that can result if power is not turned off and pressure is not relieved.

Classroom

Review the procedures for repairing and overhauling a pneumatic system. Ask whether trainees have any questions.

Safety

Stress the importance of verifying that the pneumatic system is properly prepared for maintenance.

Classroom

Explain how to remove a pneumatic actuator/operator component from the system.

Demonstration

Demonstrate how to disassemble a pneumatic actuator/operator following the maintenance procedures in the manufacturer's instruction manual.

Step 6 Turn off the electric power and the air supply to the system.

> **WARNING!**
> Failure to turn off the electric power or air supply, or to relieve all pressure can result in serious personal injury.

Step 7 Relieve pressure throughout the system.

Step 8 Lock out and tag the system.

7.2.0 Repairing and Overhauling

Repairing and overhauling pneumatic actuators/operators and their system components can be technically involved. Manufacturer's instructions and recommendations should be followed and adhered to closely for such procedures to be successful. Repairing and overhauling pneumatic actuators/operators and the related systems and components involves performing the following procedures:

- Removing pneumatic actuators/operators and components from the system
- Disassembling pneumatic actuators/operators and components
- Reassembling pneumatic actuators/operators and components
- Installing pneumatic actuators/operators and components into the system

7.2.1 Removing Actuators/Operators and Components from the System

When preparing to remove a pneumatic actuator/operator or one of its components from a system, you must observe all the procedures listed earlier in this module. As maintenance on a pneumatic system affects the pipeline, all company safety policies and procedures must be observed and followed closely.

> **WARNING!**
> Verify the pneumatic system has been prepared for maintenance and that it is locked out and tagged before performing this procedure. To avoid personal injury, verify that all pressure has been relieved before disconnecting any piping from the actuators/operators or their components.

The types, brands, and installations of pneumatic actuators/operators and their systems are numerous and varied. It is not possible to include procedures that will accurately apply to every brand and model. However, there are general procedures that apply to these types of units. After completing all the procedures given earlier in this module, you should follow these general steps to remove a pneumatic actuator/operator or component from the system:

Step 1 Clean the pneumatic actuator/operator or component thoroughly around the connections.

Step 2 Disconnect any control piping or wiring from the pneumatic actuator/operator or component, and mark or tag them for reassembly.

> **NOTE**
> Many components have few connections, and it is simple to reassemble control piping and wiring, but it is good practice to mark these connections to make reassembly easier and more accurate.

Step 3 Disconnect the air piping from the pneumatic actuator/operator or component, and cap or seal the ends of the piping.

Step 4 Take note of the pneumatic actuator/operator or component mounting arrangement to aid in reassembly after the repair is made.

Step 5 Remove the pneumatic actuator/operator or component from the system, and take it to a clean work area if practical.

7.2.2 Disassembling Actuator/Operator Components

Components should be disassembled according to the manufacturer's instruction manual. Extreme care should be exercised to prevent parts from being lost or damaged during disassembly. Follow these general steps to disassemble a pneumatic actuator/operator or component:

Step 1 Clean the actuator/operator or component thoroughly.

Step 2 Disassemble it, following the manufacturer's maintenance procedures and suggestions.

Step 3 Clean and inspect each part thoroughly for wear and damage as it is disassembled, and note any parts that need to be replaced.

Instructor's Notes:

Step 4 Place the parts in a clean, dry place, arranging them in a manner that will make reassembly easier.

> **CAUTION**
> It is important to protect the parts from dust, dirt, grit, and moisture because they are easily damaged. Parts that are to be left out for some time should be covered.

7.2.3 Reassembling Actuators/Operators and Components

Reassembly of pneumatic actuator/operator parts should be performed according to the manufacturer's instructions and recommendations.

> **CAUTION**
> Failure to follow the manufacturer's instruction manual when reassembling a pneumatic valve actuator/operator may result in valve damage, actuator/operator damage, early component failure, or a pipeline upset.

Reassembly is a reversal of disassembly. Follow these general guidelines:

- Inspect and clean all actuator/operator or component parts again as they are reassembled.
- Apply a light coat of lubricant to each part to facilitate assembly.
- Handle easily damaged parts, such as O-rings and seals, carefully to prevent nicking or damaging them.
- Inspect each part carefully as it is assembled to ensure that it fits properly and does not bind or stick.
- Set and verify all clearances using the proper measuring tools.
- Inspect all replacement parts against the manufacturer's specifications for the type and model number of the component being assembled.
- Tighten all bolts and nuts to the manufacturer's torque specifications using the proper tools.
- Operate the unit by hand, if applicable, after assembly is complete to ensure smooth operation.

7.2.4 Installing Actuators/Operators and Components

Pneumatic actuators/operators and their systems vary widely in configuration. One type and brand of actuator/operator may be used on different valve applications. Many types of actuators/operators may be compatible with many different pneumatic supply systems. However, basic procedures should be followed that are applicable to all pneumatic systems. Follow these general steps:

Step 1 Uncap and clean the ends of the air piping to be connected.

Step 2 Following manufacturer's instructions and recommendations for the unit, place the pneumatic actuator/operator or component in its proper location and connect the air piping, taking care to orient the unit to its original configuration.

Step 3 Reconnect any control piping or wiring into its original configuration.

Step 4 Follow the manufacturer's instructions and procedures for testing and servicing the component.

Step 5 Remove all locks and tags.

Step 6 Return the system to service.

Step 7 Notify the control center and any other affected personnel that the unit is back in service.

Step 8 Inspect all connections for leaks.

Step 9 Monitor the components and system operation to ensure that they are working properly.

Step10 Complete all required documentation.

7.3.0 Replacing Gauges

Pneumatic gauges are important for monitoring air pressure on pneumatic actuators/operators and their system components. When pneumatic gauges fail, they are normally replaced instead of being repaired. Gauges come in many different styles. When gauges need to be replaced, it is important to select a replacement gauge that meets system requirements according to company policy and procedure.

> **NOTE**
> Most gauges are installed in conjunction with an isolation valve so that the system does not have to be shut down or locked out and tagged in order to replace a gauge. You should verify that such isolation valves are in good operating condition before attempting to replace a gauge without locking out and tagging the system. If the gauge does not have such an isolation valve, one should be installed before a new gauge is installed in the system.

Safety

Explain how to protect parts from dust, dirt, and moisture.

Emphasize the importance of following the manufacturer's instruction manual when reassembling pneumatic devices.

Laboratory

Ask a trainee to reassemble the disassembled pneumatic actuator/operator. Have trainees observe the procedure and comment on what is done correctly and/or incorrectly.

Classroom

Using the reassembled actuator/operator as a model, explain how to install the actuator/operator. Go over any questions trainees have.

Laboratory

Bring in a variety of damaged and new gauges. Have trainees examine the gauges and determine which need to be replaced.

Safety

Ask a trainee to explain how to prepare a pneumatic system for maintenance.

Explain how to properly open the isolation valve to prevent damage.

Teaching Tip

See the Teaching Tip for Sections 7.0.0–7.3.0 at the end of this module.

Performance Testing

Have each trainee perform Performance Task 4 to your satisfaction. Fill out a Performance Profile Sheet for each trainee.

Homework

Have trainees review Sections 8.0.0–8.5.0 and answer the Review Questions for the next session.

Materials

Ensure that you have everything required for the laboratory during this session.

Audiovisual

Show Transparency 8 (Figure 13). Review the components in a spring-loaded pneumatic valve actuator.

> **WARNING!**
> Ensure that the pneumatic system has been prepared for maintenance and that it is locked out and tagged. Ensure that all pressure has been relieved before disconnecting piping from the valve. Pressure left in the system when the connection is broken could cause personal injury.

Follow these general steps to replace a pneumatic gauge:

Step 1 Close the gauge isolation valve to stop airflow and pressure to the gauge.

Step 2 Remove the gauge using the proper wrench.

Step 3 Select the proper replacement gauge.

Step 4 Install the new gauge.

Step 5 Open the isolation valve to restore airflow and pressure to the gauge.

> **CAUTION**
> The isolation valve should be opened very carefully to prevent damaging the gauge with a sudden burst of pressure. The valve should be opened slightly until pressure is established to the gauge, and then it can be opened fully.

Step 6 Inspect the new gauge and isolation valve for air leaks.

Step 7 Monitor the gauge to ensure that it is operating properly.

8.0.0 ◆ PNEUMATIC VALVE ACTUATORS/OPERATORS

The pneumatic actuator/operator uses air pressure for operation. There are two basic types of pneumatic actuators. One uses a spring and diaphragm, and the other uses a piston and cylinder.

The spring and diaphragm actuator consists of a compressed air-powered device in which the compressed air acts upon a flexible member, the diaphragm, to provide linear motion to the stem. The pneumatic spring and diaphragm unit is very common and reliable, and because it has few moving parts, it is simple to maintain. An advantage of this type is its fail-safe action on loss of signal pressure. It is available for either fail-open or fail-closed action. *Figure 13* shows a spring-loaded pneumatic valve actuator.

The piston actuator/operator consists of a compressed air-powered device in which the air acts on a moveable cylindrical member, the piston, to provide linear motion to the stem. Pneumatic piston actuators/operators are normally used when thrust and torque requirements exceed the capability of spring-loaded devices. They are also well suited for use with valves that have long strokes. Piston actuators/operators normally work with supply pressures from 50 to 150 pounds per square inch (psi). Although these can be fitted with spring return, this construction has limited capabilities. Piston actuators/operators used for throttling control require a double-acting positioner that will simultaneously load and unload opposite sides of the piston. The pneumatic piston is often used when compact, high-thrust actuators are required. *Figure 14* shows a pneumatic piston actuator on a valve.

Pneumatic actuators/operators may be used with valves to control operation of a pipeline. Electro-pneumatic positioners are used in combination with either single- or double-acting actuators/operators to provide accurate throttling of control valves. They provide a valve position that is proportional to a pneumatic or direct-current input signal.

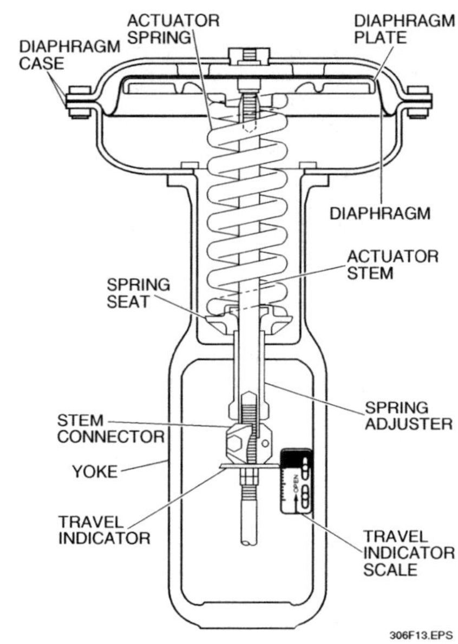

Figure 13 ◆ Spring-loaded pneumatic valve actuator.

Instructor's Notes:

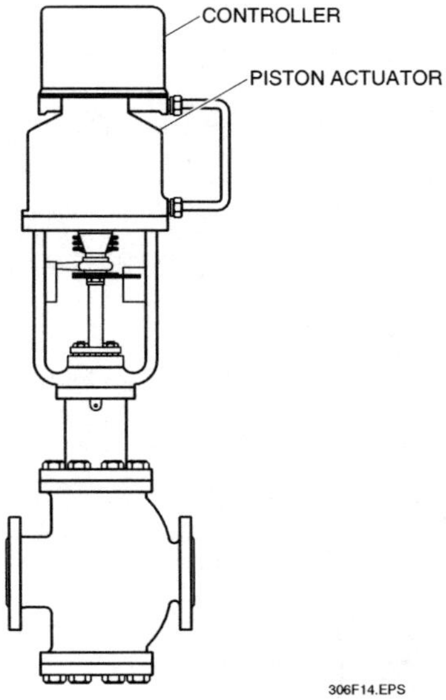

Figure 14 ◆ Pneumatic piston actuator on a valve.

8.1.0 Rack and Pinion Actuators/Operators

Pneumatic rack and pinion actuators/operators are suited for valves with 90-degree turn and designed to work off an external air supply and control system. A pneumatic rack and pinion actuator/operator is composed either of a cylinder with a piston and opposing spring to return the unit to normal position or with opposing pistons to move the actuator in both directions. The actuator/operator operates the valve stem by means of double rack and pinion gears. The pinion gear is turned as the piston pushes the rack by it. The double rack and pinion construction eliminates backlash. The pinion is connected to the valve stem by means of the machined fitting in the pinion gear.

These units (see *Figure 15*) are quick acting, powerful, compact in size, and dependable. They have very few moving parts, and action is limited to short strokes. They are available in a variety of sizes for wide application.

8.2.0 Fail-Safe Actuators/Operators

Fail-safe valves are important safety factors on the pipeline. In emergencies, the actuators/operators are set up to close the valve and shut in the system at various strategic points. In case of pipeline damage or rupture, these valves control the loss of product. Many fail-safe valves are operated by pneumatic pressure.

In case of an abnormal operating condition, the pressure is bled off the piston area, and the valve is closed by valve body and spring pressure. These units are designed to operate with a reverse ported gate valve. The valve is open in the down position. In an emergency, the actuator/operator forces the gate up, and the valve is closed. When the situation is corrected and the line is ready for operation, air pressure is supplied to the actuator/operator, opening the valve. *Figure 16* shows a pneumatic safety valve.

8.3.0 Adjusting Valve Actuators/Operators

Normal operational usage, wear, operational changes, and damage cause pneumatic valve actuators/operators to require occasional adjustment. In order to make adjustments, the valve must be operated, directly affecting the pipeline. The control center must be notified. Before beginning to adjust a pneumatic valve actuator/operator, you must complete all the preliminary preparation procedures listed earlier in this module. All cautions, warnings, and company safety policy and procedure should be meticulously observed.

Due to the many brands and models of pneumatic actuators/operators in use, it is not possible to prepare a single procedure that would apply exactly to every one. However, general procedures apply to most pneumatic equipment. Follow these general steps to adjust a pneumatic valve actuator/operator:

Step 1 Verify the location of the valve with the pneumatic actuator/operator to be adjusted.

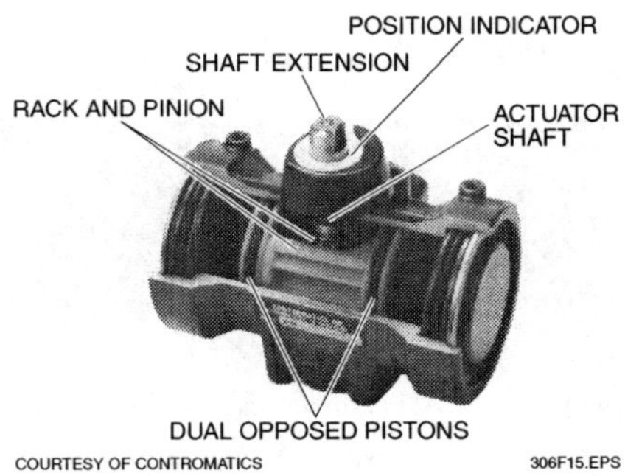

Figure 15 ◆ Pneumatic rack and pinion actuator.

Audiovisual

Show Transparency 9 (Figure 14). Discuss the role of a pneumatic piston in a pneumatic actuator/operator.

Classroom

Refer to Figure 15. Discuss the advantages of using rack and pinion actuators/operators.

Audiovisual

Show Transparency 10 (Figure 16). Review the components in a pneumatic safety valve. Explain how the valves are used in emergencies.

Classroom

Discuss the procedure to adjust a pneumatic valve actuator/operator. Ask whether trainees have any questions.

Classroom

Stress the importance of accurately completing the verification process.

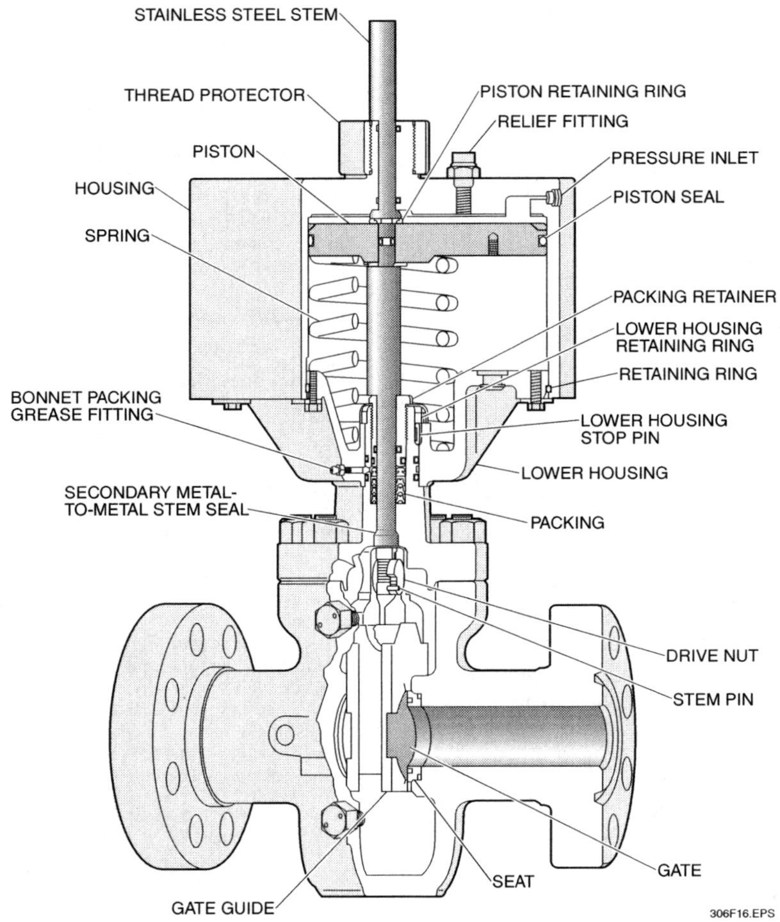

Figure 16 ◆ Pneumatic safety valve.

> **NOTE**
> Ensuring that the correct valve and pneumatic actuator/operator is located, especially before operating the valve, precludes any upsets.

Step 2 Verify its accessibility.

Step 3 Verify its number and nameplate data.

Step 4 Verify its type and manufacturer in order to verify documentation.

> **NOTE**
> If the nameplate is missing, it must be replaced according to company policy and procedure.

Step 5 Notify the control center and any other affected personnel that the work is beginning.

Instructor's Notes:

WARNING!
You must maintain an ongoing awareness of the valve and pneumatic actuator/operator activities being performed to ensure safety during the performance of the task.

Step 6 All applicable task-specific procedures must be followed. Following a task-specific procedure helps to ensure job safety, efficient task performance, standardization, and appropriate documentation. Procedures should also include work permit information and notification requirements (control center, both the upstream and downstream pump stations for main line valves, and any other personnel possibly being affected).

CAUTION
You must follow the manufacturer's instructions and industry recommendations in order to ensure personal safety, prevent mechanical damage to the pneumatic actuator/operator being adjusted, avoid internal damage to the valve, and preserve the service life of the equipment.

Step 7 Verify the correct valve position to determine the appropriate actuator/operator operational cycle.

Step 8 Set the mechanical stops, as applicable, according to manufacturer's instructions to ensure the proper operation of the valve.

Step 9 Set the pneumatic actuator/operator controlling and limiting devices, as applicable, according to manufacturer's instructions to ensure that the actuator/operator movement is correct and that neither the valve nor the actuator/operator is damaged.

Step 10 Notify the control center and/or affected personnel to verify the integrity of the system and return it to normal operation.

Step 11 Complete all required documentation.

NOTE
Documenting data for compliance, historical review, and trending is an important element of valve maintenance.

CAUTION
You must identify any potential abnormal operating conditions and initiate appropriate response.

8.4.0 Repairing Valve Actuators/Operators

Valve actuators/operators must be inspected, repaired, and maintained regularly in order to perform safely and efficiently. Improper valve operation can result in a pipeline disaster. Each operator is required by federal law to prepare and follow a manual of written procedures for regular maintenance of valve actuators/operators. This manual must be updated at least every fifteen months. Applicable portions of the manual must be kept at all locations where work is to be performed and must provide for safe maintenance operations. A valve maintenance log is used to record information about work done on actuators. *Figure 17* shows a typical valve maintenance log.

Pneumatic actuator/operator repair procedures will vary depending on company policy and on the manufacturer's recommended practices. Many companies perform only minor repairs or component swaps on pneumatic actuators/operators. Some companies prefer to remove and replace the faulty actuator/operator and send the faulty unit back to the manufacturer to make repairs. If repairs to a pneumatic actuator/operator are required to be performed by E&I technicians, all company policies and procedures and manufacturer's instructions must be followed.

Before attempting to repair a pneumatic actuator/operator, you must complete the preliminary procedures listed earlier in this module. All warnings, cautions, and company safety policy and procedure must be meticulously observed. It is not possible to prepare a single procedure that would apply exactly to every pneumatic actuator/operator. However, you should follow these basic steps to perform a general overhaul of a pneumatic valve actuator/operator:

Step 1 Verify the location of the pneumatic actuator/operator to be repaired.

NOTE
Ensuring that the correct valve and actuator/operator is located, especially before operating the valve, precludes any upsets.

Safety

Discuss the consequences of failing to stay alert while performing pneumatic actuator/operator activities.

Ask a trainee to explain why it is important to follow the manufacturer's instructions.

Emphasize the importance of identifying potential abnormal operating conditions.

Classroom

Discuss the importance of recording all work that is completed on an actuator.

Laboratory

Distribute sample valve maintenance logs. Ask trainees to identify information that you specify. Have trainees identify any trends noted on the maintenance log.

Classroom

Review the procedures for performing a general overhaul of a pneumatic valve actuator/operator. Answer any questions trainees have.

Rev. 12/97

Service Performed	Condition of Valve/Site	Deficiency	Action Required	Action Taken
VI - Visual Inspection	AC - Acceptable Condition	DO - Does Not Operate	RP - Replace Part	PR - Part Replaced
MO - Mechanical Operation		LK - Leaking	RR - Repair Required	RP - Repair Performed
LO - Local Mode Operation	NR - Needs Repair / Work	PM - Part Missing		
RO - Remote Mode Operation		DP - Defective Paint	PR - Paint Required	RE - Repainted
RM - Routine Maintenance	RR - Replacement Required	NS - Security	SS - Secure Site	SP - Security Provided
AS - All Services Listed		WO - Wash Out	DW - Dirt Work Required	DP - Dirt Work Performed
	OS - Out of Service	SU - Site Upkeep	SC - Site Clean Up	SM - Site Cleaned Up
		OT - Other (describe)	OT - Other (describe)	OT - Other [illegible]

Date	Print Order	Valve Tag No.	Serial Number	Location / Service	Service Performed	Condition of Valve/Site	Deficiency	Action Required	Action Taken
11/01/01	111	42-37B	BR549	#3 Pump Room	VI	NR	DP	PR	RE
11/13/01	112	54-18C	XJ242	Pump Station 14	MO	RR	LK	RP	PR

REMARKS: _____

DIVISION: _____ STATION: _____ INSPECTED BY: _____

306F17.EPS

Figure 17 ◆ Typical valve maintenance log.

Step 2 Verify its accessibility.

> **NOTE**
> Verification of accessibility is important, especially where large actuators/operators may necessitate the use of hoisting equipment.

Step 3 Verify its type and manufacturer.

Step 4 Verify its number and the nameplate data in order to verify documentation.

> **NOTE**
> If the nameplate is missing, it must be replaced according to company policy and procedure.

Step 5 Notify the control center and any other affected personnel that work is beginning.

> **WARNING!**
> You should maintain an ongoing awareness of the repair activities being performed to ensure safety during the performance of the task.

Step 6 All applicable task-specific procedures must be followed, which helps to ensure job safety, efficient task performance, standardization, and appropriate documentation. Procedures should also include work permit information and notification requirements (control center and both the upstream and downstream pump stations for main line valves, and others possibly being affected).

Instructor's Notes:

CAUTION

You must follow manufacturer's instructions and industry recommendations in order to ensure personal safety, prevent mechanical damage to the electric actuator/operator, avoid internal damage to the valve being repaired, and preserve the service life of the equipment.

Step 7 Using all of the precautionary steps required by company policy and procedure and those recommended by the manufacturer's instructions, as applicable, to prevent damage and ensure personal safety, remove the actuator/operator from the system.

Step 8 Following the manufacturer's instruction manual to ensure proper procedures and safety, disassemble the actuator/operator.

WARNING!

Some gasket materials may contain asbestos. Company safety policy and procedure for working with asbestos must be followed.

Step 9 Inspect, repair, and/or replace worn or damaged parts according to manufacturer's instructions, recommendations, and specifications for the unit.

Step 10 Reassemble the pneumatic actuator/operator according to the manufacturer's instructions and specifications for safe operation.

Step 11 Set the pneumatic actuator/operator stop devices, as applicable, according to the manufacturer's instructions for safe and efficient operation of the valve.

Step 12 Set the pneumatic actuator/operator controlling devices, if applicable, according to the manufacturer's specifications for safe and efficient operation.

Step 13 Assemble the gear section of the pneumatic actuator/operator, if applicable, according to the manufacturer's specifications.

Step 14 Complete the assembly and installation of the components.

Step 15 Notify the control center and any affected personnel to verify the integrity of the system and return the system to normal operation.

Step 16 Complete all required documentation of the repair and results.

NOTE

Documenting data for compliance, historical review, and trending is an important element of valve maintenance.

8.5.0 Abnormal Operating Conditions

When performing maintenance on pneumatic actuator/operators, situations may be encountered that require you to immediately respond in order to maintain proper operation of pipeline equipment. The following abnormal operating conditions could be encountered.

If a component failure or automated pneumatic valve system failure allows uncontrolled valve travel, de-energize the valve at the power source and/or operate the valve manually.

Should you be unexpectedly confronted with a hazardous liquid or gas, such as a stem leak or piping connection leak, attempt to isolate the release. If the release cannot be isolated safely at the site, the control center should be notified to isolate the site or to shut down the line as quickly as possible. You must evacuate the area and only take defensive action using the proper personal protective equipment.

If an unexpected status change occurs, attempt to stop valve travel and return the valve to the proper position. If you are unable to immediately return the valve to the proper position, take action to avoid an emergency condition. Notify the control center as quickly as possible.

In case of fire and/or explosion or ignition of hazardous, flammable liquids or gases, you should evacuate the area and only take defensive action using proper personal protective equipment. Contact the control center to isolate or to shut down the line as quickly as possible.

Safety

Explain how to prevent internal damage to the valve being repaired.

Discuss the consequences of working with materials that may contain asbestos.

Classroom

Review a variety of abnormal operating conditions that trainees may encounter while performing maintenance on pneumatic operators/actuators.

On the whiteboard/chalkboard, list several abnormal operating conditions. Ask trainees to provide an appropriate response for each condition.

Homework

Ask trainees to brainstorm questions for the guest speaker. Have trainees prepare for testing during the next session.

Materials

Ensure that you have everything required for testing during this session.

Teaching Tip

See the Teaching Tip for Sections 8.0.0–8.5.0 at the end of this module.

Summary

Pneumatic systems are used on the pipeline to operate and control actuators/operators and systems as well as to operate many tools used to maintain pipeline components. By withdrawing air from the atmosphere, compressing, processing, and directing it to the appropriate equipment in a controlled manner, much work is accomplished in a safe, economic, and efficient manner. However, the creation and control of compressed air and the pneumatic actuators/operators and systems energized by it often involve technically complicated procedures and installations.

Using pneumatic systems to operate pipeline relief valves and other safety equipment requires special care to ensure that the mechanisms are maintained and meet the required standards of operation. You must be trained and qualified before attempting repairs on actuators/operators and systems so that maintenance and repairs on this equipment will ensure the safety of the pipeline. Strict discipline must be exercised in following manufacturer's recommendations and instructions.

A strong understanding of schematic diagrams for pneumatic systems and an understanding of the operation of the various components involved in pneumatic systems being examined are essential to making efficient repairs. You are expected to be proficient in reading schematics. Being able to quickly understand and trace the components of pneumatic systems is essential for a technician to quickly pinpoint a problem and repair it.

Efficient troubleshooting procedures are necessary to reduce downtime, labor, and disruption of pipeline shipments. Although numerous troubleshooting guides and charts are available to suggest possible solutions to problems, you must also rely on their training, experience, problem-solving abilities, and logic to maintain and repair pneumatic systems. With the understanding of pneumatic systems provided in this module, you can work to achieve the level of a trained and qualified mechanical technician.

Review Questions

1. PM stands for _____.
   a. preventive maintenance
   b. purchase matrix
   c. pipeline manager
   d. prevention manual

2. The word whose letters are used to describe the basic methods for implementing PM is _____.
   a. action
   b. work
   c. control
   d. care

3. A maintenance schedule is usually formed _____.
   a. by the manufacturer
   b. in the company
   c. through state law
   d. through engineering specifications

4. Dirt and moisture are prevented from entering a compressor by the _____.
   a. intercooler
   b. in-line filter
   c. intake filter
   d. receiver

5. The air in a multistage air compressor is _____ each time it is compressed.
   a. heated
   b. cooled
   c. oiled
   d. expanded

6. The two types of intercoolers that are used in multistage air compressors are air intercoolers and _____ intercoolers.
   a. r-22
   b. r-12
   c. liquid
   d. solid

7. The discharge temperature of compressed air is reduced by a(n) _____.
   a. intercooler
   b. aftercooler
   c. separator
   d. lubricator

Instructor's Notes:

8. The type of aftercooler that is sometimes used as a compressor belt guard is the _____ aftercooler.
   a. air-cooled
   b. water-cooled
   c. refrigeration
   d. oil-cooled

9. Water vapor and oil droplets are removed from the air after it leaves the aftercooler by the _____.
   a. pressure regulator
   b. separator
   c. lubricator
   d. receiver

10. You must be able to read and understand a _____ diagram of a pneumatic system.
    a. schematic
    b. layout
    c. electrical
    d. line drawing

11. Troubleshooting the pneumatic system that serves a machine that fails can be accomplished by starting at the actuator/operator and working toward the _____.
    a. oil supply
    b. separator
    c. accumulator
    d. air supply

12. Although flow is restricted, a clogged filter in the air supply may allow enough air through to indicate adequate _____.
    a. pressure
    b. lubrication
    c. volume
    d. cooling

13. One of the first things to examine in a failing system is to verify that there is an appropriate _____ of air.
    a. leak
    b. supply
    c. cooling
    d. loss

14. An air supply saturated with _____ can cause severe damage to a pneumatic actuator/operator.
    a. oil
    b. grease
    c. lubricants
    d. water

15. To determine if a problem is in the pneumatic actuator/operator or the machine it operates, disconnect and manually operate the _____.
    a. filter
    b. actuator/operator
    c. control valve
    d. interlocks

16. A pneumatic actuator/operator may be in the form of a _____ or a motor.
    a. valve
    b. filter
    c. cylinder
    d. aftercooler

17. Machinery in a pneumatic system is protected from being operated at the wrong time by the _____.
    a. actuator
    b. interlocks
    c. control valves
    d. filters

18. A good pneumatic troubleshooting _____ is a useful tool to aid in troubleshooting pneumatic actuator/operator systems.
    a. schematic
    b. layout
    c. chart
    d. diagram

19. It is critical to close the air supply and relieve pressure from the system before performing _____ on any component.
    a. operations
    b. inventory
    c. maintenance
    d. setup

20. A basic step in the repair of a pneumatic actuator/operator is to verify the _____ of the unit.
    a. size
    b. location
    c. brand
    d. pressure

Classroom

Go over the Review Questions for this module, and answer any questions trainees have.

Examination

Administer the Module Examination. Be sure to record the results of the Exam on Craft Training Report Form 200 and submit the results to the Training Program Sponsor.

Performance Testing

Have each trainee perform Performance Tasks 5 and 6 to your satisfaction. Fill out a Performance Profile Sheet for each trainee.

Ensure that all Performance Tests have been completed and Performance Profile Sheets for each trainee are filled out. Be sure to record the results of the testing on Craft Training Report Form 200 and submit the results to the Training Program Sponsor.

Instructor's Notes:

GLOSSARY

Trade Terms Introduced in This Module

Intercooler: A heat exchanger used for cooling fluid between the stages of a multi-stage compressor, which saves power.

Receiver: An apparatus placed near the compressor to equalize the pulsations of the air as it comes from the compressor to cause a more uniform flow of air through the pipeline and to collect moisture and oil carried in the air.

MODULE 63306-02 — TEACHING TIPS

The following are suggested activities or instructional methods to help you teach the material in this AIG.

General

When you call on someone to answer a question, the rest of the class relaxes or even tunes out because they expect that the question and answer will take place only between you and the trainee you called on. Instead, use this technique to involve more trainees in answering questions and to keep them on their toes.

1. Ask trainees to define a term or explain a concept.
2. After one trainee has answered, ask a trainee seated nearby if the answer is right. Then ask whether a trainee in the back of the room agrees.
3. Ask trainees to explain why they think an answer is right or wrong.
4. Use the session to clear up incorrect ideas and encourage trainees to learn from their mistakes.

Sections 2.0.0–2.5.0

Preventive Maintenance

This exercise will familiarize trainees with the records and documentation used for preventive maintenance. For this exercise, you will need equipment records and sample preventive maintenance records and schedules. Trainees will need their Trainee Guides, pencils, and paper. Allow 30 minutes for this exercise.

1. Distribute copies of the sample records and documentation. Divide the trainees into small groups according to the number of samples.
2. Ask trainees to review the documentation and identify information that you specify. Ensure that trainees understand how to interpret the information on the records.
3. Review procedures to complete the forms, explaining the types of information to record and how to document it.
4. Answer any questions trainees have about preventive maintenance records and documentation.

Sections 3.0.0–3.8.0

Inspecting Pneumatic System Components

This Quick Quiz will familiarize trainees with the pneumatic system components. You will need photocopies of the quiz provided on the following page. Trainees will need pencils. If you allow trainees to use the Trainee Guide, decrease the amount of time you give them to complete the quiz.

1. Make as many photocopies as you will need of the quiz.
2. Give trainees 5 to 10 minutes to complete the quiz.
3. Go over the answers to the quiz.
4. Ask trainees whether they have questions about pneumatic system components.

Answers to Quick Quiz

1. d. Air-cooled aftercooler
2. h. Receiver
3. j. Pressure regulator
4. b. Air intercooler
5. i. In-line filter
6. e. Water-cooled aftercooler
7. k. Lubricator
8. a. Intake filter
9. g. Separator
10. c. Liquid intercooler

Quick Quiz ***Inspecting Pneumatic System Components***

For each description listed, identify the pneumatic system component that the text best describes. Write the corresponding letter in the blank provided.

_____ **1.** Uses finned tubing to cool compressed air on its way to the air receiver.

_____ **2.** Dissipates heat from compressed air, collects moisture, and provides smooth airflow.

_____ **3.** Regulates the pressure of the air to the user.

_____ **4.** Cools air by moving it through finned tubing between the stages of the compressor.

_____ **5.** Removes most solids and liquid particles from the compressed air.

_____ **6.** Uses headers to provide connections for water, air, and condensate piping.

_____ **7.** Injects a fine mist of oil into the air to control wear of the components.

_____ **8.** Prevents dirt and moisture from entering the compressor.

_____ **9.** Removes water vapor or oil droplets carried from the aftercooler.

_____ **10.** Cools air by circulating water through tubing.

a. Intake filter
b. Air intercooler
c. Liquid intercooler
d. Air-cooled aftercooler
e. Water-cooled aftercooler
f. Refrigeration dryer
g. Separator
h. Receiver
i. In-line filter
j. Pressure regulator
k. Lubricator

Sections 7.0.0–7.3.0

Repairing Pneumatic System Components

This exercise will give trainees the opportunity to review the procedures for repairing pneumatic system components. For this exercise, trainees will need their Trainee Guides, pencils, and paper. Allow 30 minutes for this exercise.

1. Present a variety of problems that can occur with pneumatic system components. Divide the class into small groups and assign several problems to each group.
2. Ask each group to diagnose the assigned problems and review the procedures to correct each problem.
3. Have each group share its findings with the other trainees. Provide any relevant solutions, recommendations, and/or corrections. Share examples from your own experience repairing pneumatic system components.
4. Answer any questions trainees have about repairing pneumatic system components.

Sections 8.0.0–8.5.0

Pneumatic Valve Actuators/Operators

This presentation will give trainees the opportunity to speak with a mechanical technician. For this presentation, invite a mechanical technician to speak to the class about repairing and adjusting pneumatic valve actuators/operators. Trainees need their Trainee Guides, pencils, and paper. Allow 45 to 60 minutes for this exercise.

1. Tell the trainees that a guest speaker will be presenting information on repairing and adjusting pneumatic valve actuators/operators.
2. Have trainees prepare questions for the guest speaker in advance. Suggested questions could address the following:
   - Preparation procedures
   - Safety precautions
   - Manufacturer's instructions
   - Company policies
   - Appropriate documentation
   - Abnormal operating conditions
3. Introduce the guest speaker. Remember to include the speaker's name, title, company, and relevant credentials.
4. Allow 30 to 45 minutes for the speaker and 15 minutes for questions.
5. Following the presentation, ask trainees to share their questions with the guest speaker.

MODULE 63306-02 — ANSWERS TO REVIEW QUESTIONS

Answer	Section
1. a	2.0.0
2. d	2.1.0
3. b	2.2.0
4. c	3.1.0
5. a	3.2.0
6. c	3.2.0
7. b	3.3.0
8. a	3.3.0
9. b	3.4.0
10. a	4.0.0
11. d	5.0.0
12. a	5.1.0
13. b	5.1.0
14. d	5.1.0
15. b	5.2.0
16. c	5.2.0
17. b	5.5.0
18. c	6.0.0
19. c	7.1.0
20. b	7.1.0

CONTREN™ LEARNING SERIES — USER UPDATES

The NCCER makes every effort to keep these textbooks up-to-date and free of technical errors. We appreciate your help in this process. If you have an idea for improving this textbook, or if you find an error, a typographical mistake, or an inaccuracy in NCCER's Contren™ textbooks, please write us, using this form or a photocopy. Be sure to include the exact module number, page number, a detailed description, and the correction, if applicable. Your input will be brought to the attention of the Technical Review Committee. Thank you for your assistance.

Instructors – If you found that additional materials were necessary in order to teach this module effectively, please let us know so that we may include them in the Equipment/Materials list in the Instructor's Guide.

Write: Curriculum Revision and Development Department
National Center for Construction Education and Research
P.O. Box 141104, Gainesville, FL 32614-1104

Fax: 352-334-0932

E-mail: curriculum@nccer.org

Craft ______ Module Name ______

Copyright Date ______ Module Number ______ Page Number(s) ______

Description ______

(Optional) Correction ______

(Optional) Your Name and Address ______

Module 63307-02

Maintain, Troubleshoot, and Repair Hydraulic Valve Actuators/Operators and Systems

(CT 19.7 and 21.4)

Maintain, Troubleshoot, and Repair Hydraulic Valve Actuators/Operators and Systems Annotated Instructor's Guide

Module 63307-02

MODULE OVERVIEW

This module explains the basic hydraulic principles, hydraulic system equipment, and hydraulic system schematic diagrams. Trainees will learn how to inspect, troubleshoot, repair, and adjust hydraulic valve actuators/operators and systems.

PREREQUISITES

Please refer to the Course Map in the Trainee Module. Prior to training with this module, it is recommended that the trainee shall have successfully completed the following:

Core Curriculum; Pipeline Core, Modules 66101-02 and 66102-02; Pipeline Mechanical Levels One and Two; Pipeline Mechanical Level Three, Modules 63301-02 through 63306-02

OBJECTIVES

Upon completion of this module, the trainee will be able to:

1. Inspect hydraulic system equipment.
2. Read hydraulic system schematic diagrams.
3. Explain the basic hydraulic principles that must be considered before troubleshooting.
4. Troubleshoot hydraulic systems.
5. Repair hydraulic system components.
6. Adjust hydraulic valve actuators/operators (CT 19.7).
7. Repair hydraulic valve actuators/operators (CT 21.4).

PERFORMANCE TASKS

Under the supervision of the instructor, the trainee should be able to:

1. Inspect hydraulic system equipment.
2. Troubleshoot hydraulic system components.
3. Repair hydraulic system components.
4. Adjust hydraulic valve actuators/operators (CT 19.7).
5. Repair hydraulic valve actuators/operators (CT 21.4).

NCCER STANDARDIZED CRAFT TRAINING PROGRAM

The National Center for Construction Education and Research (NCCER) provides a standardized national program of accredited craft training. Key features of the program include instructor certification, competency-based training, and performance testing. The program provides trainees, instructors, and companies with a standard form of recognition through a National Craft Training Registry. The program is described in full in the *Guidelines for Accreditation*, published by the NCCER. For more information on standardized craft training, contact the NCCER by writing us at P.O. Box 141104, Gainesville, FL 32614-1104; calling 352-334-0911; or e-mailing info@nccer.org. More information may be found at our Web site, www.nccer.org.

HOW TO USE THIS ANNOTATED INSTRUCTOR'S GUIDE

Each page presents two sections of information. The larger section displays each page exactly as it appears in the Trainee Module. The narrow column ties suggested trainee and instructor actions to each page and provides icons to call your attention to material, safety, audiovisual, or testing requirements. The bottom of each page includes space for your notes.

Teaching Tip

If you see the Teaching Tip icon, that means there is a teaching tip associated with this section. Also refer to any suggested teaching tips at the end of the module.

SAFETY CONSIDERATIONS

Ensure that the trainees are equipped with appropriate personal protective equipment. Emphasize the importance of following all safety precautions and procedures when working with power tools.

PREPARATION

Before teaching this module, you should review the Module Outline, Objectives, Performance Tasks, and the Materials and Equipment List. Be sure to allow ample time to prepare your own training or lesson plan and gather all required equipment and materials.

MATERIALS AND EQUIPMENT LIST

Transparencies
Markers/chalk
Blank acetate sheets
Transparency pens
Pencils and scratch paper
Module Examinations*
Performance Profile Sheets*
Copies of Performance Verifications for Covered Tasks 19.7 and 21.4**
Overhead projector and screen
Whiteboard/chalkboard
Appropriate personal protective equipment
Copies of *49 CFR Part 192* (Gas) and/or *49 CFR Part 195* (Liquid)
Copies of your company policy and procedures manual
Sample preventive maintenance schedules
Sample maintenance records and equipment logs
Pyrometer
Clamp-on ammeter
Sample hydraulic pump with the necessary tools for disassembly and assembly
Manufacturers' instruction manuals for the sample hydraulic pumps
Hydraulic system control valves and the necessary tools for disassembly and assembly
A variety of replacement parts for a hydraulic actuator/operator and the corresponding manufacturers' specifications
A variety of hydraulic hoses and the necessary tools to cut them
Hydraulic schematic drawings
Copies of manufacturers' instruction manuals for a variety of hydraulic system components

* Located in the Test Booklet.
**Available for download from www.nccer.org.

TEACHING TIME FOR THIS MODULE

An outline for use in developing your lesson plan is presented below. Note that each Roman numeral in the outline equates to one session of instruction. Each session has a suggested time period of 2½ hours. This includes 10 minutes at the beginning of each session for administrative tasks and one 10-minute break during the session. Approximately 15 hours are suggested to cover *Maintain, Troubleshoot, and Repair Hydraulic Valve Actuators/Operators and Systems (CT 19.7 and 21.4)*. You will need to adjust the time required for hands-on activity and testing based on your class size and resources.

Topic	Planned Time
Session I. Preventive Maintenance	
A. PM Methods	________
B. Scheduling	________
1. Laboratory	________
C. Manufacturer's Instructions and Warranties	________
D. Training	________
E. Documentation	________
1. Laboratory	________
Session II. Inspecting Hydraulic System Components	
A. Inspecting Hydraulic Fluid Reservoirs	________
B. Inspecting Filters and Strainers	________
C. Inspecting Hydraulic Pumps	________
D. Inspecting Hydraulic Motors	________
E. Inspecting Control Valves	________
F. Inspecting Check Valves	________
G. Inspecting Cylinders	________
H. Inspecting Hoses, Fluid Lines, and Gauges	________
1. Performance Testing	________
Session III. Performing Hydraulic System Troubleshooting Procedures	
A. Reading Hydraulic Schematic Diagrams	________
B. Preliminary Diagnoses	________
C. Troubleshooting Hydraulic Systems	________
D. Repair Options	________
1. Performance Testing	________
Session IV. Adjusting and Repairing Hydraulic System Components, Part One	
A. Preparing the System for Shutdown and Repair	________
B. Repairing and Overhauling Pumps	________
1. Laboratory	________
C. Repairing and Overhauling Control Valves	________
1. Laboratory	________
Session V. Adjusting and Repairing Hydraulic System Components, Part Two	
A. Repairing and Overhauling Actuators/Operators	________
1. Laboratory	________
B. Replacing Hoses	________
C. Replacing Gauges	________
1. Performance Testing	________

Session VI. Adjusting and Repairing Hydraulic System Components, Part Three

A. Adjusting and Repairing Hydraulic Valve Actuators/Operators ____________

B. Summary

1. Summarize module ____________
2. Answer questions ____________

C. Module Examination ____________

1. Trainees must score 70% or higher to receive recognition from the NCCER.
2. Record the testing results on Craft Training Report Form 200 and submit the results to the Training Program Sponsor.

D. Performance Testing ____________

1. Trainees must perform each task to the satisfaction of the instructor to receive recognition from the NCCER.
2. Record the testing results on Craft Training Report Form 200 and submit the results to the Training Program Sponsor.

Module 63307-02

Maintain, Troubleshoot, and Repair Hydraulic Valve Actuators/Operators and Systems

(CT 19.7 and 21.4)

Instructor's Notes:

Instructor's Notes:

Introduction to the Pipeline Operator Qualification Rule, API, and NCCER

It could be said that petroleum products keep America running. Petroleum products like oil and natural gas are the principal sources of energy in the United States and are used in many manufacturing and chemical processes. Petroleum products are primarily transported by pipeline. There are more than a million miles of liquid and natural gas pipeline in this country, ranging in diameter from two inches to four feet. These pipelines bring crude oil and natural gas from production fields to processing plants and refineries—and ultimately to the consumer.

Pipeline Operator Qualification

Petroleum products are considered hazardous, so pipeline personnel must take extra precautions and follow specific procedures when working with them. To help ensure the safe handling and transportation of petroleum products, the United States Department of Transportation (DOT) mandates safe practices as federal regulations.

One such regulation, the Operator Qualification (OQ) rule, requires pipeline operators to develop a written qualification plan to evaluate their personnel and their contract personnel (see *49 CFR Part 192 Subpart N—Gas,* and *49 CFR Part 195 Subpart G—Liquid*). The evaluation hinges on the operator's ability to perform certain covered tasks and to be able to recognize and respond to abnormal operating conditions that may be encountered while performing these covered tasks.

Covered Tasks

A pipeline performance task is deemed *covered* if the activity passes the following four-part test:

1. Is it performed on a pipeline facility?
2. Is it an operations or maintenance task?
3. Is it performed as a requirement of *49 CFR Part 195 (Liquid)* or *49 CFR Part 192 (Gas)*?
4. Does it affect the operation or integrity of the pipeline facility?

If the answer to all four questions is *yes,* then the activity is considered *covered* and must figure into the evaluation process of pipeline personnel.

American Petroleum Institute

The OQ rule and its related covered tasks provided the catalyst to develop a unified pipeline training and assessment program—so that pipeline operators and owners have ready access to the assessment and training tools they need. This effort was the direct result of the vision and leadership provided by The American Petroleum Institute (API) and API members.

Efforts to unify the industry are nothing new to API. Indeed, the impetus for forming the Institute in 1919 was the need to standardize engineering specifications for drilling and production equipment. Today, the most pressing issues revolve around public perceptions and government policies toward the industry. Speaking with one voice on these issues has become as essential as having interchangeable parts in the field. API's members determine the petroleum industry's positions on public policy as well as the standards that govern its day-to-day operations worldwide.

Consortium on Operator Qualification

API assembled a focus group whose primary purpose was to identify covered tasks and provide compliance recommendations to pipeline operators.

The Consortium on Operator Qualification (COOQ) became the guiding force for the identification and development of covered tasks for the liquid pipeline industry. Their work product, API publication 1161, *Guidance Document for the Qualification of Liquid Pipeline Personnel,* formed the basis for the resulting Pipeliner Training and Assessment Program.

The covered task list created in API publication 1161 has now evolved to include gas covered tasks and is maintained by NCCER. The most current covered task list can be downloaded at www.nccer.org.

National Center for Construction Education and Research

With the guidance and support of API, its members, and the Consortium on Operator Qualification, the National Center for Construction Education and Research (NCCER) was able to design and implement the Pipeliner Training and Assessment Program.

NCCER is a not-for-profit 501(c)(3) education foundation established in 1995 by the world's largest and most progressive construction and maintenance contractors and national contractor associations. NCCER was founded to address the

severe workforce shortage facing the construction and maintenance industries and to develop a standardized training process and curricula. Today, NCCER is supported by hundreds of leading construction and maintenance companies, manufacturers, and national associations and developed more than 35 construction and maintenance titles.

NCCER also maintains a National Registry, which contains the records of trainees and instructors who have received NCCER training. All completion information for every trainee is sent to the NCCER and filed within the Registry. The Registry can then confirm training and skills for workers as they move from company to company, state to state, or even within their own company.

Standardized Curriculum (NCCER's Contren™ Learning Series)

NCCER curricula are developed using the expertise of industry subject matter experts (SMEs). The use of SMEs as an integral part of every new development or revision ensures that the curriculum is current, industry-driven, and indicative of industry practices. The development and/or revision process is facilitated from start to finish by NCCER project managers, who work closely with the SMEs and a select team of technical writer(s) on the development and presentation of the curriculum.

Pipeline: Five Training Paths

The pipeline curriculum focuses on essential skills and competencies within five skill areas:

- Corrosion Control
- Electrical and Instrumentation
- Maintenance
- Mechanical
- Operations (Field and Control Center)

SMEs from the pipeline industry have determined the skills and competencies needed based on API's covered tasks and the consensus of the group/industry. Specific company information is generally not incorporated into the curriculum unless the SMEs determine that the information is the industry standard.

Training Modules and Covered Tasks

The smallest unit of instruction in the curriculum is a *module*. Each curriculum is composed of modules (the equivalent of stand-alone chapters) that are logically sequenced into complete *levels* of instruction. Every module is task-oriented and requires the successful completion of a written (knowledge-based) examination and a performance (hands-on) test. Modules may be individually selected and/or combined for efficient task training. Modules in the pipeline curricula, to the extent possible, will align with API-developed covered tasks. While not all modules will relate to covered tasks, all covered tasks will relate to at least one module.

Benefits of Training to the Pipeline Marketplace

Applying NCCER's standardized training process allows pipeline owners and contractors in the United States to establish the benchmark for acceptable skill and competency levels within the pipeline industry. This process will assist training organizations in developing and/or improving standardized pipeline training programs that effectively match their resources and operating conditions while responding to regulatory demands and to the needs of the industry. The process will also recognize training organizations that meet the minimum standards of competence, warranting regulatory, public, and professional confidence. This in turn will foster national unity within the pipeline industry for the mutual benefit of individual technicians, owners, contractors, and ultimately pipeline-users and society.

Additional Resources

To review the Pipeliner Training and Assessment Program Implementation and Resource Guide and the latest version of pipeline covered tasks, visit NCCER's Web site at www.nccer.org, or contact:

National Center for Construction Education and Research
PO Box 141104
Gainesville, FL 32614-1104
(888) 622-3720
Web site: www.nccer.org

For more information on API, please contact:

American Petroleum Institute
1220 L Street, Northwest
Washington, D.C. 20005
(202) 682-8000
Web site: www.api.org

To order *API Pub 1161*, contact Global Engineering Documents by calling 1-800-854-7149 or visiting their Web site at www.global.his.com.

Additional sources of information on the pipeline operator qualification regulation include:

DOT – Office of Pipeline Safety
www.ops.dot.gov

Transportation Safety Institute
www.tsi.dot.gov

Instructor's Notes:

COURSE MAP

This course map shows all of the modules in the third level of the Pipeline Mechanical curriculum. The suggested training order begins at the bottom and proceeds up. Skill levels increase as you advance on the course map. The local Training Program Sponsor may adjust the training order.

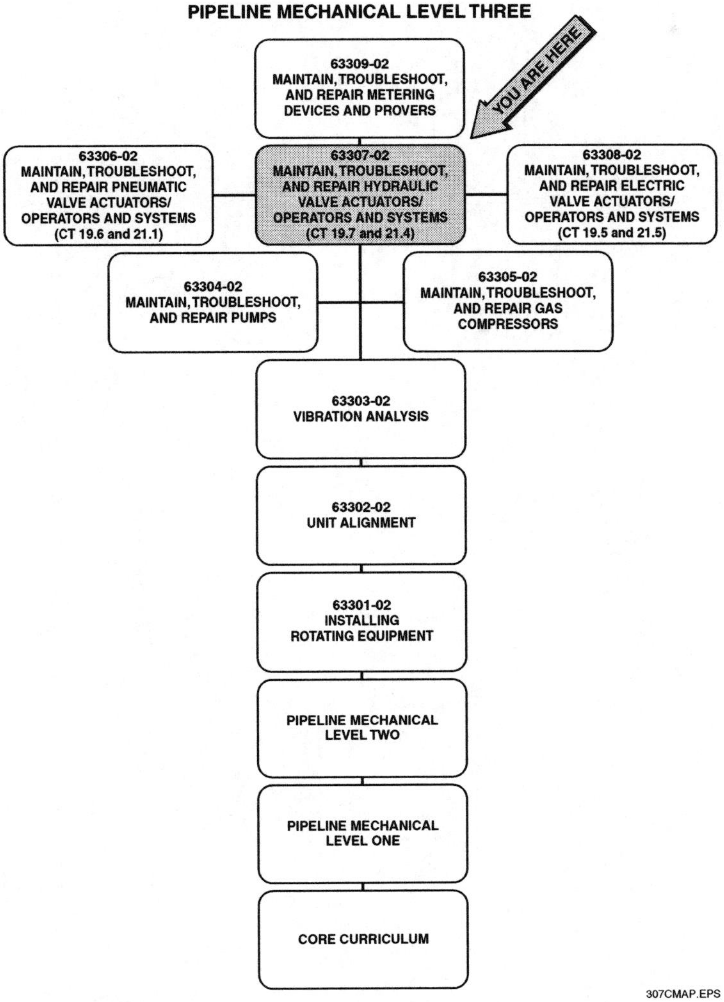

307CMAP.EPS

Homework

Assign reading of Module 63307-02.

Instructor's Notes:

MODULE 63307-02 CONTENTS

Figures

Table

Instructor's Notes:

MODULE 63307-02

Maintain, Troubleshoot, and Repair Hydraulic Valve Actuators/Operators and Systems (CT 19.7 and 21.4)

Objectives

When you have completed this module, you will be able to do the following:

1. Inspect hydraulic system equipment.
2. Read hydraulic system schematic diagrams.
3. Explain the basic hydraulic principles that must be considered before troubleshooting.
4. Troubleshoot hydraulic systems.
5. Repair hydraulic system components.
6. Adjust hydraulic valve actuators/operators (CT 19.7).
7. Repair hydraulic valve actuators/operators (CT 21.4).

Prerequisites

Before you begin this module, it is recommended that you successfully complete the following: Core Curriculum; Pipeline Core, Modules 66101-02 and 66102-02; Pipeline Mechanical Level One; Pipeline Mechanical Level Two; Pipeline Mechanical Level Three, Modules 63301-02 through 63305-02.

Required Trainee Materials

1. Pencil and paper
2. Appropriate personal protective equipment

1.0.0 ◆ INTRODUCTION

Hydraulic systems (see *Figure 1*) are used extensively to operate and control pipeline equipment. These systems operate within a closed-loop system to accomplish a specific task, such as operating a valve actuator/operator to open and close a valve. By the use of hydraulic equipment, work can be accomplished in a safe, economic, and efficient manner. Using hydraulic equipment to operate pipeline valves and other equipment requires special care to ensure that the mechanisms are maintained and meet the required standards of operation.

A valve actuator/operator is a mechanical device that is used to open, close, or position a valve manually or automatically. Manual actuators/operators come in the form of handwheels, levers, and wrenches that are attached to the valve stem, either directly or by means of a gearbox. Automatic actuators/operators are used when the valve is remote from the main work area, when the frequency of operation would require unreasonable human effort, or if rapid opening and closing of the valve is required. In some installations, automatic remote control actuator/operators are installed as a matter of convenience, reducing the number of employees required for facility operation. Large valves that require great effort to turn must be equipped with some type of powered mechanism, such as electric, pneumatic, or hydraulic motors.

Note that in some locations the gearbox is called an actuator. In other locations, it is called an actuator/operator. Technically speaking, the operator is the gearbox unit or hand device that actually moves the valve. The actuator is the power unit that is added to turn the operator automatically. In this module, the term actuator/operator will be used collectively to refer to the entire unit.

You must also understand that many actuators/operators are devices that experience heavy

Materials

Ensure that you have everything required to teach the course. Check the Materials and Equipment List at the front of this module.

Classroom

Explain that terms shown in bold are defined in the Glossary at the back of this module.

Audiovisual

Show Transparency 1, Course Objectives.

Show Transparency 2, Performance Tasks.

Teaching Tip

See the general Teaching Tip at the end of this module.

Classroom

Explain how hydraulic systems and valve actuators/operators are used on pipeline equipment. Review the various applications for each.

Audiovisual

Show Transparency 3 (Figure 1). Explain how hydraulic systems operate.

Classroom

Review the activities typically included in a preventive maintenance program.

Stress the importance of obtaining a detailed maintenance history of the system's equipment.

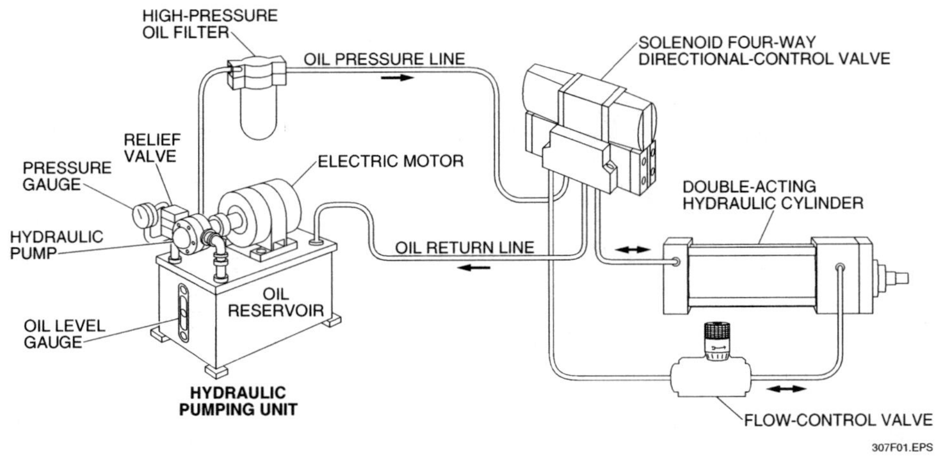

Figure 1 ◈ Hydraulic system.

loads although they may be idle for many days at a time. Some valves, if turned manually, would require so much time and effort that an opening or closing operation would be self-defeating if it were not an emergency. For example, some large valves require more than 10,000 handwheel revolutions from open to close. The time that would be required for a power unit to open or close such a valve would be extensive. For that reason, some mechanical technicians and crews carry gasoline engine, air, or electric powered portable valve actuators/operators to assist them with cycling valves.

Such valve equipment demonstrates the fact that an actuator/operator must be durable and in good condition to carry such loads. You must be able to maintain, troubleshoot, and repair the actuators/operators so that they perform in a reliable and efficient manner. With proper maintenance and repair, valve actuators/operators can give long, reliable service with minimal downtime.

2.0.0 ◈ PREVENTIVE MAINTENANCE

Preventive maintenance (PM) is the periodic inspection of facility assets and equipment to uncover conditions that may lead to failure or shorten anticipated life. The long-term goal for a good PM program is to increase reliability, decrease unplanned equipment downtime, reduce overall maintenance costs, and support improvements in quality and production. The duties involved in PM are generally grouped into the following activities:

- Inspecting components
- Lubricating components
- Replacing components
- Cleaning components
- Adjusting components
- Documenting work accomplished

The start of any PM program involves performing a detailed analysis of past equipment failures, including their costs, their causes, any remedial actions, and the changing demands on the equipment. A detailed maintenance history of all of the equipment within the system must be developed. After this information is gathered, it is divided and prioritized according to the maintenance item's effect, ranging from maximum to minimum. The criteria used to determine effect includes safety, product costs, overhead costs, and equipment replacement costs.

A good PM program targets the items that cause economic damage as first priority. The second priorities are the maintenance items that need to be done but can wait until the first priorities have been dealt with. The lowest priority maintenance items are those that have little monetary and absolutely no safety impact; these can be attended to after they have failed.

In a general application of *49 CFR 192.605*, each pipeline operator is required by federal law to

Instructor's Notes:

prepare and follow a manual of written procedures for regular maintenance of valve actuator/operators. This manual must be updated at least once every fifteen months. Applicable portions of it must be kept at locations where work is performed. Such manuals must provide for safety of operations during these maintenance procedures. You must follow company policies and procedures during maintenance activities.

The law only applies to maintenance activities on the pipeline that could result in danger to personnel, the public, facilities, or the environment. Inspections, diagnostics, repair of units located in areas isolated from the pipeline, bench work in a shop, and how information gathered about such units is managed is not included.

However, sound management practice requires that all equipment be maintained in good working condition. It is not only a matter of safe and efficient operation but also economic operation. Valve actuators/operators represent a sizeable investment that must be protected.

2.1.0 PM Methods

PM methods are systematic maintenance procedures that reduce emergencies, downtime, and costly production delays. PM procedures vary from one piece of pneumatic equipment to another, but the basic principles are the same. The basic methods for implementing PM procedures can be remembered using the word CARE:

- C = *Cleaning* the system and equipment
- A = *Adjusting* the components
- R = *Replacing* disposable components
- E = *Examining* or inspecting equipment

2.2.0 Scheduling

Scheduling PM is an important element in the PM program. PM performed too frequently increases costs, as does PM that is performed too infrequently. Companies use computerized systems, card files, and maintenance logs to assign individuals to manage the responsibilities of maintenance. Some company computers generate work orders at regular intervals for the equipment to make certain that legally required schedules are met and that equipment in the most isolated locations is included. Schedules must be reviewed on a continual basis to check their accuracy and effectiveness. The schedule is always changing and depends on consistent documentation.

The maintenance schedule, formed within the company, specifies all the recommended maintenance required to keep all equipment in good operating condition.

2.3.0 Manufacturer's Instructions and Warranties

Preventive maintenance must be performed according to manufacturer's instructions and warranty requirements. Many actuators/operators are supplied with warranties, and in order to retain that warranty, they must be maintained according to the manufacturer's instructions. If it is still under warranty, the work will need to be carried out only as authorized by the manufacturer. In some cases, the manufacturer may require that the actuator/operator to be returned to the factory for warranty service. The record for the particular actuator/operator should be reviewed to see if the warranty is still in effect before any work is attempted. The specified maintenance procedures should be followed to ensure proper operation and to maintain warranty requirements.

2.4.0 Training

You should become completely familiar with the types of actuators/operators being used and the types of maintenance products required for their job assignments. An ongoing training program with regular updates and reviews is considered essential by some companies. Training should also include the use of proper methods of documentation.

2.5.0 Documentation

Records are also a very important element of the PM system. Records are maintained to build a detailed and accurate historical file of the equipment in relation to hours of operation, parts usage, and cost of maintenance. The records system must store accurate information in the following categories:

- Listing of each piece of equipment requiring PM
- Location of each piece of equipment
- Running history of maintenance actions on the equipment, including the parts replaced
- Time expended on PM
- Cost of replacement parts

3.0.0 ◆ INSPECTING HYDRAULIC SYSTEM COMPONENTS

Hydraulic system preventive maintenance (PM) inspections should be performed regularly to ensure that the system components are in proper working order and to maintain a long working life for components. Company policy and procedures usually specify how often PM inspections should

Classroom

Discuss the procedures involved in a basic preventive maintenance program. Answer any questions trainees have.

Laboratory

Distribute sample preventive maintenance schedules. Have trainees review the schedules and determine whether any of the activities have been performed too frequently or infrequently.

Classroom

Emphasize the importance of adhering to the manufacturer's warranty.

Laboratory

Distribute sample maintenance records and equipment logs. Ask trainees to complete the forms using hypothetical information you provide. Stress the need to record accurate information.

Homework

Have trainees review Sections 3.0.0–3.8.0 for the next session.

Materials

Ensure that you have everything required for the demonstrations and testing during this session.

Classroom

Review the components that require inspection as part of a hydraulic system preventive maintenance program.

Discuss the procedures to inspect a hydraulic fluid reservoir.

Audiovisual

Show Transparency 4 (Figure 2). Explain how to check the sight glass. Ask a trainee to explain what may cause a false line to appear.

Classroom

Refer to Figure 3. Discuss problems that can result if contaminants build up in a hydraulic system.

be performed. The hydraulic system PM inspection includes inspecting the following components:

- Hydraulic fluid reservoirs
- Filters and strainers
- Hydraulic pumps
- Hydraulic motors
- Control valves
- Check valves
- **Cylinders**
- Hoses, fluid lines, and gauges
- Relief valve settings
- Nitrogen supply
- Controls for automatic operations
- Alarms

3.1.0 Inspecting Hydraulic Fluid Reservoirs

A reservoir in a hydraulic system stores the fluid used in the system. In the reservoir, air can separate from the fluid, and particles of contamination can settle. The working fluid is cooled by passing through the reservoir, where the heat dissipates from the fluid into the atmosphere, keeping the operating temperature of the fluid within allowable limits. Reservoirs are usually made of welded steel plate. A baffle plate in the center of the reservoir helps returning fluid circulate and cool before being pumped back into the system. Drain plugs and removable covers help drain and clean the reservoir. Perform the following procedures to inspect a hydraulic fluid reservoir:

- Inspect the reservoir tank to ensure that there are no leaks and that it is clean.

> **NOTE**
> The reservoir of the hydraulic system is the point where heat is dissipated from the fluid. If the tank is dirty, heat is held in and the fluid cannot be efficiently cooled.

- Inspect the reservoir air breather to ensure that it is clean.
- Inspect all line connections to and from the reservoir to ensure that they are tight and that there is no fluid leaking from the system or air leaking into the system.
- Examine the maintenance records to determine if the reservoir is due for a scheduled fluid change and cleaning.
- Look at the reservoir sight glass to ensure that the fluid is at the proper operating level, and add fluid as necessary. When checking the sight glass, be aware that a false line may appear due to oil stains or lines from previous oil levels. *Figure 2* shows a hydraulic fluid reservoir.
- Inspect the reservoir for water contamination.
- Inspect the reservoir tank for evidence of corrosion.

3.2.0 Inspecting Filters and Strainers

As hydraulic fluid moves through the system, it picks up contaminants, or particles of unwanted materials. If contaminants are allowed to build up in the system, they shorten the working life of the fluid and damage the system parts. Strainers and filters (see *Figure 3*) trap and remove contaminants from the fluid as it flows through them.

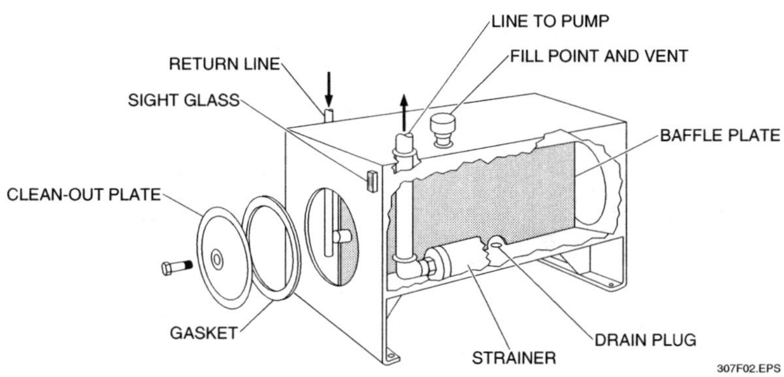

Figure 2 ◆ Hydraulic fluid reservoir.

Instructor's Notes:

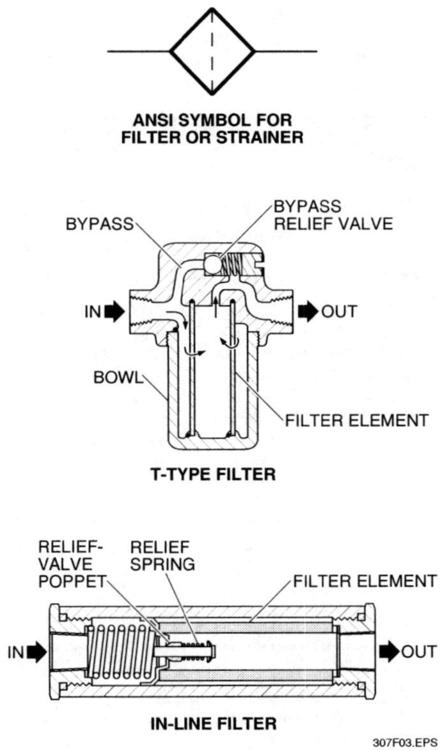

Figure 3 ◆ Filters.

Strainers remove the larger particles from the fluid, and filters remove smaller particles that slip through the strainers. Perform the following procedures to inspect filters and strainers:

- Inspect the filter or strainer housing for leaks.
- Inspect the filter or strainer housing for cleanliness.
- Examine the filter or strainer maintenance records to determine if it is due for scheduled cleaning or replacement.
- Clean and/or replace the filter or strainer as necessary.

3.3.0 Inspecting Hydraulic Pumps

Pumps in hydraulic systems change the mechanical energy produced by the pump drive motor into hydraulic energy. The pumps accomplish this by exerting force on the fluid and creating flow in the hydraulic system. Perform the following procedures to inspect a hydraulic pump:

- Inspect the pump casing for leaks.
- Inspect the shaft seal for leakage.
- Inspect all fluid line connections to ensure that they are tight and that there is no fluid leaking from the system or air leaking into the system.
- Listen for unusual noises from the pump.
- Inspect the pump spring if the pump is pressure-compensated.

> **NOTE**
> A noisy pump may indicate wear or cavitation. Cavitation occurs when the pump does not receive adequate fluid because of low fluid level in the reservoir or a plugged strainer or filter.

- Examine the pump and drive motor for an unusually high temperature using a pyrometer or other temperature-measuring device. *Figure 4* shows a pyrometer.

> **NOTE**
> An excessively high temperature may indicate bad pump or motor bearings or binding due to worn parts or misalignment. It may also indicate that the fluid in the system is not being properly cooled or that the flow is restricted.

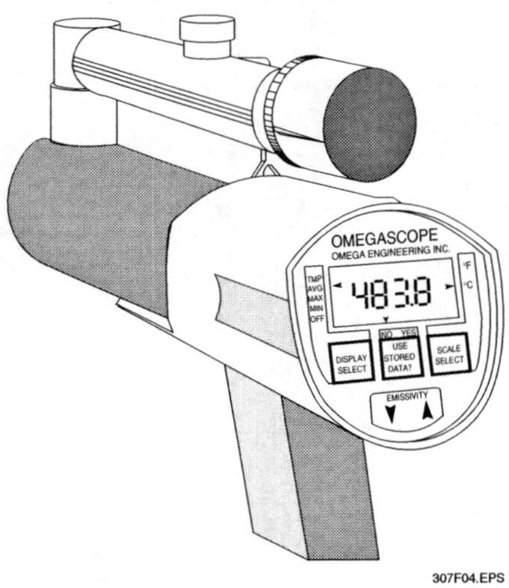

Figure 4 ◆ Pyrometer.

Classroom

Explain how to inspect filters and strainers for leaks and cleanliness.

Review the procedures to inspect a hydraulic pump for leaks, unusual noises, and proper fluid line connections. Go over any questions trainees have.

Demonstration

Bring in a pyrometer. Demonstrate how the meter is used to examine the pump and drive motor for excessive temperature.

Safety

Discuss the consequences of failing to turn off the electrical power during an inspection.

Stress the importance of turning off the drive motor before working on belts and pulleys.

Demonstration

Bring in a clamp-on ammeter. Demonstrate how to test the drive motor running amperes using the ammeter.

Classroom

Discuss the inspection procedures for a hydraulic motor.

- Examine the pump and drive motor mounting bolts to ensure that they are tight.
- Examine the pump and drive motor for excessive vibration.

> **NOTE**
> Excessive vibration in a pump or drive motor may indicate misalignment of the pump-to-motor coupling, worn bearings, or restricted suction.

- Ensure that the pump and drive motor are clean.

The following inspections should be performed by a qualified electrician:

> **WARNING!**
> To prevent electrical shock, ensure that the electrical power is turned off and locked out before touching the connections.

- Test the drive motor running **amperes** (amps), using an ammeter, to ensure that the motor is not exceeding the maximum amps specified on the motor nameplate. *Figure 5* shows a clamp-on ammeter.

> **NOTE**
> A high amp reading on the drive motor may indicate that the pump or motor is binding due to bad bearings, misalignment, or worn pump parts. It may also indicate that the load on the pump exceeds specifications.

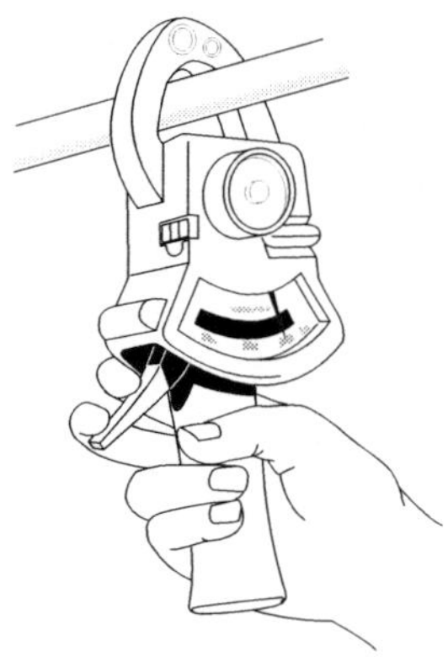

307F05.EPS

Figure 5 ◆ Clamp-on ammeter.

- Examine the belts and pulleys on belt-driven pumps closely to ensure that they are in good condition, are properly aligned, and that the belt tension is properly adjusted.

> **WARNING!**
> To prevent personal injury, ensure that the pump drive motor is turned off and locked out before working on the belts and pulleys.

3.4.0 Inspecting Hydraulic Motors

Hydraulic motors transform hydraulic working energy into rotary mechanical energy that is applied to a resisting load. Fluid is supplied under pressure to the motor by the hydraulic pump. The fluid turns the internal parts of the motor that turn the motor shaft. The motor shaft is connected to the resisting load. Perform the following procedures to inspect a hydraulic motor:

- Inspect the motor housing to ensure that there are no leaks and it is clean. Clean it as necessary.
- Inspect all line connections to the motor to ensure that they are tight and there is no fluid leaking from the system or air leaking into the system.
- Listen for unusual noises from the motor.

> **NOTE**
> A noisy motor may indicate worn or damaged parts or misalignment between the motor drive shaft and the load.

- Test the motor for an unusually high temperature using a pyrometer or other temperature-measuring device.

> **NOTE**
> An excessively high motor temperature may indicate worn or damaged parts or misalignment between the motor drive shaft and the load. It may also indicate that the load has become too great for the motor or that the flow is restricted.

- Inspect the motor mounting bolts to ensure that they are tight.
- Examine the motor for excessive vibration.

> **NOTE**
> Excessive vibration may indicate misalignment of the motor-to-load coupling or worn bearings.

Instructor's Notes:

3.5.0 Inspecting Control Valves

The correct fluid pressure must be maintained in a hydraulic system for the system to function properly and to prevent damage to system components. Pressure-control valves limit and regulate the pressure in the system. Perform the following procedures to inspect control valves:

- Inspect the valve body and connections to ensure that there are no leaks and that they are clean. Clean them as necessary.
- Operate the valve controls to ensure that they operate smoothly and that there is a smooth, quick response.

> **WARNING!**
> To prevent injuries, warn personnel working in the area of the equipment startup before operating the controls.

- Have a qualified electrician inspect all connections to electrically controlled valves to ensure that the connections are tight.

> **WARNING!**
> To prevent personal injury from electrical shock, ensure that the power to the valve is turned off and locked out before working on the connections.

- Inspect any pneumatic or hydraulic control lines to the valve to ensure that they are tight and that there are no leaks.

3.6.0 Inspecting Check Valves

Check valves control the direction of fluid flow in a hydraulic system. Different types of check valves control the fluid in different ways. These valves can start, stop, or change the direction of fluid flow. Many types of check valves are used in hydraulic systems. You should perform the following general procedures to inspect check valves:

- Inspect the valve body to ensure that there are no leaks.
- Inspect the valve and connections to ensure that they are clean, and clean them as necessary.
- Listen for unusual noises from the valve.

3.7.0 Inspecting Cylinders

Cylinders in a hydraulic system are actuators/operators. They change hydraulic energy into mechanical energy. The pressure and movement of the hydraulic fluid move the cylinder rod to perform work. Cylinders are called linear actuators/operators because they produce motion and force in a straight line. Perform the following procedures to perform an external inspection of cylinders:

- Inspect the cylinder to ensure that there are no leaks and that it is clean. Clean it as necessary.
- Inspect the cylinder shaft to ensure that it has no scratches or burrs, if applicable.
- Examine the cylinder mounting bracket to ensure that the bolts are in good condition and properly adjusted.
- Listen for unusual noises from the cylinder.
- Monitor the cylinder for jerky, rough, sluggish, or otherwise irregular operation.

> **NOTE**
> A noisy cylinder or the irregular operation of a cylinder may indicate bad seals, a bent cylinder shaft, clogged ports, or scratched cylinder walls.

- Test the cylinder for excessively high temperature using a pyrometer or other temperature-measuring device.

> **NOTE**
> An excessively high cylinder temperature may indicate worn internal parts, an overloaded cylinder, or cylinder binding.

3.8.0 Inspecting Hoses, Fluid Lines, and Gauges

Hoses, fluid lines, and gauges make up the part of a hydraulic system that carries the fluid to the different system components. The type and size of the hoses, fluid lines, and gauges used in a system depend on the requirements of the system. Perform the following procedures to inspect hoses, fluid lines, and gauges:

- Inspect all hose, fluid line, and gauge connections to ensure that they are tight and that there is no leaking.
- Inspect all gauges to ensure that they are in good operating condition and are calibrated according to company policy.
- Examine all flexible hoses closely to ensure that there are no areas of dry rot, worn or thin spots, or kinks.
- Examine all fluid lines to ensure that they are properly supported by hangers or clamps.

Safety

Ask a trainee to explain how to prevent injuries while inspecting control valves.

Explain how to avoid electrical shock.

Classroom

Review the guidelines for inspecting control valves, check valves, and cylinders. Explain how to inspect hoses, fluid lines, and gauges. Ask whether trainees have any questions.

Teaching Tip

See the Teaching Tip for Sections 3.0.0–3.8.0 at the end of this module.

Performance Testing

Have each trainee perform Performance Task 1 to your satisfaction. Fill out a Performance Profile Sheet for each trainee.

Homework

Have trainees review Sections 4.0.0–4.4.0 for the next session.

Materials

Ensure that you have everything required for testing during this session.

Classroom

Explain why the process of elimination is an essential part of successful troubleshooting.

Review the importance of using troubleshooting charts and schematic diagrams to efficiently diagnose problems.

Refer to Table 1. Present problems that may be encountered on a hydraulic system. Ask trainees to diagnose the problem and determine appropriate solutions.

4.0.0 ◆ PERFORMING HYDRAULIC SYSTEM TROUBLESHOOTING PROCEDURES

Troubleshooting consists of locating and eliminating the cause of problems or malfunctions in a machine or system. To effectively troubleshoot a hydraulic system, you must have a good understanding of how the system and it components operate.

The best method for troubleshooting is the process of elimination. After the problem and possible causes are identified, the causes should be eliminated one by one, starting with the one that requires the least amount of downtime or major disassembly. Tests and inspections that involve extensive downtime or major disassembly should be saved until last.

There is no substitute for a thorough knowledge of hydraulic principles and a good understanding of how a machine works. However, a good troubleshooting chart is a useful tool for troubleshooting hydraulic systems (see *Table 1*). Troubleshooting charts describe some of the problems, possible causes, and possible solutions for a common system. Once a problem has been identified, the troubleshooting chart can be used to help determine the cause and the solution that should be taken.

4.1.0 Reading Hydraulic Schematic Diagrams

The ability to read a schematic diagram is essential in troubleshooting a hydraulic system. Using the schematic diagram, you can locate the various components to help determine where the problem is occurring.

Schematic diagrams are made up of symbols that represent the components and fluid lines of the system. Reading a schematic diagram consists of recognizing the standard symbols used and being able to trace the system lines from one component to another. *Figure 6* shows common hydraulic symbols.

When tracing a system, begin at the reservoir and trace the line through the system until the entire system has been traced and all of the components identified. Being able to trace the system and identify the components and their functions helps to pinpoint problems when troubleshooting.

Table 1 Hydraulic System Troubleshooting Chart (1 of 2)

Problem	Possible Cause	Possible Solution
The actuator is not moving.	The drive motor is not running.	Repair or replace the drive motor.
	The directional-control valve is not shifting.	Repair or replace the directional-control valve.
	The drive motor-to-pump coupling is loose or sheared.	Tighten or replace the coupling.
	The reservoir fluid level is low.	Fill the reservoir to the proper level.
	The pressure-compensated pump spring is broken.	Replace the spring.
	The relief or unloading valve is stuck open.	Free or replace the unloading valve.
	The actuator is binding.	Realign the actuator.
The actuator is moving too slowly.	The pump drive motor speed is low.	Adjust the pump drive motor speed.
	The line is restricted.	Locate and clear the restriction.
	The directional-control valve is not shifting completely.	Repair or replace the directional-control valve.
	The flow-control valve is defective or set wrong.	Adjust or replace the valve.
	The pressure-compensated or variable-volume pump is out of adjustment.	Adjust the pump.
The actuator is slowing or stopping under load.	The relief valve is set too low.	Adjust the relief valve.
	The pump or hydraulic motor is worn.	Replace the pump or motor.
	The actuator is leaking.	Repair or replace the actuator.
	The pressure-compensated pump is set too low.	Adjust the pump.
	The actuator is binding.	Align the actuator.
The actuator is creeping.	The directional-control valve is leaking excessively.	Repair or replace the directional-control valve.

307T01A.EPS

Instructor's Notes:

Table 1 Hydraulic System Troubleshooting Chart (2 of 2)

Problem	Possible Cause	Possible Solution
The actuator is moving too fast.	The flow-control valve is out of adjustment. The pump speed is too high.	Adjust the flow-control valve. Adjust the pump speed.
The actuator is moving in a jerky or erratic manner.	The fluid is contaminated. The directional control valve is contaminated. The accumulator is defective. The solenoid is defective. The system pressure or control valve is out of adjustment.	Change the system fluid. Remove, clean, and replace the directional-control valve. Repair or replace the accumulator. Replace the solenoid. Adjust the valve.
The load is settling.	The load lock valve is defective. The directional control valve, rod seal, or piston seal is leaking.	Repair or replace the valve. Replace as necessary.
The hydraulic pump is noisy.	The fluid level is low. The system pressure is too high. The variable-volume or pressure-compensated pump is adjusted for too much flow. The suction line is leaking. The pump speed is too high. The motor and pump are out of alignment. The suction line is restricted. The intake strainer is plugged. The pump is worn.	Fill the reservoir to the proper level. Adjust the pressure-relief valve to low pressure. Adjust the pump. Locate and correct the leaks. Adjust the pump speed. Align the pump and motor. Locate and clear the restriction. Clean or replace the strainer. Repair or replace the pump.
The pump is too hot.	The unloading valve is not working properly. The pump is worn. The relief valve is adjusted too high.	Repair or replace the valve. Repair or replace the pump. Adjust the relief valve.
The relief valve is too hot or noisy.	The unloading valve is defective. The relief valve seats are worn or leaking. The flow through the valve is excessive.	Replace the unloading valve. Replace the relief valve. Adjust the flow rate of the pump.
The hydraulic motor is too hot.	The motor is worn. The lubricity or viscosity of the fluid is broken down. The motor is leaking internally.	Repair or replace the motor. Drain and replace the system fluid. Repair or replace the motor.
The hydraulic motor is noisy.	Motor parts are worn or broken. The motor coupling is misaligned or defective. The system pressure is too high.	Repair or replace the motor. Align or replace the coupling. Adjust the pressure-relief valve.
There are hot spots in the cylinder housing.	The piston seals are leaking.	Replace the piston seals.
The pump is not delivering.	There is an air leak in the suction line. The suction line strainer is clogged. The fluid level in the reservoir is low.	Locate and correct the leak. Clean or replace the strainer. Add fluid to the reservoir.
The system pressure is low.	The relief valve is malfunctioning. There are leaks in the system. The pump is worn or damaged.	Clean, repair, or replace the relief valve. Locate and correct the leaks. Repair or replace the pump.

307T01B.EPS

Classroom

On the whiteboard/ chalkboard, draw a variety of hydraulic symbols. Ask trainees to interpret the symbols. Answer any questions trainees have about the symbols' meanings.

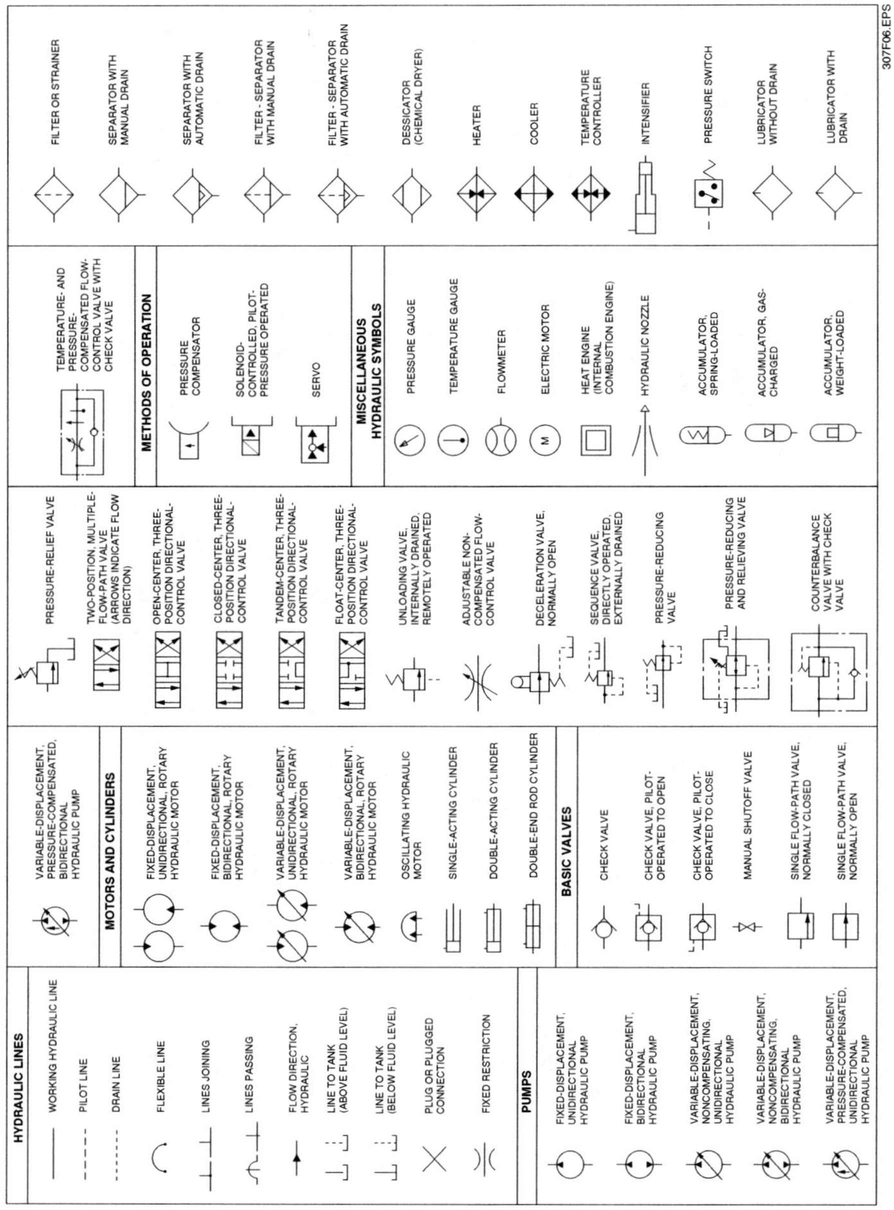

Figure 6 ◆ Hydraulic symbols.

Instructor's Notes:

4.2.0 Preliminary Diagnoses

Typically, problems in valve actuators/operators are initially misdiagnosed as problems with valves because the operation of the valves is the demonstration of the problem. Field personnel first notice the problem when a valve fails to operate properly. You might be told that the valve is:

- *Hung* – Actuator/operator not functioning
- *Seized* – Actuator/operator not functioning
- *Out of control* – Actuator/operator overactive
- *Vibrating* – Could be in pipeline
- *Making strange noises* – Squeaking, hammering, or banging
- *Not powerful enough for the job* – Suggests the unit is slow or erratic in operation

One or more such conditions might be reported when the only observable problem is that the valve is not operating properly. It is up to you to pinpoint the actual problem and repair it safely and efficiently. That is not to say that you should ignore the input from others because other personnel can be a valuable resource. At the same time, such input should be weighed with other solid evidence collected before deciding how to proceed. For example, if a valve is actually hung, you can determine this by simply attempting to turn the handwheel. Most likely, the problem is in the actuator/operator.

4.3.0 Troubleshooting Hydraulic Systems

Three basic hydraulic principles must be considered before troubleshooting a hydraulic system:

- Movement depends on flow. There must be fluid flow to the actuators/operators of a hydraulic system before anything can move.
- Flow rate determines speed. The speed at which an actuator/operator moves depends on the rate of fluid flow. The faster a fluid fills the cylinder of an actuator/operator, the faster the actuator/operator moves.
- Load determines pressure. The amount of pressure that is developed in an actuator/operator depends on the load that it must move. The greater the load, the higher the pressure rises.

If an actuator/operator does not move and no fluid flow is present, you must determine the reason for the lack of fluid flow. Possible causes include:

- Low or empty fluid reservoir
- Plugged input strainer or filter
- Nonfunctioning pump
- Worn-out or damaged seals
- Open relief valve
- Malfunctioning directional-control valve
- Malfunctioning solenoids
- Blockage in a fluid line to or from the flow
- Excessive internal clearances

If an actuator/operator is not working at the proper speed and the flow rate is inadequate, you must determine what is reducing the flow rate to the actuator/operator. Some of the possible causes of reduction in fluid flow rate are:

- Low pump flow rate
- Directional-control valve that is not completely shifted
- Contaminated fluid

If fluid flow is adequate and the actuator/operator still does not move the load, you must determine why the load does not move. Some of the reasons an actuator/operator does not move the load are:

- Excessive load for the system
- Inadequate system pressure due to a slipping pump or low relief valve setting
- Leaks in the system

With so many possible causes, you must be methodical when troubleshooting. The following are basic steps to take when diagnosing a problem with a hydraulic valve actuator/operator:

> **WARNING!**
> The control center must be notified before any work is begun. Company safety policy must be followed when troubleshooting a valve actuator/operator.

Step 1 Understand how the valve and its actuator/operator function within the system.

Step 2 Learn about the problem by investigating the symptoms. Include input from local operations personnel, the unit's historical information, and personal assessment.

Step 3 If it is a repeat problem, contact the manufacturer's technical assistance department for information and suggestions about the particular actuator/operator.

Step 4 Take note of all attached equipment influencing the operation of the unit. The operation of limiting devices and other related units are especially important in diagnosing problems.

Classroom

Review the possible ways in which valve actuators/operators are misdiagnosed.

Discuss the basic hydraulic principles. Review the possible causes of problems with the fluid flow in an actuator/operator.

Safety

Emphasize the importance of notifying the control center before performing any work on a valve actuator/operator.

Classroom

Explain how to diagnose problems with hydraulic valve actuators/operators. Go over any questions trainees have.

Safety

Ask a trainee to explain why cleanliness is essential when disassembling actuators/operators.

Teaching Tip

See the Teaching Tip for Sections 4.0.0–4.4.0 at the end of this module.

Performance Testing

Have each trainee perform Performance Task 2 to your satisfaction. Fill out a Performance Profile Sheet for each trainee.

Homework

Have trainees review Sections 5.0.0–5.3.4 for the next session.

Materials

Ensure that you have everything required for the laboratories and demonstration during this session.

Classroom

Explain how to prepare a hydraulic system for shutdown.

Safety

Explain how to ensure that all loads are blocked.

Step 5 Notify the control center, upstream and downstream pumping stations, and any other affected personnel on the pipeline.

Step 6 Switch the actuator/operator to manual control and run the unit through its cycle completely several times to observe the actual problem.

Step 7 Form a personal theory, supported by facts, about the cause of the problem.

Step 8 Form a plan of action for the most efficient solution to the problem and the return of the actuator/operator to normal operation.

Step 9 Initiate all applicable lockout/tagout procedures and follow company policy for PPE and safety procedures.

Step 10 Perform repairs/adjustments according to company policy and manufacturer's recommendations.

Step 11 Inspect the unit for worn or damaged parts. They should be repaired or replaced while the unit is disassembled.

> **CAUTION**
> When disassembling an actuator/operator for inspection and repairs, clean the area, keep it clean, and keep the actuator/operator parts clean.

Step 12 Test the valve for proper operation.

Step 13 Return the unit to normal use.

Step 14 Notify the control center and other affected personnel, and complete all required documentation per company policy.

4.4.0 Repair Options

Common options for repairing an actuator/operator include but are not limited to:

- Repair the problem on site using manufacturer's repair manuals and recommendations.
- Replace the unit, and send the failed unit to the company repair shop.
- Replace the unit and return the failed unit to the manufacturer. With some types of sealed units, this is the only recourse.

5.0.0 ◆ ADJUSTING AND REPAIRING HYDRAULIC SYSTEM COMPONENTS

When a hydraulic system component becomes worn past the point of efficiently performing the function for which it was intended, it must be repaired. There are numerous types and brands of hydraulic components, and it is essential to refer to the specific manufacturer's instruction manual when repairing or overhauling. This module provides general procedures that must be performed in conjunction with the service manual from the manufacturer of the components being used in the repair. Repairing hydraulic system components includes performing the following procedures:

- Preparing the system for shutdown and repair
- Repairing and overhauling pumps
- Repairing and overhauling valves
- Repairing and overhauling actuator/operators
- Replacing hoses
- Replacing gauges

5.1.0 Preparing the System for Shutdown and Repair

Before repairing or replacing any hydraulic system component, steps must be taken to prepare for maintenance. It is critical to relieve pressure from the system before performing maintenance on any component. Follow these steps to prepare the system to be shut down and repaired:

Step 1 Gather all the necessary tools and hoisting equipment if applicable.

Step 2 Obtain the manufacturer's service manual for the actuator/operator to be serviced, and review the repair and installation procedures.

Step 3 Lower all vertical cylinders.

Step 4 Discharge all accumulators.

Step 5 Block all loads.

> **WARNING!**
> All loads must be blocked to prevent them from falling and causing equipment damage or personal injury when pressure is relieved.

Step 6 Turn off the electrical power to the system.

Step 7 Relieve all pressure throughout the system.

Step 8 Lock out and tag out all the points of energy control in the system.

Instructor's Notes:

5.2.0 Repairing and Overhauling Pumps

Repairing and overhauling hydraulic pumps (see *Figure 7*) involves performing the following procedures:

- Removing a pump from the system
- Disassembling a pump
- Reassembling a pump
- Installing a pump

5.2.1 Removing a Pump from the System

The number of different configurations of hydraulic systems is endless. However, most hydraulic systems are much alike, and all use the same principles of operation. Although it is not possible to give an accurate procedure for each model and installation, most pump removal procedures follow the same general steps as those that follow, with variations for the particular model. You should follow these general steps to remove a pump from the system:

WARNING!
Ensure that the hydraulic system has been prepared for maintenance and that it is locked out and tagged before performing this procedure. Verify that all pressure has been relieved in the system before disconnecting piping from the pump. Pressure left in the system when the pipe connection is broken could cause personal injury.

Figure 7 ◆ Hydraulic system installed on a pipeline.

Step 1 Clean the pump thoroughly around the connections.

Step 2 Disconnect the coupling between the motor and the pump.

Step 3 Disconnect the piping from the pump, and cap or seal the ends of the pipe.

Step 4 Make notes of the pump mounting arrangement to aid in reinstallation after the repair is made.

Step 5 Remove the mounting bolts.

Step 6 Remove the shims from under the base.

NOTE
Note the number of shims at each mounting bolt so that the same number of shims can be reinstalled at each point later. Measure the thickness of each shim pack and number them according to each foot location.

Step 7 Remove the pump from the mounting base, and take it to a clean work area.

5.2.2 Disassembling a Pump

Pump disassembly should be performed according to the manufacturer's instruction manual. Special care should be taken to ensure that no parts are misplaced or damaged during disassembly. Observe the following general guidelines during disassembly:

- Ensure that the work area is clean and free of dirt and other foreign materials.
- Lay out the parts in an orderly manner as the pump is disassembled to facilitate reassembly.
- Inspect all parts for wear or damage as they are disassembled, and replace them as necessary.
- Clean all parts except bearings thoroughly using a safe solvent.

NOTE
Do not use solvent on bearings because it would flush out the internal lubricant. Wipe bearings lightly with a clean rag to remove any foreign materials or old grease; then apply fresh lubricant.

- Disassembled parts that are to be left out for an extended period should be covered with a clean covering and carefully placed to prevent contamination or loss.

Classroom

Review the general procedures to remove a pump from a hydraulic system.

Safety

Discuss the consequences of failing to properly prepare a hydraulic system for maintenance.

Laboratory

Ask a trainee to demonstrate the procedures to disassemble a pump, following the manufacturer's instructions. Have trainees observe the procedure and explain what was done correctly and/or incorrectly.

Demonstration

Using the disassembled pump, demonstrate reassembly. Identify differences between the general guidelines and the manufacturer's instructions.

Classroom

Explain how to install the reassembled hydraulic pump in the system. Ask whether trainees have any questions.

Safety

Explain problems that can result if the pump and motor are not perfectly aligned.

Ask a trainee to explain how to properly lock out and tag a hydraulic system.

5.2.3 Reassembling a Pump

Pump reassembly should also be performed specifically according to the manufacturer's instruction manual. Reassembly is a reversal of disassembly. The following guidelines should be observed:

- Inspect and clean all parts again as they are reassembled.
- Apply a light coat of lubricant to each part to facilitate assembly.
- Take care with parts, such as O-rings and seals, to prevent nicking or damaging them when the pump is reassembled.
- Check and measure all replacement parts against the manufacturer's specifications for the type and model of pump being assembled.
- Check each part carefully as it is assembled to ensure that it fits properly and does not bind or stick.
- Set and check all clearances using the proper measuring tools.
- Tighten all bolts and nuts to the required torque using the proper tools.
- Turn the pump by hand after assembly is complete to ensure smooth operation.

5.2.4 Installing a Pump

Follow these steps to install a pump in the system:

Step 1 Clean the pump and mounting area.

Step 2 Place the pump on the mounting base.

Step 3 Install the pump mounting bolts, but leave them loose.

Step 4 Install the same number of shims that were removed when the pump was removed.

Step 5 Tighten the pump mounting bolts.

Step 6 Ensure that the piping connections are clean and free of any foreign materials.

Step 7 Connect the piping, and properly tighten the connections.

Step 8 Align the pump and the motor.

> **CAUTION**
> The pump and motor must be perfectly aligned to prevent binding and premature wear-out or breakdown.

Step 9 Remove the locks and tags.

Step 10 Restore power to the pump motor.

Step 11 Start the pump, adhering strictly to the manufacturer's startup instructions.

Step 12 Monitor the pump for unusual noises or excessive heat buildup.

Step 13 Monitor the system operation for proper flow and pressures.

Step 14 Return all tools and manuals to their proper locations.

5.3.0 Repairing and Overhauling Control Valves

Repairing and overhauling hydraulic system control valves involves performing the following procedures:

- Removing a valve from the system
- Disassembling a valve
- Reassembling a valve
- Installing a valve

5.3.1 Removing a Valve from the System

Hydraulic system control valves can frequently be repaired without being removed from the system. If a valve cannot be repaired while it is in the system, it must be removed. Follow these steps to remove a hydraulic valve from the system:

> **WARNING!**
> Ensure that the hydraulic system has been prepared for maintenance and that it is locked out and tagged before performing this procedure. Verify that all pressure has been relieved in the system before disconnecting piping from the valve. Pressure left in the system when the pipe connection is broken could cause personal injury.

Step 1 Clean the valve thoroughly around the end connections.

Step 2 Have an electrician disconnect the electrical leads from the pilot section of the valve.

> **NOTE**
> Disconnect the air lines to the pilot valve if applicable.

Instructor's Notes:

Step 3 Have an electrician disconnect electrical wiring from the solenoid, if applicable, and label the wires for reassembly.

Step 4 Disconnect piping from the valve, and cap or seal the ends of the pipe.

Step 5 Take note of the valve mounting arrangement to facilitate reinstallation after repair.

Step 6 Disconnect the nearest union, unscrew the valve from the piping, and remove the valve from the system, taking care not to drop it.

> **NOTE**
> On some larger hydraulic systems, valves are installed with flanges instead of threaded connections. In this case, the flange bolts must be removed to remove the valve from the system.

Step 7 After removing the valve from the system, take it to a clean work area.

5.3.2 Disassembling a Valve

Valve disassembly should be performed according to the manufacturer's instruction manual. Special care should be taken to ensure that no parts are misplaced or damaged during disassembly. The following general guidelines should be observed during disassembly:

- The work area should be clean and free of dirt and other foreign materials.
- The parts should be laid out in an orderly manner to facilitate reassembly.
- All parts should be inspected for wear or damage as they are disassembled and should be replaced as necessary.
- All parts should be thoroughly cleaned with a suitable solvent.
- Disassembled parts that will be left out for an extended time should be covered with a clean covering and carefully placed to prevent contamination and loss.

5.3.3 Reassembling a Valve

Valve reassembly should also be performed specifically according to the manufacturer's instruction manual. Reassembly is a reversal of disassembly. The following general guidelines should be observed, as applicable, during most reassembly procedures:

- Inspect and clean all parts again as they are reassembled.
- Apply a light coat of lubricant to each part to facilitate assembly.
- Handle easily damaged parts, such as O-rings and seals, with care to prevent nicking or damaging them before the valve is reassembled.
- Examine and measure all replacement parts against the manufacturer's specifications for the type and model of the component being assembled.
- Examine each part carefully as it is assembled to ensure that it fits properly and that it does not bind or stick.
- Set and verify all clearances using the proper measuring tools.
- Tighten all bolts and nuts to the manufacturer's torque specifications using the proper tools.

5.3.4 Installing a Valve

Follow these general steps to install a valve in a hydraulic system:

Step 1 Clean the ends of the pipe connections.

Step 2 Screw the valve into the system, taking care not to overtighten or otherwise damage the valve.

> **NOTE**
> If the valve is flange-mounted, it is placed in the system with gaskets, and the flange bolts are installed and tightened.

Step 3 Connect and tighten the pipe union.

Step 4 Have a qualified electrician connect the electrical leads to the solenoid and the valve pilot section if applicable.

Step 5 Remove locks and tags.

Step 6 Start up the system.

Step 7 Jog the equipment through all positions to test the valve operation.

Step 8 Monitor the valve to verify correct operation.

Step 9 Monitor the system operation for proper pressure and flow.

Step 10 Return all tools and manuals to their proper locations.

Classroom

Review the procedures to remove and disassemble a control valve.

Laboratory

Set up several stations with disassembled valves and the necessary tools for reassembly. Ask trainees to follow the general guidelines for reassembly. Review and assess trainees' work.

Classroom

Ask a trainee to explain how to install a valve into a hydraulic system. Go over any questions trainees have.

Homework

Have trainees review Sections 5.4.0–5.6.0 for the next session.

Materials

Ensure that you have everything required for the laboratory, demonstration, and testing during this session.

Classroom

Discuss the difference between preparing for repairs on small actuators/ operators and on large pipeline actuators/operators.

Safety

Stress the importance of relieving all of the pressure from the system before disconnecting the actuator/ operator.

Classroom

Review the procedures for removing an actuator/ operator from the system.

Discuss the general guidelines for disassembling and reassembling an actuator/operator.

5.4.0 Repairing and Overhauling Actuators/Operators

Hydraulic system actuators/operators all depend on the same principles of hydraulics for their operation as do other closed loop or machine hydraulic systems. However, there is a big difference between working with small actuators/ operators in small systems and some of the large applications on the pipeline. Small actuators/ operators on machine systems can often be held in one hand. Some large pipeline actuators/operators will require the use of heavy mobile hoisting equipment to move some of the components. Repairs to large pipeline units often must be approached from an entirely different perspective. In many cases, a great deal of preparation must be accomplished before you can begin a repair on a large actuator/operator. In all cases, repairing and overhauling hydraulic actuators/ operators will involve performing the following procedures:

- Removing an actuator/operator from the system
- Disassembling an actuator/operator
- Reassembling an actuator/operator
- Installing an actuator/operator

5.4.1 Removing an Actuator/Operator from the System

Prior to removing an actuator/operator from a hydraulic system, have a qualified electrician remove any electrical connections from the actuator/operator. The extensive number of models and brands of hydraulic actuators/operators prevent them all from being covered in a module. However, you should follow these general steps to remove an actuator/operator from the system:

> **WARNING!**
> Ensure that the hydraulic system has been prepared for maintenance and that it is locked out and tagged before performing this procedure. Verify that all pressure has been relieved in the system before disconnecting piping from the actuator/operator. Pressure existing in the system when the pipe connection is broken could cause personal injury.

Step 1 Clean the actuator/operator thoroughly around the fluid line connections.

Step 2 Disconnect the inlet and outlet fluid lines from the actuator/operator and cap or seal the ends of the lines.

Step 3 Disconnect the actuator/operator from the actuator/operator mount.

Step 4 Note the mounting arrangement, placement, and total thickness of the shim pack and brackets to facilitate reinstallation.

Step 5 Remove the actuator/operator from the system and take it to a clean work area.

5.4.2 Disassembling an Actuator/Operator

Hydraulic actuator/operator disassembly should be performed according to the manufacturer's instruction manual. Special care should be taken to ensure that no parts are misplaced or damaged during disassembly. Follow these general guidelines during disassembly:

- The work area should be clean and free of dirt and other foreign materials.
- Parts should be laid out in an orderly manner to facilitate reassembly as the actuator/operator is disassembled.
- All parts should be inspected for wear or damage as they are disassembled and should be replaced as necessary.
- All parts should be thoroughly cleaned with a suitable solvent.
- Disassembled parts to be left out for an extended time should be covered with a clean covering and carefully placed to prevent contamination and loss.

5.4.3 Reassembling an Actuator/Operator

Hydraulic actuator/operator reassembly should also be performed specifically according to the manufacturer's instruction manual. Reassembly is a reversal of disassembly. Follow these general guidelines during reassembly:

- Inspect and clean all parts again as they are reassembled.
- Apply a light coat of lubricant to each part to facilitate assembly.
- Take care of easily damaged parts, such as O-rings and seals, to prevent nicking or damaging them when the actuator/operator is reassembled.
- Examine and measure all replacement parts against the manufacturer's specifications for the type and model of actuator/operator being assembled.
- Examine each part carefully as it is assembled to ensure that it fits properly and does not bind or stick.

Instructor's Notes:

- Set and verify all clearances using the proper measuring tools.
- Tighten all bolts and nuts to the manufacturer's torque specifications using the proper tools.

5.4.4 Installing an Actuator/Operator

Every system and installation is different. However, certain steps and procedures (with variations) are common to most locations. You should follow these general steps to install an actuator/operator in the system:

Step 1 Clean all fluid line connections thoroughly.

Step 2 Mount the actuator/operator on the actuator/operator mount using the shims and other components as they were arranged before disassembly.

Step 3 Attach accessories to the actuator/operator rod, taking care not to scratch the rod.

Step 4 Lubricate the actuator/operator pivot points.

Step 5 Connect the inlet and outlet port fluid piping.

Step 6 Remove locks and tags.

Step 7 Adjust the system pressure-control valves to a reduced pressure.

Step 8 Start up the system.

Step 9 Adjust the cushion action of the actuator/operator until a smooth stop action is obtained.

Step 10 Return the pressure-control valves to operating pressure.

Step 11 Monitor the actuator/operator for proper operation.

Step 12 Return all tools and manuals to their proper locations.

5.5.0 Replacing Hoses

When a hydraulic hose fails, it is usually replaced instead of repaired. Hose comes in many different lengths with fittings already attached. If the proper length is not available, the hose must be cut to length, and the fittings must be attached. Replacing hydraulic hoses involves performing the following activities:

- Removing a hose from the system
- Assembling a hose
- Installing a hose

5.5.1 Removing a Hose from the System

Follow these steps to remove a hose from the system:

> **WARNING!**
> Ensure that the hydraulic system has been prepared for maintenance and that it is locked out and tagged before performing this procedure. Verify that all pressure has been relieved from the system before loosening the connections. Pressure existing in the system when the pipe connection is broken could cause personal injury.

Step 1 Clean the hose thoroughly around the hose connections.

Step 2 Loosen the hose connections.

Step 3 Remove the hose from the system, and take it to a clean work area.

5.5.2 Assembling a Hose

Special care must be taken to select the proper replacement hose for a hydraulic system. A hydraulic hose must be selected according to the temperature range and pressure range of the system and the type of fluid used in the system. Selecting the wrong hose can result in premature hose failure and equipment damage. A hose of the wrong temperature range may become too soft or even melt. A hose of the wrong pressure range may blow out. A hose not compatible with the type of fluid in the system may deteriorate, and its material may then contaminate the fluid in the system.

Hose fittings are selected according to the pressure range of the system. There are many types of hydraulic hose fittings. This section explains how to assemble a hydraulic hose using medium- and high-pressure fittings. Follow these steps to assemble a hose:

Step 1 Select the proper hose for the system.

Step 2 Cut the hose to the required length.

Step 3 Select the proper type of fitting for the hose application according to the pressure rating of the system.

> **NOTE**
> Many types of hose fittings are reusable. If the existing fittings are reusable, they should be removed from the old hose and cleaned for reuse on the new hose.

Laboratory

Bring in a variety of replacement parts for a hydraulic actuator/operator and the corresponding manufacturer's specifications. Ask trainees to examine and measure the parts against the specifications.

Classroom

Explain how to install a hydraulic actuator/operator. Emphasize that every system and installation is different. Ask whether trainees have any questions.

Safety

Discuss the consequences of disconnecting piping while pressure exists in the system.

Demonstration

Bring in a variety of hoses appropriate for hydraulic systems. Demonstrate how to inspect the hoses for damage and how to cut them for installation.

Classroom

Explain how to select the appropriate hose for the system. Review the procedures to assemble and install a hose.

Safety

Ensure that trainees understand how to verify that all of the pressure has been relieved in the system.

Emphasize the importance of carefully opening the isolation valve.

Teaching Tip

See the Teaching Tip for Sections 5.0.0–5.6.0 at the end of this module.

Performance Testing

Have each trainee perform Performance Task 3 to your satisfaction. Fill out a Performance Profile Sheet for each trainee.

Homework

Have trainees review Sections 5.7.0–5.7.3 and answer the Review Questions for the next session.

Step 4 Place the fitting socket in a vise, being careful not to crush or deform the socket.

Step 5 Dip the end of the hose in hose oil or other heavy oil.

Step 6 Thread the hose into the socket counterclockwise until it bottoms out, and then back it off half a turn.

Step 7 Position the hose and socket in the vise to install the nipple.

Step 8 Dip the hose end of the nipple in hose oil.

Step 9 Thread the nipple into the socket until the nipple hex shoulders touch the socket.

Step 10 Repeat Steps 4 through 9 to assemble a fitting onto the opposite end of the hose.

5.5.3 Installing a Hose

Follow these steps to install a hose in the system:

Step 1 Clean the hose connections thoroughly.

Step 2 Place the hose in the proper position.

Step 3 Connect the hose fittings to the system components.

Step 4 Tighten the connections using the proper size and type of wrench.

Step 5 Remove locks and tags.

Step 6 Start up the system.

Step 7 Check the hose connections for leaks.

Step 8 Return all tools to their proper locations.

5.6.0 Replacing Gauges

When a hydraulic gauge fails, it is usually replaced instead of repaired. Gauges come in many different styles. When gauges need to be replaced, it is important to select a replacement gauge that is an exact match of the gauge being replaced. Follow these steps to replace a gauge:

> **WARNING!**
> Ensure that the hydraulic system has been prepared for maintenance and that it is locked out and tagged before performing this procedure. Verify that all pressure has been relieved in the system before disconnecting piping from the valves. Pressure left in the system when the pipe connection is broken could cause personal injury.

Step 1 Close the gauge isolation valve to stop fluid flow and pressure to the gauge.

> **NOTE**
> All gauges should have an isolation valve that provides a means for gauge maintenance without shutting down the system. If the gauge being removed does not have an isolation valve, a valve should be installed on the gauge before the new gauge is installed in the system.

Step 2 Remove the gauge using the proper wrench.

Step 3 Select the correct replacement gauge.

Step 4 Install the new gauge.

Step 5 Open the isolation valve to restore flow and pressure to the gauge.

> **CAUTION**
> The isolation valve should be opened very carefully to prevent damaging the gauge with a sudden burst of pressure. The valve should be opened slightly until pressure is established to the gauge, and then it can be opened fully.

Step 6 Check the gauge and isolation valve for fluid leaks.

Step 7 Monitor the gauge to ensure that it is operating properly.

5.7.0 Hydraulic Valve Actuators/Operators

The hydraulic actuator/operator (see *Figure 8*) operates by the use of fluid pressure. The fluid used is normally some type of oil specified for the unit. Fluid pressure acts on a duplex piston that is connected to the valve stem through a cylinder. The required length of stem travel and seating thrust is obtained by varying the cylinder size. Fluid pressure above and below the piston moves the stem up and down, opening and closing the valve. Many have spring-loaded pistons in either the open or closed position that return the valve to its safe position in case of pressure failure. Units usually have hand pumps for manual operation when necessary.

Hydraulic systems provide exceptional stiffness, fast stroking speeds, very high thrust, and dynamic response. Very large valves are often

Materials

Ensure that you have everything required for testing during this session.

Audiovisual

Show Transparency 5 (Figure 8). Explain how fluid pressure moves the piston in a hydraulic actuator/operator.

Instructor's Notes:

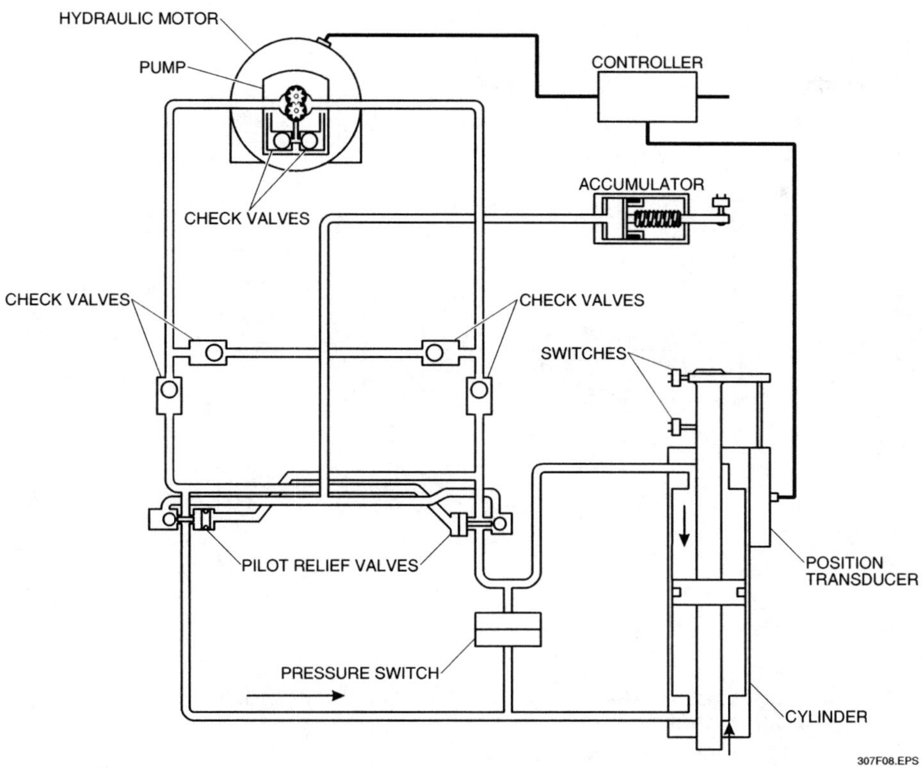

Figure 8 ◆ Hydraulic actuator/operator.

hydraulically operated because of the large amount of force available from a relatively small actuator/operator.

Electro-hydraulic actuators/operators are units that operate off a self-contained hydraulic system. The unit employs a hydraulic pump, a reservoir, control valves, and other accessories needed to operate the system. The actuator/operator is normally remote controlled but can be equipped with a handwheel. These actuators/operators are quick acting and reasonably safe for explosive atmospheres as long as the manufacturer's recommendations are followed.

When the unit is activated, the pump supplies hydraulic pressure to the designated inlet valve. The fluid is then directed into a cylinder to the appropriate side of the piston. The piston is attached to the valve stem by means of a stem connector and crank arm. When a signal is received, the electric-powered pump is started. Pressure supplied to the manifold connection is diverted to the top or bottom side of the piston to open or close the valve as required.

These actuators/operators must be installed in a vertical position. As with all hydraulic and electrical units, the connections, switches, and assemblies must be kept free of dirt and corrosion. The hydraulic system must be kept free of dirt, sludge, and air. Excessive noise from the pump may indicate air in the system, which must be bled off as quickly as possible to avoid damage. *Figure 9* shows a schematic of an electro-hydraulic actuator/operator.

Audiovisual

Show Transparency 6 (Figure 9). Explain how an electro-hydraulic actuator/operator is activated.

Classroom

Stress that trainees adhere to company policies and procedures when beginning or adjusting hydraulic valve actuators/operators.

Safety

Emphasize the importance of maintaining an ongoing awareness while performing actuator/operator activities.

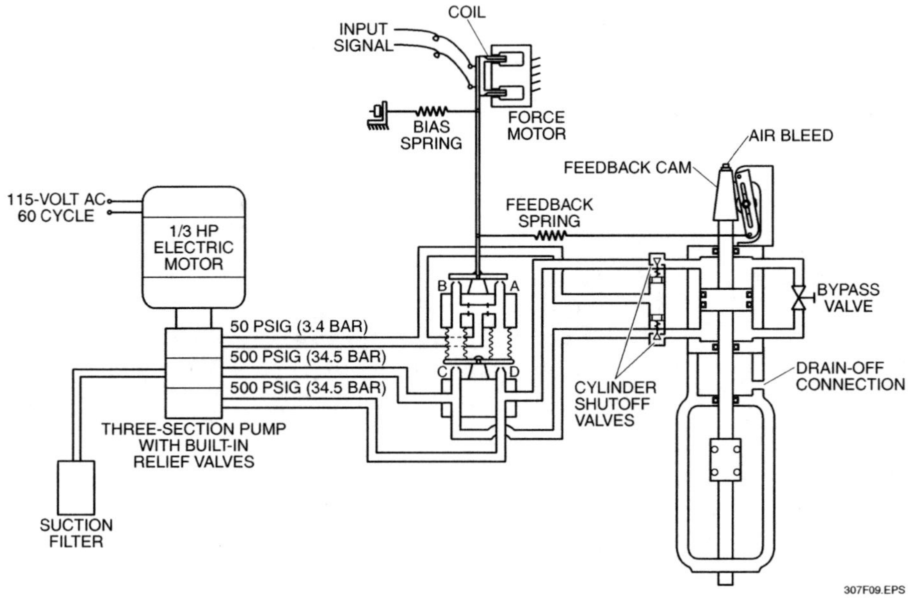

Figure 9 ◆ Schematic of an electro-hydraulic actuator/operator.

5.7.1 Adjusting Hydraulic Valve Actuators/Operators

Before beginning to adjust a hydraulic valve actuator/operator, you should complete the preliminary procedures listed earlier in this module. All cautions, warnings, and company safety policy and procedures should be meticulously observed. Due to the many brands and models of hydraulic actuators/operators in use, it is not possible to prepare a single procedure that would apply exactly to every one. However, you should follow these general steps to adjust a hydraulic valve actuator/operator.

Step 1 Ensure that the correct valve and hydraulic actuator/operator are located, especially before operating the valve, to preclude any upsets.

> **NOTE**
> If the nameplate is missing, it must be replaced according to company policy and procedure.

Step 2 Ensure that there are no obstructions blocking accessibility to the valve.

Step 3 Verify the documentation.

Step 4 Verify its type and manufacturer.

Step 5 Notify the control center and any other affected personnel that the work is beginning.

> **WARNING!**
> You must maintain an ongoing awareness of the valve and actuator/operator activities being performed to ensure safety during the performance of the task.

Step 6 All applicable task-specific procedures must be followed. Following a task-specific procedure helps to ensure job safety, efficient task performance, standardization, and appropriate documentation. Procedures should also include work permit information and notification requirements (control center and both the upstream and downstream pump stations for main line valves, any others possibly being affected).

Instructor's Notes:

CAUTION
You must take care to follow the manufacturer's instructions and industry recommendations in order to ensure personal safety, prevent mechanical damage to the actuator/operator being adjusted, avoid internal damage to the valve, and preserve the service life of the equipment.

Step 7 Verify the correct valve position to determine the appropriate limit switch setting.

Step 8 Properly set the **limit switches**, as applicable, to ensure that the limit switch contacts operate at the correct valve positions.

Step 9 Properly set the actuator/operator travel if applicable.

Step 10 Notify the control center and/or affected personnel to verify the integrity of the system and return the system to normal operation.

Step 11 Complete all required documentation and results of the adjustment.

NOTE
Documenting data for compliance, historical review, and trending is an important element of valve maintenance.

5.7.2 Repairing Hydraulic Valve Actuators/Operators

Before attempting to repair a hydraulic valve actuator/operator, (see *Figure 10*) you should complete the preliminary procedures listed earlier in this module. All cautions, warnings, and company safety policy and procedure must be meticulously observed and followed. Due to the many brands and models of hydraulic actuators/operators in use, it is not possible to prepare a single procedure that would apply exactly to every one. However, follow these general steps to repair a hydraulic valve actuator/operator.

NOTE
If the nameplate is missing, it must be replaced according to company policy and procedure.

Step 1 Ensure that the correct valve and hydraulic actuator/operator is located, especially before operating the valve to preclude any upsets.

Step 2 Ensure that there are no obstructions blocking accessibility to the valve actuator/operator.

NOTE
Verification of accessibility is very important, especially where large actuators/operators may necessitate the use of mobile hoisting equipment.

Step 3 Verify the hydraulic valve actuator/operator documentation.

Step 4 Verify the valve number and nameplate data of the hydraulic actuator/operator to be adjusted.

Step 5 Notify the operations center and any other affected personnel that the work is beginning.

WARNING!
You must maintain an ongoing awareness of the valve and actuator/operator repair activities being performed to ensure safety during the performance of the task.

ELECTRO-HYDRAULIC ACTUATOR/OPERATOR

307F10.EPS

Figure 10 ◆ Hydraulic valve actuator/operator.

Safety

Ask a trainee to explain problems that can result if manufacturer's instructions are not followed.

Classroom

Ensure that trainees understand how to complete all of the required documentation. Answer any questions trainees have.

Refer to Figure 10. Review the general procedures to repair a hydraulic valve actuator/operator.

Safety

Discuss the consequences of failing to maintain the proper level of awareness while repairing actuators/ operators.

Safety

Stress that trainees adhere to the industry recommendations to ensure personal safety.

Explain the hazards associated with materials that contain asbestos.

Classroom

Review the abnormal operating conditions that trainees may encounter while performing maintenance on hydraulic actuators/operators.

Present a variety of abnormal operating conditions that can occur on hydraulic actuators/operators. Ask trainees to identify the source of the problem and determine the appropriate safety response.

Teaching Tip

See the Teaching Tip for Sections 5.7.0–5.7.3 at the end of this module.

Step 6 All applicable task-specific procedures must be followed. Following a task-specific procedure helps to ensure job safety, efficient task performance, standardization, and appropriate documentation. Procedures should also include work permit information and notification requirements (control center and both the upstream and downstream pump stations for main line valves, and others possibly being affected).

> **CAUTION**
> You must take care to follow manufacturer's instructions and industry recommendations in order to ensure personal safety, prevent mechanical damage to the actuator/operator, avoid internal damage to the valve being repaired, and preserve the service life of the equipment.

Step 7 Using all of the precautionary steps recommended by the manufacturer's instructions as applicable, remove the actuator/operator.

Step 8 Disassemble the hydraulic actuator/operator, following the manufacturer's instruction manual, to ensure proper procedures and safety.

> **WARNING!**
> Some gasket materials may contain asbestos. Company safety policy and procedures for working with asbestos must be followed.

Step 9 Inspect all gears, bearings, shafts, and other components for wear and damage, comparing them to the manufacturer's specifications for the unit.

Step 10 Reassemble the actuator per the manufacturer's specifications for safe and proper operation.

Step 11 Install the actuator per the manufacturer's or industry recommendations.

Step 12 Properly adjust the actuator/operator per the manufacturer's or company specifications.

Step 13 Notify the control center and any affected personnel to verify the integrity of the system, and return it to normal operation.

Step 14 Complete all required documentation of the repair and results.

> **NOTE**
> Documenting data for compliance, historical review, and trending is an important element of valve maintenance.

5.7.3 Abnormal Operating Conditions

When performing maintenance on hydraulic actuator/operators, situations may be encountered that require you to immediately respond in order to maintain proper operation of pipeline equipment. The following abnormal operating conditions could be encountered.

If a component failure or automated hydraulic valve system failure allows uncontrolled valve travel, de-energize the valve at the power source and/or operate the valve manually.

Should you be unexpectedly confronted with a hazardous liquid or gas such as a stem leak or piping connection leak, attempt to isolate the release. If the release cannot be isolated safely at the site, the control center should be notified to isolate the site or to shut down the line as quickly as possible. You must evacuate the area and only take defensive action using the proper personal protective equipment.

If an unexpected status change occurs, attempt to stop valve travel and return the valve to the proper position. If you are unable to immediately return the valve to the proper position, take action to avoid an emergency condition. Notify the control center as quickly as possible.

In case of fire and/or explosion or ignition of hazardous, flammable liquids or gases, you should evacuate the area and only take defensive action using proper personal protective equipment. Contact the control center to isolate or to shut down the line as quickly as possible.

Instructor's Notes:

Summary

Mechanical technicians must be trained and qualified for repairs on hydraulic equipment so that maintenance and repairs on the equipment will help to ensure safe and efficient operation of the pipeline. Strict discipline must be exercised in following manufacturer's recommendations and instructions.

To perform mechanical work on actuator/operators, you must be trained and qualified and have a strong mechanical aptitude, as well as a certain amount of problem-solving ability. Troubleshooting a valve actuator/operator can be a long, tedious process requiring much patience. Like any other machine, an actuator/operator can be complex. The ability to stick to the task enables a technician to make a well thought-out diagnosis. You can then complete a repair or adjustment efficiently and return the pipeline to normal. Discipline is required to follow the many complex procedures required by manufacturer's instructions and company policies and procedures.

The valve actuators/operators govern the pipeline, maintaining safe and efficient operation of valves. In order to do that, actuators/operators must be maintained in good operating condition. It is up to you to learn, study, and train to become qualified to maintain such important equipment.

Classroom

Go over the Review Questions for this module, and answer any questions trainees have.

Review Questions

1. The reservoir of a hydraulic system is the point where heat is _____ the fluid.
   a. dissipated from
   b. gauged in
   c. equalized with
   d. applied to

2. A _____ removes large particles from hydraulic fluid.
   a. control valve
   b. regulator
   c. strainer
   d. filter

3. The _____ in a hydraulic system converts the mechanical energy of the drive motor into hydraulic energy.
   a. accumulator
   b. pump
   c. fluid
   d. dissipator

4. Unusual noises from a pump may indicate _____.
   a. dissipation
   b. actuation
   c. isolation
   d. cavitation

5. All of the following procedures should be completed when inspecting a control valve *except* _____.
   a. ensuring that there are no leaks from the valve body
   b. operating the valve controls
   c. disassembling the valve to check for wear
   d. examining the valve controls to see if they are clean

6. Cylinders are called _____ because they produce force and motion in a straight line.
   a. pushers
   b. rotational motivators
   c. workhorses
   d. linear actuators/operators

7. Using a _____ diagram, you can trace the system and its components to determine where a problem exists.
   a. schematic
   b. three-dimensional
   c. quality-control
   d. pressure-flow

8. When tracing a system, begin at the _____.
   a. relief valve
   b. actuator/operator
   c. flow-control valve
   d. reservoir

Examination

Administer the Module Examination. Be sure to record the results of the Exam on Craft Training Report Form 200 and submit the results to the Training Program Sponsor.

Performance Testing

Have each trainee perform Performance Tasks 4 and 5 to your satisfaction. Fill out a Performance Profile Sheet for each trainee.

Ensure that all Performance Tests have been completed and Performance Profile Sheets for each trainee are filled out. Be sure to record the results of the testing on Craft Training Report Form 200 and submit the results to the Training Program Sponsor.

9. In a hydraulic system, movement depends on _____.

a. flow
b. pressure
c. temperature
d. drive

10. The speed at which an actuator moves depends on the _____.

a. flow rate
b. pressure
c. temperature
d. drive

11. Hydraulic hose fittings are selected according to the _____ range of the equipment.

a. fluid
b. pressure
c. flow
d. temperature

12. All gauges should have a(n) _____ valve installed to provide means of maintenance without shutting down the system.

a. control
b. hydraulic
c. regulator
d. isolation

13. Before beginning the repair of a hydraulic actuator/operator, you should notify the _____ and any other affected personnel.

a. control center
b. E&I department
c. company management
d. maintenance shop

14. The first step in adjusting a hydraulic valve actuator/operator should be to verify the _____.

a. location
b. number
c. nameplate data
d. brand

15. The final step in repairing a hydraulic actuator/operator is to complete the _____.

a. cleaning
b. documentation
c. notification
d. reinstalling

Instructor's Notes:

GLOSSARY

Trade Terms Introduced in This Module

Ampere (A or amps): A unit of measurement of electrical current. For example, one volt across one ohm of resistance causes a current flow of one ampere.

Cylinder: A device that converts fluid force into linear mechanical force by a piston and rod being pushed and pulled within a capped tube.

Limit switch: A switch designed to cut off power at or near the limit of travel of a moving object controlled by electrical means.

ACKNOWLEDGMENTS

Figure Credits

John Copeland, Industrial Development Corporation	307F10
Bill Coulter	307F07

Instructor's Notes:

MODULE 63307-02 — TEACHING TIPS

The following are suggested activities or instructional methods to help you teach the material in this AIG.

General

When you call on someone to answer a question, the rest of the class relaxes or even tunes out because they expect that the question and answer will take place only between you and the trainee you called on. Instead, use this technique to involve more trainees in answering questions and to keep them on their toes.

1. Ask trainees to define a term or explain a concept.
2. After one trainee has answered, ask a trainee seated nearby if the answer is right. Then ask whether a trainee in the back of the room agrees.
3. Ask trainees to explain why they think an answer is right or wrong.
4. Use the session to clear up incorrect ideas and encourage trainees to learn from their mistakes.

Sections 3.0.0–3.8.0

Inspecting Hydraulic System Components

Trainees have reviewed the components that are part of a hydraulic system. This exercise will familiarize trainees with the procedures to inspect these components. For this exercise, trainees will need their Trainee Guides and pencils and paper to take notes for discussion. Allow 20 minutes for this exercise.

1. Ask trainees to list the components to include in a hydraulic system preventive maintenance inspection.
2. Have trainees identify the role of each component in the system and explain the procedures for inspecting each component.
3. Suggest techniques for efficient inspections from your own experience working on hydraulic systems.
4. Answer any questions trainees have about inspecting hydraulic system components.

Sections 4.0.0–4.4.0

Performing Hydraulic System Troubleshooting Procedures

This exercise will help trainees develop their troubleshooting skills. For this exercise, you will need a variety of hydraulic schematic diagrams. Trainees will need their Trainee Guides and pencils and paper to take notes for discussion. Allow 30 minutes for this exercise.

1. Distribute schematic diagrams among the trainees. Divide the class into small groups according to the number of diagrams.
2. Assign each group a set of hypothetical problems and any relevant information that trainees need to diagnose the problems.
3. Have each group make a preliminary diagnosis. Review each group's troubleshooting techniques.
4. Have trainees identify the potential cause of the problem using the basic steps to diagnose a problem with a hydraulic valve actuator/operator. Review and discuss repair options for each problem.
5. Answer any questions trainees may have about performing hydraulic system troubleshooting procedures.

Sections 5.0.0–5.6.0 *Adjusting and Repairing Hydraulic System Components*

Trainees have reviewed the different types of hydraulic system components and the basic steps to repair them. This exercise will emphasize the importance of following the manufacturers' instruction manuals. For this exercise, you will need instruction manuals for a variety of hydraulic system components. Trainees will need their Trainee Guides and pencils and paper to take notes for discussion. Allow 30 minutes for this exercise.

1. Distribute copies of manufacturers' instruction manuals for a variety of hydraulic system components. Divide the class into small groups according to the number of manuals.
2. Ask each group to review the manufacturers' instructions for one of the procedures covered to this point:
   - Preparing the system for shutdown and repair
   - Repairing and overhauling pumps
   - Repairing and overhauling control valves
   - Repairing and overhauling actuators/operators
   - Replacing hoses
   - Replacing gauges
3. Have each group compare the manufacturers' instructions with the general guidelines. Identify major differences, similarities, additional precautions, and/or specific requirements. Review and discuss trainees' findings.
4. Ensure that trainees understand how to perform the procedures for adjusting and repairing hydraulic system components. Answer any questions trainees have.

Sections 5.7.0–5.7.3 *Hydraulic Valve Actuators/Operators*

This exercise will give trainees opportunities to observe repair and/or adjustment procedures for a hydraulic valve actuator/operator. For this exercise, you will need to locate a nearby site or control center where a repair or adjustment procedure is scheduled. Trainees will need their Trainee Guides and pencils and paper to take notes for discussion. Allow at least 60 minutes for this exercise.

1. Schedule a visit to a local site where a repair or adjustment procedure is scheduled.
2. Ask the operations personnel to explain the scheduled procedure to the trainees.
3. Ask trainees to observe the adjustment and/or repair procedures and to assist with the procedures, as appropriate.
4. Following the site visit, review trainees' observations.
5. Answer any questions trainees have about adjusting and repairing hydraulic valve actuators/operators.

MODULE 63307-02 — ANSWERS TO REVIEW QUESTIONS

Answer	Section
1. a	3.1.0
2. c	3.2.0
3. b	3.3.0
4. d	3.3.0
5. c	3.5.0
6. d	3.7.0
7. a	4.1.0
8. d	4.1.0
9. a	4.3.0
10. a	4.3.0
11. b	5.5.2
12. d	5.6.0
13. a	5.7.1
14. a	5.7.1
15. b	5.7.2

CONTREN™ LEARNING SERIES — USER UPDATES

The NCCER makes every effort to keep these textbooks up-to-date and free of technical errors. We appreciate your help in this process. If you have an idea for improving this textbook, or if you find an error, a typographical mistake, or an inaccuracy in NCCER's Contren™ textbooks, please write us, using this form or a photocopy. Be sure to include the exact module number, page number, a detailed description, and the correction, if applicable. Your input will be brought to the attention of the Technical Review Committee. Thank you for your assistance.

Instructors – If you found that additional materials were necessary in order to teach this module effectively, please let us know so that we may include them in the Equipment/Materials list in the Instructor's Guide.

Write: Curriculum Revision and Development Department
National Center for Construction Education and Research
P.O. Box 141104, Gainesville, FL 32614-1104

Fax: 352-334-0932

E-mail: curriculum@nccer.org

Craft ______________________ Module Name ______________________

Copyright Date __________ Module Number __________ Page Number(s) __________

Description ______________________

____

____

____

(Optional) Correction ______________________

____

____

(Optional) Your Name and Address ______________________

____

____

Module 63308-02

Maintain, Troubleshoot, and Repair Electric Valve Actuators/Operators and Systems

(CT 19.5 and 21.5)

Maintain, Troubleshoot, and Repair Electric Valve Actuators/Operators and Systems Annotated Instructor's Guide

Module 63308-02

MODULE OVERVIEW

This module discusses preventative maintenance procedures on electric valve actuators/operators and systems. Trainees will learn to inspect, troubleshoot, adjust, and repair electric actuator/operator components.

PREREQUISITES

Please refer to the Course Map in the Trainee Module. Prior to training with this module, it is recommended that the trainee shall have successfully completed the following:

Core Curriculum; Pipeline Core, Modules 66101-02 and 66102-02; Pipeline Mechanical Levels One and Two; Pipeline Mechanical Level Three, Modules 63301-02 through 63307-02

OBJECTIVES

Upon completion of this module, the trainee will be able to:

1. Perform preventative maintenance procedures on electric actuators/operators.
2. Inspect electric actuator/operator components.
3. Troubleshoot problems with electric actuators/operators.
4. Adjust electric actuator/operator components (CT 19.5).
5. Repair electric actuator/operator components (CT 21.5).

PERFORMANCE TASKS

Under the supervision of the instructor, the trainee should be able to:

1. Perform preventative maintenance procedures on electric actuators/operators.
2. Inspect electric actuator/operator components.
3. Troubleshoot problems with electric actuators/operators.
4. Adjust electric actuator/operator components (CT 19.5).
5. Repair electric actuator/operator components (CT 21.5).

NCCER STANDARDIZED CRAFT TRAINING PROGRAM

The National Center for Construction Education and Research (NCCER) provides a standardized national program of accredited craft training. Key features of the program include instructor certification, competency-based training, and performance testing. The program provides trainees, instructors, and companies with a standard form of recognition through a National Craft Training Registry. The program is described in full in the *Guidelines for Accreditation*, published by the NCCER. For more information on standardized craft training, contact the NCCER by writing us at P.O. Box 141104, Gainesville, FL 32614-1104; calling 352-334-0911; or e-mailing info@nccer.org. More information may be found at our Web site, www.nccer.org.

HOW TO USE THIS ANNOTATED INSTRUCTOR'S GUIDE

Each page presents two sections of information. The larger section displays each page exactly as it appears in the Trainee Module. The narrow column ties suggested trainee and instructor actions to each page and provides icons to call your attention to material, safety, audiovisual, or testing requirements. The bottom of each page includes space for your notes.

Teaching Tip

If you see the Teaching Tip icon, that means there is a teaching tip associated with this section. Also refer to any suggested teaching tips at the end of the module.

SAFETY CONSIDERATIONS

Ensure that the trainees are equipped with appropriate personal protective equipment. Emphasize the importance of following all safety precautions and procedures when working with power tools.

PREPARATION

Before teaching this module, you should review the Module Outline, Objectives, Performance Tasks, and the Materials and Equipment List. Be sure to allow ample time to prepare your own training or lesson plan and gather all required equipment and materials.

MATERIALS AND EQUIPMENT LIST

- Transparencies
- Markers/chalk
- Blank acetate sheets
- Transparency pens
- Pencils and scratch paper
- Module Examinations*
- Performance Profile Sheets*
- Copies of Performance Verifications for Covered Tasks 19.5 and 21.5**
- Overhead projector and screen
- Whiteboard/chalkboard
- Appropriate personal protective equipment
- Copies of *49 CFR Part 192* (Gas) and/or *49 CFR Part 195* (Liquid)
- Copies of your company policy and procedures manual
- Sample valve maintenance logs
- Computer-generated work orders
- Grease guns
- Valve stems
- Copies of *Figure 10* with the legend covered
- Tools and materials to
  - Apply external and internal lubricants
  - Disassemble bolt-on and machine-fitted spur-gear attachments
  - Remove worm gear assemblies
- Tools and materials to disassemble and assemble
  - Single reduction worm gears
  - Actuator/operator bearing assemblies
  - Bevel gear actuator/operators
- Lubricants
- Brushes
- Cleaning solvents
- Shop cloths
- Sample P&IDs
- Inspection checklists
- Bolt-on and machine-fitted spur gear attachments
- Manufacturer's instruction manuals for limit switches
- Torque switch
- Stem nut assembly

* Located in the Test Booklet.
**Available for download from www.nccer.org.

TEACHING TIME FOR THIS MODULE

An outline for use in developing your lesson plan is presented below. Note that each Roman numeral in the outline equates to one session of instruction. Each session has a suggested time period of 2½ hours. This includes 10 minutes at the beginning of each session for administrative tasks and one 10-minute break during the session. Approximately 15 hours are suggested to cover *Maintain, Troubleshoot, and Repair Electric Valve Actuators/Operators and Systems (CT 19.5 and 21.5).* You will need to adjust the time required for hands-on activity and testing based on your class size and resources.

Topic	Planned Time
Session I. Preventive Maintenance	
A. PM Methods	________
B. Scheduling	________
1. Laboratory	________
C. Manufacturer's Instructions and Warranties	________
D. Training	________
E. Documentation	________
1. Performance Testing	________
Session II. Lubricating Electric Actuators/Operators	
A. Lubrication Methods	________
B. Applying External Lubricants	________
1. Laboratory	________
C. Applying Lubrication Internally	________
Session III. Inspecting Electric Actuator/Operator Components	
A. Locating the Actuator/Operator	________
B. Inspection Checklists	________
C. Verifying Actuator/Operator Control	________
D. Painting	________
E. Caring for Out-of-Service Equipment	________
1. Performance Testing	________
Session IV. Troubleshooting Electric Valve Actuators/Operators	
A. Preliminary Diagnoses	________
B. Non-Actuator/Operator Problems	________
C. Basic Procedures for Troubleshooting	________
D. Repair Options	________
1. Performance Testing	________
Session V. Adjusting and Repairing Electric Actuator/Operator Components, Part One	
A. Removing an Electric Actuator/Operator from the Valve Stem	________
B. Removing the Spur Gear Attachment	________
C. Removing the Worm Gear Assembly	________
1. Laboratory	________
D. Adjusting the Valve Travel	________
1. Laboratory	________

Session VI. Adjusting and Repairing Electric Actuator/Operator Components, Part Two

A. Replacing the Stem Nut ____________
   1. Laboratory ____________
B. Replacing the Motor ____________
C. Installing Electric Actuators/Operators ____________
   1. Laboratory ____________
D. Performing Adjustments and Repairs ____________
E. Summary
   1. Summarize module ____________
   2. Answer questions ____________
F. Module Examination ____________
   1. Trainees must score 70% or higher to receive recognition from the NCCER.
   2. Record the testing results on Craft Training Report Form 200 and submit the results to the Training Program Sponsor.
G. Performance Testing ____________
   1. Trainees must perform each task to the satisfaction of the instructor to receive recognition from the NCCER.
   2. Record the testing results on Craft Training Report Form 200 and submit the results to the Training Program Sponsor.

Module 63308-02

Maintain, Troubleshoot, and Repair Electric Valve Actuators/Operators and Systems

(CT 19.5 and 21.5)

Instructor's Notes:

Instructor's Notes:

Introduction to the Pipeline Operator Qualification Rule, API, and NCCER

It could be said that petroleum products keep America running. Petroleum products like oil and natural gas are the principal sources of energy in the United States and are used in many manufacturing and chemical processes. Petroleum products are primarily transported by pipeline. There are more than a million miles of liquid and natural gas pipeline in this country, ranging in diameter from two inches to four feet. These pipelines bring crude oil and natural gas from production fields to processing plants and refineries—and ultimately to the consumer.

Pipeline Operator Qualification

Petroleum products are considered hazardous, so pipeline personnel must take extra precautions and follow specific procedures when working with them. To help ensure the safe handling and transportation of petroleum products, the United States Department of Transportation (DOT) mandates safe practices as federal regulations.

One such regulation, the Operator Qualification (OQ) rule, requires pipeline operators to develop a written qualification plan to evaluate their personnel and their contract personnel (see *49 CFR Part 192 Subpart N—Gas*, and *49 CFR Part 195 Subpart G—Liquid*). The evaluation hinges on the operator's ability to perform certain covered tasks and to be able to recognize and respond to abnormal operating conditions that may be encountered while performing these covered tasks.

Covered Tasks

A pipeline performance task is deemed *covered* if the activity passes the following four-part test:

1. Is it performed on a pipeline facility?
2. Is it an operations or maintenance task?
3. Is it performed as a requirement of *49 CFR Part 195 (Liquid)* or *49 CFR Part 192 (Gas)*?
4. Does it affect the operation or integrity of the pipeline facility?

If the answer to all four questions is *yes*, then the activity is considered *covered* and must figure into the evaluation process of pipeline personnel.

American Petroleum Institute

The OQ rule and its related covered tasks provided the catalyst to develop a unified pipeline training and assessment program—so that pipeline operators and owners have ready access to the assessment and training tools they need. This effort was the direct result of the vision and leadership provided by The American Petroleum Institute (API) and API members.

Efforts to unify the industry are nothing new to API. Indeed, the impetus for forming the Institute in 1919 was the need to standardize engineering specifications for drilling and production equipment. Today, the most pressing issues revolve around public perceptions and government policies toward the industry. Speaking with one voice on these issues has become as essential as having interchangeable parts in the field. API's members determine the petroleum industry's positions on public policy as well as the standards that govern its day-to-day operations worldwide.

Consortium on Operator Qualification

API assembled a focus group whose primary purpose was to identify covered tasks and provide compliance recommendations to pipeline operators.

The Consortium on Operator Qualification (COOQ) became the guiding force for the identification and development of covered tasks for the liquid pipeline industry. Their work product, API publication 1161, *Guidance Document for the Qualification of Liquid Pipeline Personnel*, formed the basis for the resulting Pipeliner Training and Assessment Program.

The covered task list created in API publication 1161 has now evolved to include gas covered tasks and is maintained by NCCER. The most current covered task list can be downloaded at www.nccer.org.

National Center for Construction Education and Research

With the guidance and support of API, its members, and the Consortium on Operator Qualification, the National Center for Construction Education and Research (NCCER) was able to design and implement the Pipeliner Training and Assessment Program.

NCCER is a not-for-profit 501(c)(3) education foundation established in 1995 by the world's largest and most progressive construction and maintenance contractors and national contractor associations. NCCER was founded to address the

severe workforce shortage facing the construction and maintenance industries and to develop a standardized training process and curricula. Today, NCCER is supported by hundreds of leading construction and maintenance companies, manufacturers, and national associations and developed more than 35 construction and maintenance titles.

NCCER also maintains a National Registry, which contains the records of trainees and instructors who have received NCCER training. All completion information for every trainee is sent to the NCCER and filed within the Registry. The Registry can then confirm training and skills for workers as they move from company to company, state to state, or even within their own company.

Standardized Curriculum (NCCER's Contren™ Learning Series)

NCCER curricula are developed using the expertise of industry subject matter experts (SMEs). The use of SMEs as an integral part of every new development or revision ensures that the curriculum is current, industry-driven, and indicative of industry practices. The development and/or revision process is facilitated from start to finish by NCCER project managers, who work closely with the SMEs and a select team of technical writer(s) on the development and presentation of the curriculum.

Pipeline: Five Training Paths

The pipeline curriculum focuses on essential skills and competencies within five skill areas:

- Corrosion Control
- Electrical and Instrumentation
- Maintenance
- Mechanical
- Operations (Field and Control Center)

SMEs from the pipeline industry have determined the skills and competencies needed based on API's covered tasks and the consensus of the group/industry. Specific company information is generally not incorporated into the curriculum unless the SMEs determine that the information is the industry standard.

Training Modules and Covered Tasks

The smallest unit of instruction in the curriculum is a *module*. Each curriculum is composed of modules (the equivalent of stand-alone chapters) that are logically sequenced into complete *levels* of instruction. Every module is task-oriented and requires the successful completion of a written (knowledge-based) examination and a performance (hands-on) test. Modules may be individually selected and/or combined for efficient task training. Modules in the pipeline curricula, to the extent possible, will align with API-developed covered tasks. While not all modules will relate to covered tasks, all covered tasks will relate to at least one module.

Benefits of Training to the Pipeline Marketplace

Applying NCCER's standardized training process allows pipeline owners and contractors in the United States to establish the benchmark for acceptable skill and competency levels within the pipeline industry. This process will assist training organizations in developing and/or improving standardized pipeline training programs that effectively match their resources and operating conditions while responding to regulatory demands and to the needs of the industry. The process will also recognize training organizations that meet the minimum standards of competence, warranting regulatory, public, and professional confidence. This in turn will foster national unity within the pipeline industry for the mutual benefit of individual technicians, owners, contractors, and ultimately pipeline-users and society.

Additional Resources

To review the Pipeliner Training and Assessment Program Implementation and Resource Guide and the latest version of pipeline covered tasks, visit NCCER's Web site at www.nccer.org, or contact:

National Center for Construction Education and Research
PO Box 141104
Gainesville, FL 32614-1104
(888) 622-3720
Web site: www.nccer.org

For more information on API, please contact:

American Petroleum Institute
1220 L Street, Northwest
Washington, D.C. 20005
(202) 682-8000
Web site: www.api.org

To order *API Pub 1161*, contact Global Engineering Documents by calling 1-800-854-7149 or visiting their Web site at www.global.his.com.

Additional sources of information on the pipeline operator qualification regulation include:

DOT – Office of Pipeline Safety
www.ops.dot.gov

Transportation Safety Institute
www.tsi.dot.gov

Instructor's Notes:

COURSE MAP

This course map shows all of the modules in the third level of the Pipeline Mechanical curriculum. The suggested training order begins at the bottom and proceeds up. Skill levels increase as you advance on the course map. The local Training Program Sponsor may adjust the training order.

PIPELINE MECHANICAL LEVEL THREE

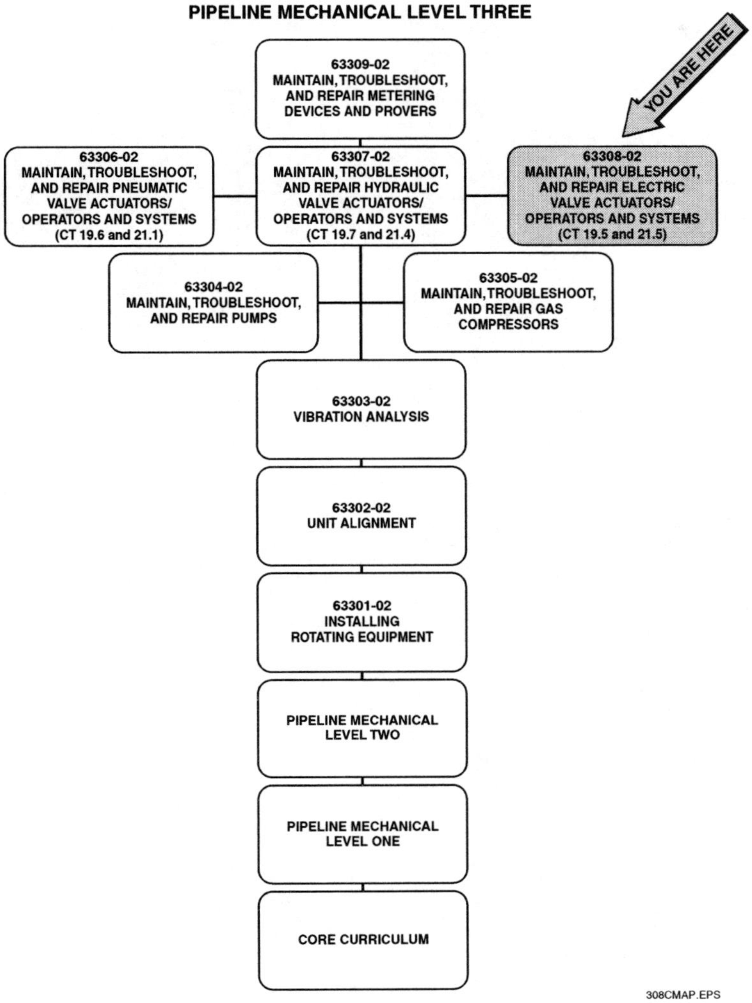

308CMAP.EPS

Homework

Assign reading of Module 63308-02.

Instructor's Notes:

MODULE 63308-02 CONTENTS

MODULE 63308-02 CONTENTS (Continued)

Instructor's Notes:

Figures

Instructor's Notes:

MODULE 63308-02

Maintain, Troubleshoot, and Repair Electric Valve Actuators/Operators and Systems (CT 19.5 and 21.5)

Objectives

When you have completed this module, you will be able to do the following:

1. Perform preventative maintenance procedures on electric actuators/operators.
2. Inspect electric actuator/operator components.
3. Troubleshoot problems with electric actuators/operators.
4. Adjust electric actuator/operator components (CT 19.5).
5. Repair electric actuator/operator components (CT 21.5).

Prerequisites

Before you begin this module, it is recommended that you successfully complete the following: Core Curriculum; Pipeline Core, Modules 66101-02 and 66102-02; Pipeline Mechanical Level One; Pipeline Mechanical Level Two; Pipeline Mechanical Level Three, Modules 63301-02 through 63305-02.

Required Trainee Materials

1. Pencil and paper
2. Appropriate personal protective equipment

1.0.0 ◆ INTRODUCTION

A valve actuator/operator is a mechanical device that is used to open, close, or position a valve manually or automatically. Manual actuators/operators come in the form of handwheels, levers, and wrenches that are attached to the valve stem, either directly or by means of a gearbox. Automatic actuators/operators are used when the valve is remote from the main work area, when the frequency of operation would require unreasonable human effort, or if rapid opening and closing of the valve is required. In some installations, automatic remote control actuator/operators are installed as a matter of convenience, reducing the number of employees required for facility operation. Large valves that require great effort to turn must be equipped with some type of powered mechanism, such as electric, pneumatic, or hydraulic motors.

In discussions in the field, power units are often included and referred to as an integral part of the actuator/operator assembly. However, this is technically incorrect. The power units are not actually a part of the operator. They are the power supply that moves the operator.

Note that in some locations the gearbox is called an actuator. In other locations, it is called an actuator/operator. Technically speaking, the operator is the gearbox unit or hand device that actually moves the valve. The actuator is the power unit that is added to turn the operator automatically. In this module, the term actuator/operator will be used collectively to refer to the entire unit.

You must also understand that many actuators/operators are devices that experience heavy loads although they may be idle for many days at a time. Some valves, if turned manually, would require so much time and effort that an opening or closing operation would be self-defeating if it were not an emergency. For example, some large valves require more than 10,000 handwheel revolutions from open to close. The time that would be required for a power unit to open or close such a valve would be extensive. For that reason, some mechanical

Materials

Ensure that you have everything required to teach the course. Check the Materials and Equipment List at the front of this module.

Classroom

Explain that terms shown in bold are defined in the Glossary at the back of this module.

Audiovisual

Show Transparency 1, Course Objectives.

Show Transparency 2, Performance Tasks.

Teaching Tip

See the general Teaching Tip at the end of this module.

Classroom

Explain how electric valve actuators/operators are used on a pipeline.

Review the difference between the actuator and the operator and how they both function.

Classroom

Discuss the duties involved in a preventive maintenance program. Emphasize the importance of conducting a detailed analysis of past equipment failures.

Refer to *49 CFT 192.605*. Explain how it applies to maintenance activities on the pipeline.

Laboratory

Distribute copies of valve maintenance logs. Ask trainees to identify information that you specify. Ensure that trainees understand how to document maintenance procedures.

Classroom

Review the basic methods for implementing preventive maintenance procedures. Answer any questions trainees have.

Demonstration

Distribute copies of computer-generated work orders. Demonstrate how to review the documents and how to determine whether work is being completed according to required schedules.

technicians and crews carry gasoline engine, air, or electric powered portable valve actuators/operators to assist them with cycling valves.

Such valve equipment demonstrates the fact that an actuator/operator must be durable and in good condition to carry such loads. You must be able to maintain, troubleshoot, and repair the actuators/operators so that they perform in a reliable and efficient manner. With proper maintenance and repair, valve actuators/operators can give long, reliable service with minimal downtime.

2.0.0 ◆ PREVENTIVE MAINTENANCE

Preventive maintenance (PM) is the periodic inspection of facility assets and equipment to uncover conditions that may lead to failure or shorten anticipated life. The long-term goal for a good PM program is to increase reliability, decrease unplanned equipment downtime, reduce overall maintenance costs, and support improvements in quality and production. The duties involved in PM are generally grouped into the following activities:

- Inspecting components
- Lubricating components
- Replacing components
- Cleaning components
- Adjusting components
- Documenting work accomplished

The start of any PM program involves performing a detailed analysis of past equipment failures, including their costs, their causes, any remedial actions, and the changing demands on the equipment. A detailed maintenance history of all of the equipment within the system must be developed. After this information is gathered, it is divided and prioritized according to the maintenance item's effect, ranging from maximum to minimum. The criteria used to determine effect includes safety, product costs, overhead costs, and equipment replacement costs.

A good PM program targets the items that cause economic damage as first priority. The second priorities are the maintenance items that need to be done but can wait until the first priorities have been dealt with. The lowest priority maintenance items are those that have little monetary and absolutely no safety impact; these can be attended to after they have failed. *Figure 1* shows a valve maintenance log, including actuators/operators.

In a general application of *49 CFR 192.605*, each pipeline operator is required by federal law to prepare and follow a manual of written procedures for regular maintenance of valve actuator/operators. This manual must be updated at least once every fifteen months. Applicable portions of it must be kept at locations where work is performed. Such manuals must provide for safety of operations during these maintenance procedures. You must follow company policies and procedures during maintenance activities.

The law only applies to maintenance activities on the pipeline that could result in danger to personnel, the public, facilities, or the environment. Inspections, diagnostics, repair of units located in areas isolated from the pipeline, bench work in a shop, and how information gathered about such units is managed is not included.

However, sound management practice requires that all equipment be maintained in good working condition. It is not only a matter of safe and efficient operation but also economic operation. Valve actuators/operators represent a sizeable investment that must be protected.

2.1.0 PM Methods

PM methods are systematic maintenance procedures that reduce emergencies, downtime, and costly production delays. PM procedures vary from one piece of pneumatic equipment to another, but the basic principles are the same. The basic methods for implementing PM procedures can be remembered using the word CARE:

- C = *Cleaning* the system and equipment
- A = *Adjusting* the components
- R = *Replacing* disposable components
- E = *Examining* or inspecting equipment

2.2.0 Scheduling

Scheduling PM is an important element in the PM program. PM performed too frequently increases costs, as does PM that is performed too infrequently. Companies use computerized systems, card files, and maintenance logs to assign individuals to manage the responsibilities of maintenance. Some company computers generate work orders at regular intervals for the equipment to make certain that legally required schedules are met and that equipment in the most isolated locations is included. Schedules must be reviewed on a continual basis to check their accuracy and effectiveness. The schedule is always changing and depends on consistent documentation.

The maintenance schedule, formed within the company, specifies all the recommended maintenance required to keep all equipment in good operating condition.

Instructor's Notes:

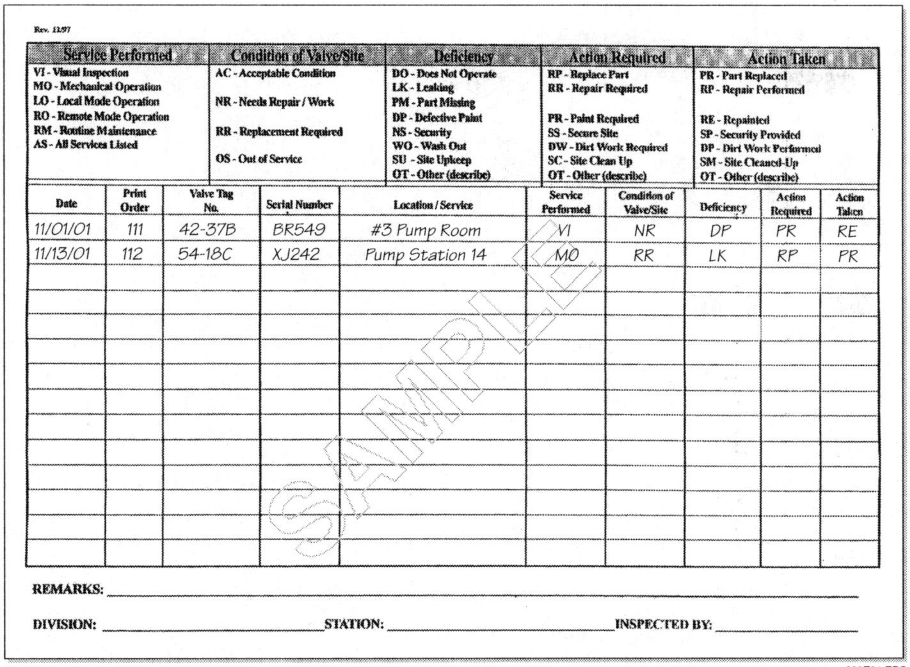

Rev. 11/97

Service Performed	Condition of Valve/Site	Deficiency	Action Required	Action Taken
VI - Visual Inspection MO - Mechanical Operation LO - Local Mode Operation RO - Remote Mode Operation RM - Routine Maintenance AS - All Services Listed	AC - Acceptable Condition NR - Needs Repair / Work RR - Replacement Required OS - Out of Service	DO - Does Not Operate LK - Leaking PM - Part Missing DP - Defective Paint NS - Security WO - Wash Out SU - Site Upkeep OT - Other (describe)	RP - Replace Part RR - Repair Required PR - Paint Required SS - Secure Site DW - Dirt Work Required SC - Site Clean Up OT - Other (describe)	PR - Part Replaced RP - Repair Performed RE - Repainted SP - Security Provided DP - Dirt Work Performed SM - Site Cleaned-Up OT - Other (describe)

Date	Print Order	Valve Tag No.	Serial Number	Location / Service	Service Performed	Condition of Valve/Site	Deficiency	Action Required	Action Taken
11/01/01	111	42-37B	BR549	#3 Pump Room	VI	NR	DP	PR	RE
11/13/01	112	54-18C	XJ242	Pump Station 14	MO	RR	LK	RP	PR

REMARKS: ______

DIVISION: ______ STATION: ______ INSPECTED BY: ______

SAMPLE

308F01.EPS

Figure 1 ◆ Valve maintenance log.

2.3.0 Manufacturer's Instructions and Warranties

Preventive maintenance must be performed according to manufacturer's instructions and warranty requirements. Many actuators/operators are supplied with warranties, and in order to retain that warranty, they must be maintained according to the manufacturer's instructions. If it is still under warranty, the work will need to be carried out only as authorized by the manufacturer. In some cases, the manufacturer may require that the actuator/operator to be returned to the factory for warranty service. The record for the particular actuator/operator should be reviewed to see if the warranty is still in effect before any work is attempted. The specified maintenance procedures should be followed to ensure proper operation and to maintain warranty requirements.

2.4.0 Training

You should become completely familiar with the types of actuators/operators being used and the types of maintenance products required for their job assignments. An ongoing training program with regular updates and reviews is considered essential by some companies. Training should also include the use of proper methods of documentation.

2.5.0 Documentation

Records are also a very important element of the PM system. Records are maintained to build a detailed and accurate historical file of the equipment in relation to hours of operation, parts usage, and cost of maintenance. The records system must store accurate information in the following categories:

- Listing of each piece of equipment requiring PM
- Location of each piece of equipment
- Running history of maintenance actions on the equipment, including the parts replaced
- Time expended on PM
- Cost of replacement parts

Classroom

Emphasize the importance of adhering to warranties and of properly documenting all maintenance activities.

Teaching Tip

See the Teaching Tip for Sections 2.0.0–2.5.0 at the end of this module.

Performance Testing

Have each trainee perform Performance Task 1 to your satisfaction. Fill out a Performance Profile Sheet for each trainee.

Homework

Have trainees review Sections 3.0.0–3.3.4 for the next session.

Materials

Ensure that you have everything required for the laboratory during this session.

Classroom

Review the problems prevented by properly lubricating electric valve actuators/operators.

Compare the methods for external lubrication with those for internal lubrication.

Audiovisual

Show Transparency 3 (Figure 2). Review the layout of a single reduction worm gear.

Classroom

Refer to Figure 3. Ask a trainee to explain why some stems may require hand lubrication with a brush.

3.0.0 ◆ LUBRICATING ELECTRIC ACTUATORS/OPERATORS

Lubrication is the major preventive maintenance task performed on electric actuators/operators. Proper lubrication protects them from rust, corrosion, damage from freezing moisture, and seizure of parts. If the electric actuator/operator is exposed to rain, snow, ice, or immersion, moisture may destroy lubrication if the wrong type of grease is used. If the proper grease level is not maintained in some actuators/operators, moisture may form in the void and cause the actuator/operator to freeze in cold weather. Moisture will contribute significantly to the formation of rust and corrosion. In areas prone to flooding or tidal activities, blowing sand or dust, or severe temperature, special lubricants or lubrication schedules may be required. Factors that should be considered when selecting lubricants include:

- Climate
- Exposure of the component
- Application of the component
- Manufacturer's recommendations
- Company policy

Many manufacturers recommend that units seldom used or in storage should be cycled on a monthly schedule. The fittings should be filled with fresh grease from a grease gun. Other lubrication must be kept up to appropriate levels for reliable operation.

3.1.0 Lubrication Methods

Properly lubricated electric actuators/operators move freely without binding, requiring less energy to turn them. Electric actuator/operator lubrication methods can be separated into two types: external lubrication, and internal lubrication. External lubrication applies mostly to manual valve actuators/operators with threads or working parts exposed to the elements. Internal lubrication applies more to those actuators/operators using gear drive or automatic drive units.

3.1.1 External Lubrication

Components of equipment exposed to the elements tend to dry out, harden, and deteriorate as time passes. Dirt and grime collect in unacceptable places and cause problems with movement, increasing wear, corrosion, and damage to machined and polished surfaces. Periodic cleaning and lubrication of certain parts of the electric actuator/operator prevents or at least slows down the process, greatly extending the life and efficiency of components.

3.1.2 Internal Lubrication

Electric valve actuators/operators employing a closed gearbox design require that lubrication be injected or packed inside the gearbox housing or case in such a way that it reaches all bearings, bushings, sleeves, and gears. Some units are designed so that the gears and shafts run in an immersion. These must be kept at manufacturer's recommended levels in the housing to displace moisture. Other units require that the lubricant or grease must be hand packed into the bearings and sleeves as the unit is assembled. Regular lubrication with a gun or repacking by hand is required to maintain lubrication levels in such designs. *Figure 2* shows a layout of a single reduction worm gear with grease fittings.

3.2.0 Applying External Lubricants

Applying grease with a brush and/or grease gun is a common procedure performed on stem-type valves on the pipeline. This procedure applies to a variety of stem-type valves with manual actuators/operators and some electric actuators/operators. Keep in mind that a grease gun uses hydraulic pressure from the small pump in the gun to move grease into the fitting. In cold weather, great pressure may be required to apply grease to the actuators/operators. Pumping the grease from the gun may require extra effort. At the same time, grease being brushed onto a stem or other part may be much more difficult to apply in cold weather. *Figure 3* shows a stem requiring lubrication with a brush.

3.2.1 Preparing for Shutdown and Repair

Every company has safety policies and procedures that must be followed when performing maintenance and repairs on electric valve actuators/operators. These procedures must be followed before every inspection and repair procedure on electric actuators/operators discussed in this module:

Step 1 Secure all required work orders, permits, and PPE.

Step 2 Follow all company safety policies and procedures.

Step 3 Verify the location of the electric valve actuator/operator to be inspected.

Instructor's Notes:

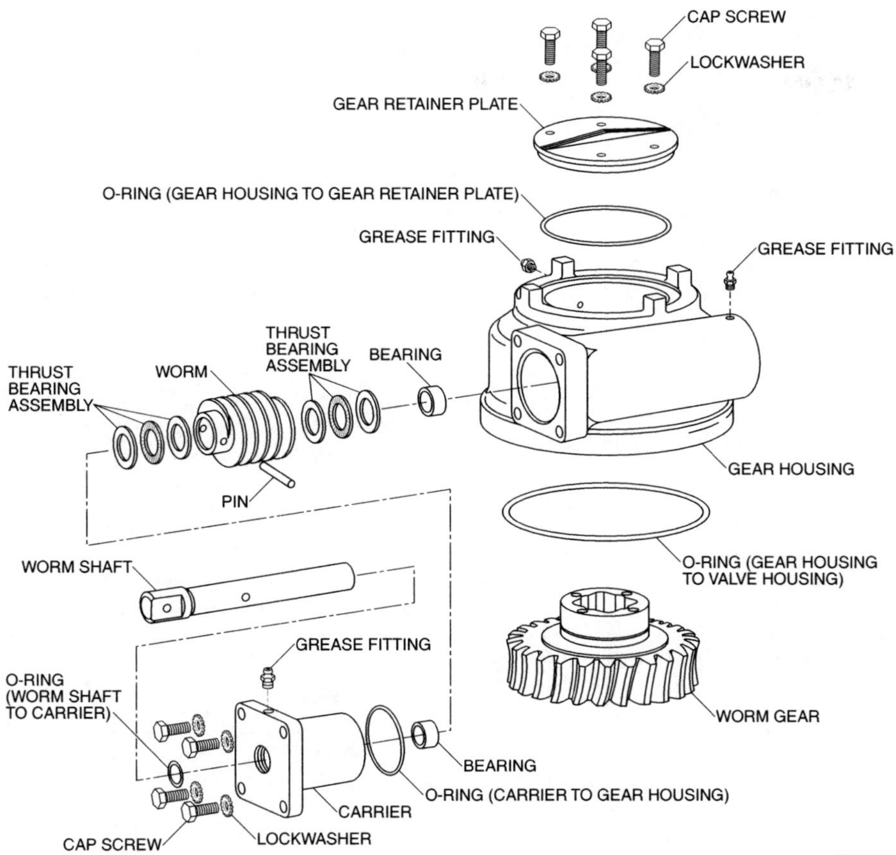

Figure 2 ◆ Layout of a single reduction worm gear with grease fittings.

Step 4 Verify the accessibility of the actuator/operator.

Step 5 Verify the actuator/operator number and nameplate data.

Step 6 If the nameplate is missing or illegible, replace it according to company policy and procedures.

Step 7 Notify the control center that work is beginning.

3.2.2 Application Procedures

Applying grease to a valve stem and electric actuator/operator includes some or all of the following steps:

Step 1 Verify that conditions are safe for personnel to perform maintenance.

Step 2 Turn the valve to the completely open position so that the stem threads are extended as far as possible.

Classroom

Review the procedures to prepare an electric actuator/operator for shutdown.

Explain how to apply grease to a valve stem using a grease gun. Go over any questions trainees have.

Audiovisual

Show Transparency 4 (Figure 4). Discuss techniques for applying grease to the grease fittings.

Classroom

Refer to Figure 5. Explain how to remove the individual bearings and races from the housing.

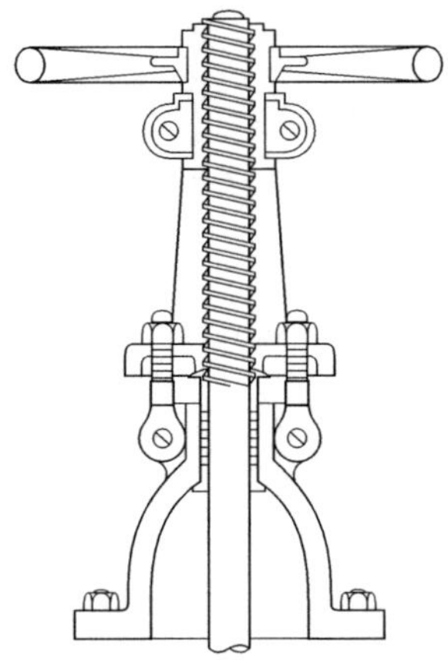

Figure 3 ◆ Stem requiring lubrication with a brush.

Step 3 Clean the threads of all rust, dirt, and grime with a wire brush, solvent, and shop cloths.

Step 4 Wipe off all residues from moving parts and seal areas with clean shop cloths.

Step 5 Locate and remove safety caps from grease fittings on the sides of the valve body if applicable.

Step 6 Pump grease gun lever several strokes and operate valve stem an inch in each direction to spread grease on seals.

Step 7 Pump grease gun lever several strokes on each fitting. *Figure 4* shows an actuator/operator with grease fittings.

Step 8 Operate the valve through several complete cycles to spread lubrication.

Step 9 Move the valve to full open position and pump grease gun several more strokes to fill cavity.

NOTE
Replace safety caps on all grease fittings.

Step 10 Brush the recommended grease heavily onto the outer threads of the stem for those valves that do not have grease fittings or do have exposed stem sections.

Step 11 Cycle the valve several times.

Step 12 Remove the handwheel nut and handwheel, exposing the bearing housing.

Step 13 Remove the setscrew holding the bearing assembly in place. *Figure 5* shows a typical actuator/operator bearing assembly.

Step 14 Remove the individual bearings and races from the housing.

Step 15 Clean and inspect the handwheel bearings and races.

Step 16 Hand-pack the handwheel bearings and races.

Step 17 Reassemble the races and bearings into the housing.

Step 18 Slide the bearing housing back over the stem.

Step 19 Tighten the assembly-housing setscrew.

Step 20 Reinstall the handwheel and nut.

Step 21 Torque the handwheel nut to the manufacturer's specification.

Step 22 Remove all lockout devices and tags.

Step 23 Return the valve to the appropriate position.

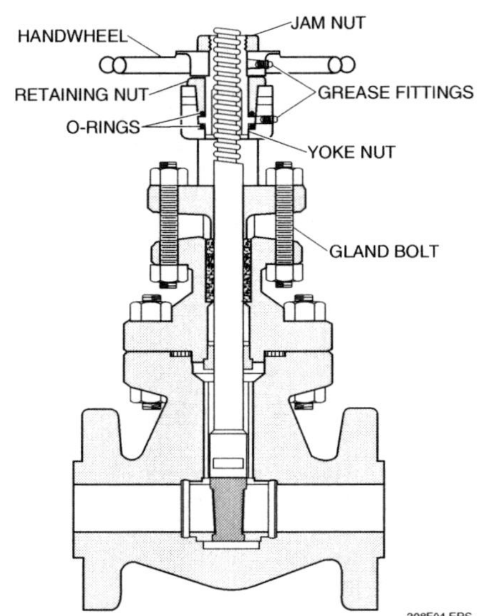

Figure 4 ◆ Actuator/operator with grease fittings.

Instructor's Notes:

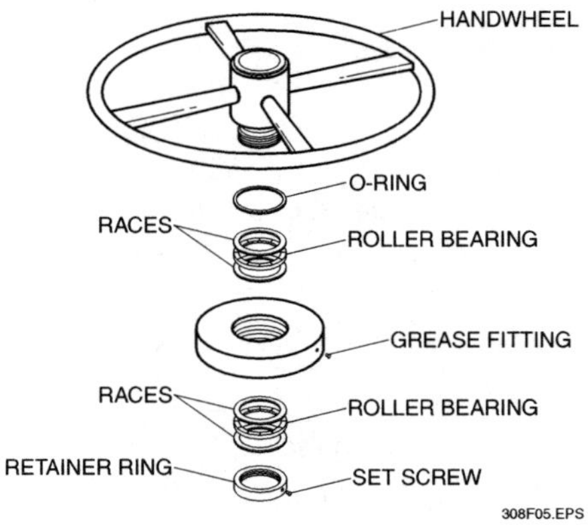

Figure 5 ◆ Actuator/operator bearing assembly.

Once the actuator/operator and valve are returned to service, the control center operator must be notified so that any affected pipeline section can be started up again. You must return the work order with all permits, lockout tags, and other documentation to the control center for filing. Appropriate signatures must be secured before filing. The work must be recorded in the control center logs, and any other required documentation must be completed.

3.3.0 Applying Lubrication Internally

Electric actuators/operators with sealed cases are designed to operate in a horizontal orientation and to maintain a standard grease level inside the case. Components turn inside this grease, and the proper level must be maintained. Levels should be checked regularly by removing the designated fill plug and verifying that the grease is at operational level.

Some electric actuators/operators can be installed in any direction. In cases where the electric actuator/operator is not installed in a normal orientation, required lubricant levels may be recommended that are higher than normal. Some electric actuators/operators require that the case be completely filled if installed in an unusual orientation. A different type of grease may also be recommended for this type installation. A different fill plug will be the designated checkpoint for orientations of units that are not horizontal. *Figure 6* shows electric valve actuators/operators mounted in positions that might require higher lubricant levels.

Manufacturer's recommendations include checking grease levels on a regular schedule. Before starting a new electric actuator/operator or one that has been out of service, you should remove the plug and check the grease level. The condition of the lubricant should also be noted for viscosity and any contamination that might be present. Actuators/operators in service should have the appropriate plug removed and the grease checked at intervals specified in company policy or manufacturer's recommendations.

Electric valve actuators/operators are usually heavy. Do not attempt to manually lift them. Manufacturers install lifting lugs in the tops or drill special holes in the housings of some electric valve actuators/operators for lifting (see *Figure 7*). Manufacturer's lifting lugs or holes must be used so that the actuator/operator is not distorted or damaged when removal is necessary. When lifting an electric actuator/operator, it must be separated from the valve body. Lugs on electric actuator/operators are not designed to lift the entire assembly. If attempted, serious damage to the actuator/operator and/or valve may result.

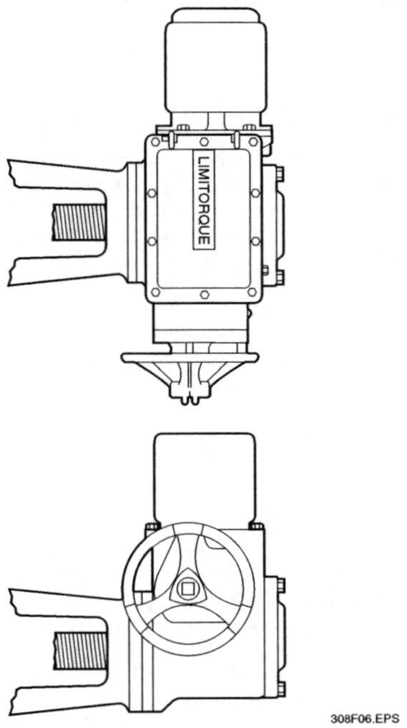

Figure 6 ◆ Possible positions requiring additional lubricant.

Laboratory

Set up several stations with grease guns and valve stems. Ask trainees to practice using a grease gun to lubricate the fittings. Review and assess trainees' work.

Classroom

Explain how to inspect the lubrication levels of internal components.

Audiovisual

Show Transparency 5 (Figure 6). Discuss the types of grease, lubrication levels, and fill plugs required for electric actuators/operators that operate in a horizontal orientation.

Classroom

Discuss problems that can result if the lugs on an electric actuator/operator are used to lift the entire assembly. Ask whether trainees have any questions.

Audiovisual

Show Transparency 6 (Figure 8). Emphasize the importance of accounting for every part when disassembling spur gear reduction units.

Classroom

Explain that add-ons for spur gear reduction units must be serviced separately.

Figure 7 ◆ Electric actuator/operator with lifting lugs.

3.3.1 Spur Gear Reduction Units

Electric actuators/operators with spur gear reduction units are used directly on valve stems. Some involve multiple gear trains with many gears, bushings, and bearings in the system. Some of these units must be disassembled and hand-packed. Others have grease fittings for grease gun use. If a unit must be disassembled, the manufacturer's instructions must be followed. Each part, seal, and gasket must be accounted for and reinstalled in its exact location for the unit to work well, be leak proof, and be weather-tight again. Each bearing must be hand-packed, and each shaft must be hand-coated with the appropriate grease. The case must be packed or filled to manufacturer's recommendations for the unit to resist moisture buildup. Some manufacturers suggest that the case be at least 80% full. *Figure 8* shows the layout of a double reduction spur gear actuator/operator with no grease fittings.

Spur gear reduction units are commonly added on to other types of actuators/operators in order to multiply mechanical advantage to enable the turning of heavy valves. Because the spur gear reduction unit is a separate unit, no lubrication passes between the actuator/operator and the reduction unit. Each unit depends on a self-contained lubricant supply. You must be careful to recognize these add-on units because they must be serviced as a separate unit; a different set of manufacturer's recommendations and instructions apply to the spur gear reduction unit.

3.3.2 Bevel Gear Actuator/Operators

Bevel gear actuators/operators sometimes require hand lubrication. Many of these actuators/operators require that they be manually disassembled, bearings packed, and shafts coated with grease.

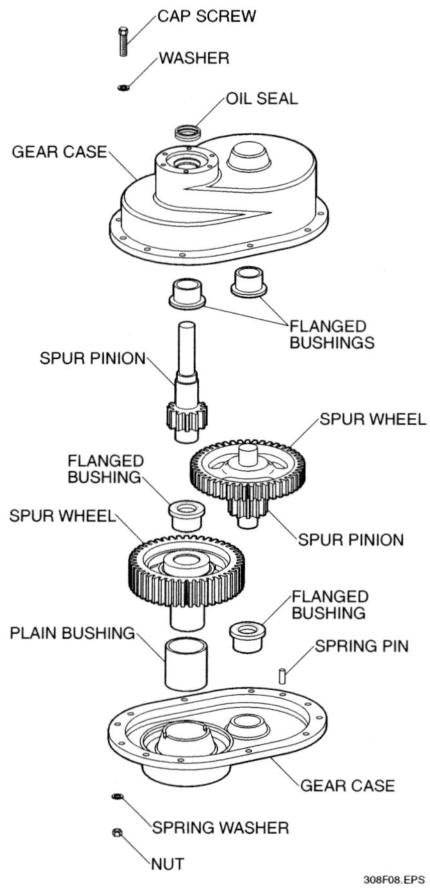

Figure 8 ◆ Double reduction spur gear actuator/operator with no grease fittings.

Instructor's Notes:

Again, it is important that you correctly identify the model being serviced so that the manufacturer's instruction manual can be followed and the correct part numbers identified. Care must be taken with seals because these units usually operate directly on the valve stem. *Figure 9* shows the layout of a bevel gear actuator/operator with no grease fittings.

3.3.3 Multiple Reduction Actuators/Operators

Many gear-type actuators/operators require multiple reduction sets in order to obtain sufficient mechanical advantage to turn the valve. Basic gear designs are used by many companies. Most companies have their own variations of placement and number of lubrication points. Manufacturer's recommendations should be consulted for the exact application and type of grease to be used. General procedures call for application by grease gun. However, you must be careful to locate every grease fitting so that those fittings, obscured by their location or that of other equipment, will not be missed. Missing a grease fitting on several occasions will eventually lead to failure of that part. Secondly, when a gear case involves larger volumes of lubricant, the gun may need to be pumped a number of times, the valve cycled, and then a few more strokes of the pump added. *Figure 10* shows the layout of a double reduction worm gear actuator/operator with several grease fittings.

3.3.4 Automatic Electric Actuators/Operators

Automatic electric actuators/operators add another dimension to the maintenance of valve actuators/operators because they operate the valves without human assistance and use a source of control and power from another location. Proper lubrication is essential to the longevity of any power transmission equipment, and electric valve actuators/operators are no different. Some electric actuators/operators are factory-packed and sealed and only need to be checked once every twelve to eighteen months under normal usage. They need to be inspected regularly for seal leakage or bearing failure; however, electric actuators/operators subjected to storage for a year or more should be observed closely after being returned to service to verify that they do not leak. (Seals tend to leak after units are unused for a time.)

Limit switches, **torque switches**, and limiting valves are integral parts of electric valve actuators/operators because they signal the unit to start and stop (see *Figure 11*). If a limit switch, torque switch, or limiting valve were not involved in the actuator/operator's action, the equipment would soon destroy itself. Limit switches and torque switches are often gear-type devices that measure the travel of the valve or the amount of torque involved when in operation from an electrical actuator/operator. The principle is that the valve must turn far enough to ensure a tight seal, but not far

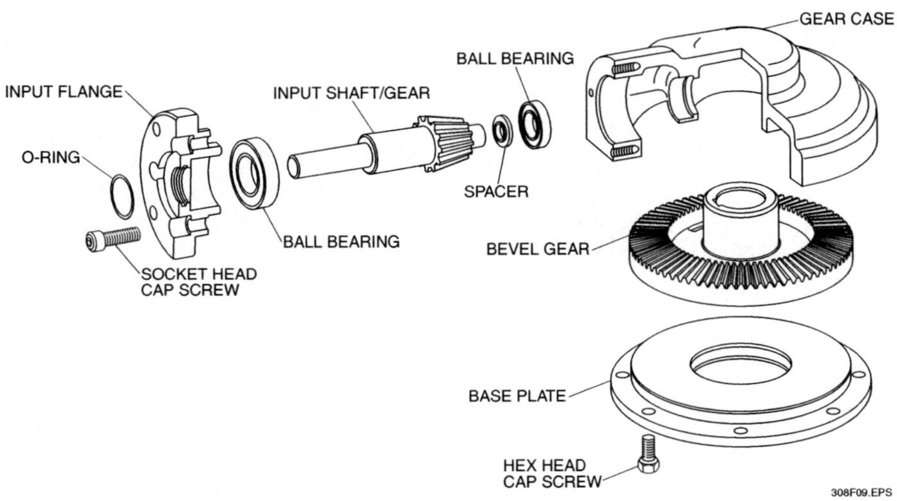

Figure 9 ◆ Bevel gear actuator/operator with no grease fittings.

Audiovisual

Show Transparency 7 (Figure 9). Explain why bevel gear actuators/operators may require hand lubrication and manual disassembly.

Classroom

Encourage trainees to consult the manufacturer's recommendations for the application and type of grease to be used on multiple reduction actuators/operators.

Distribute copies of Figure 10 with the legend covered. Ask trainees to identify the components in the double reduction worm gear actuator/operator. Have trainees review techniques to lubricate the fittings.

Explain the role of limit switches, torque switches, and limiting valves in electric valve actuators/operators.

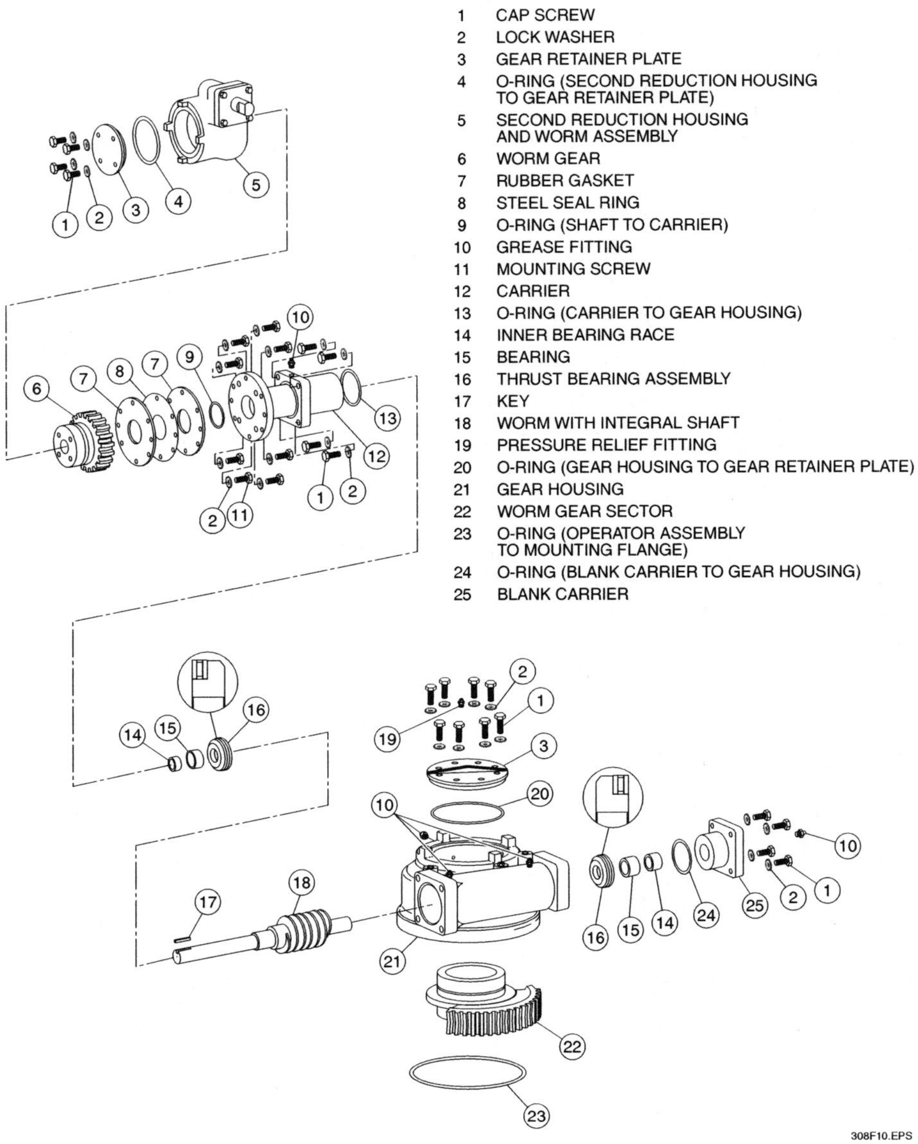

Figure 10 ◈ Double reduction worm gear actuator/operator with several grease fittings.

Instructor's Notes:

308F11.EPS

Figure 11 ◆ Double torque limit switch.

enough to damage the seals, gears, and other components. The electric actuator/operator must stop before the unit stalls and motor damage results.

Limit switches and torque switches must be lubricated according to the manufacturer's recommendations. Many utilize electronic devices and switches, and the type of lubricant used must be compatible with them. You must follow manufacturer's recommendations for compatible products to be used.

> **NOTE**
> Most internal gear lube is not compatible with the electronic movements.

A leak can enter and saturate the motor and switching compartments of some electric actuators/operators if they are installed in a nonstandard configuration. If this occurs, the units and compartments of these mechanisms must be flushed and cleaned with an electrically compatible solvent. The mechanisms should be completely dried before use.

4.0.0 ◆ INSPECTING ELECTRIC ACTUATOR/OPERATOR COMPONENTS

Only trained, qualified mechanical technicians should inspect electric actuator/operator components. Many of the parts are only accessible after the electric actuator/operator has been disassembled and the parts are separated from their adjoining components. Most technicians do not disassemble an electric actuator/operator just for a routine examination of parts, though they sometimes disassemble ones that must be repacked with grease at certain intervals in the field. Electric actuators/operators with many parts involving several systems and complicated gear trains are best disassembled and repaired in a well-equipped shop.

4.1.0 Locating the Actuator/Operator

When assigned to inspect an electric actuator/operator, you must be able to find the designated one. At times this might not be as easy as it sounds because most companies have many valves, and some are located in seemingly inaccessible places. A valve on the terminal yard may be easier to get to physically than if it were in an area of rugged terrain or swamp. However, a valve in a terminal yard might be more easily confused with other valves around it. Numbers should be checked carefully. Some valves might be located under water at times, such as in tidewater areas and flood plains. You may have to use a map along with P&IDs to trace the location of some valves if they are far out in open terrain.

When the valve is found, you must verify that it is the correct unit by comparing the nameplate data and serial number with those on the work order. If the nameplate is missing or illegible, it must be replaced immediately according to company policy and procedures.

4.2.0 Inspection Checklists

Many company procedures have official checklists for mechanical technicians to follow as they carry out the task of inspecting electric actuators/operators. The checklist becomes part of the electric actuator/operator maintenance documentation.

When the electric actuator/operator is found, you should approach the site cautiously so that all conditions may be carefully observed and evaluated, especially if there is an odor of product or signs of dead vegetation. You should also be watchful for natural dangers such as stinging insects, snakes, and other wildlife.

Teaching Tip

See the Teaching Tip for Sections 3.0.0–3.3.4 at the end of this module.

Homework

Have trainees review Sections 4.0.0–4.5.0 for the next session.

Materials

Ensure that you have everything required for testing during this session.

Classroom

Emphasize that only trained mechanical technicians are qualified to inspect electric actuators/operators.

Explain how to use P&IDs to locate an electric valve actuator/operator in a terminal yard. Answer any questions trainees have.

Classroom

Review the elements of an inspection checklist. Explain how these checklists are used as part of an electric actuator/operator inspection.

Safety

Stress the importance of completing all preparation procedures before adjusting the limiting device on an electric valve.

Audiovisual

Show Transparency 8 (Figure 12). Explain how to use solvent to clean the contacts on the switches.

Classroom

Discuss the problems prevented by painting pipeline equipment.

An inspection checklist might include but would not be limited to the following items:

- General electric actuator/operator appearance, such as the condition of finish, damaged or missing parts, and damaged housing
- Molded covers and conduits in place and watertight
- Seal leakage, indicated by slight weeping or moistness, which is normal on some models
- General tightness of the installation, indicated by a lack of loose parts or units
- Main gear case lubricant

> **NOTE**
> Do not mix lubricant types or grades. Use the manufacturer's recommendations or an approved equivalent.

- Secondary gear case lubricant, such as the reduction gearbox on a worm gear actuator/operator
- Limit switch condition and settings if applicable
- Torque switch condition and settings if applicable
- Unit wiring pattern; compare with the applicable wiring diagram
- Electric motor turning in the correct direction of rotation
- Obstructions that might hinder access to or operation of the valve, such as plant growth, flowing water, or debris

4.3.0 Verifying Actuator/Operator Control

When performing maintenance on electric actuator/operators, the limit switches and torque switches should be tested to verify proper settings. The contacts on the switches should be cleaned with a solvent designed for electrical components as directed in the manufacturer's instructions. *Figure 12* shows an external type limit switch.

> **WARNING!**
> Only trained, qualified mechanical technicians should adjust limiting devices on an electric valve actuator/operator. Before beginning this task, you should complete all the preparation procedures listed earlier in this module.

You should cycle the valve under power to verify the settings. If the valve movement is wrong, the switches should be adjusted. Manufacturer's instructions and company policy must be followed. Only trained and qualified technicians should adjust the limiting devices on an electric valve actuator/operator. If the limiting device is of questionable operation, it should be replaced as a unit. Some devices can only be repaired by sending them to a shop or the manufacturer while others must be discarded. You should follow company policy and procedure. Notification of the control center is required for work on any part of the actuator/operator.

> **NOTE**
> Work orders and permits are required for repairs on electric actuator/operator limiting devices.

Once the unit's operation is confirmed, a small amount of recommended, compatible grease should be applied to the appropriate parts of the limiting device.

4.4.0 Painting

Companies use regular painting to protect expensive pipeline equipment. Paint inspection is usually one of the points on regular inspection checklists. Electric valve actuators/operators should be painted when new and then touched up or repainted on a regular schedule.

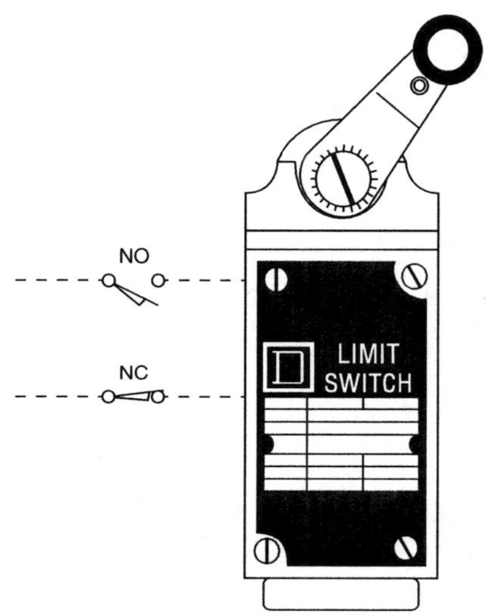

308F12.EPS

Figure 12 ◆ External type limit switch.

Instructor's Notes:

Paint is a strong defense against rust and corrosion. Manufacturers normally coat new electric actuators/operators with some kind of corrosion resistant, oxide-type primer to protect them until delivery. Some electric actuators/operators may have the final coat applied by the manufacturer. Once actuators/operators are delivered, some companies add their own paint to the factory coating, according to company policy, for resistance to the weather or for product coding. *Figure 13* shows a contrasting pipeline paint scheme with actuator/operator paint left as the factory applied it.

During inspections and maintenance work, you should pay close attention to the condition of paint and coatings on actuators/operators. Damaged paint allows rust and corrosion to begin. A chip, ding, or flaked area should be cleaned and touched up before rust and corrosion advance. You should arrange to have painting scheduled as soon as possible.

Figure 13 ◆ Actuator/operator with factory finish.

4.5.0 Caring for Out-of-Service Equipment

Out-of-service units require certain maintenance activities even though they are not being used. This applies to both new units that are stored in original crates and repaired units that are waiting to be reinstalled. Some units may be in the shop storage (shelf or floor) area for weeks or months before being needed. The following practices are generally applicable to the storage of actuator/operators:

- Store actuators/operators out of the weather.
- Units stored outside should be covered with polyethylene and a supply of silica gel or other desiccant to absorb moisture. Connect internal heaters if they are available on the unit.
- Use space heaters to control dampness in storage areas.
- Maintain all valve stems in a clean, lubricated fashion.
- Set up a periodic cycling schedule for actuators/operators not often used.
- Maintain a clean area for all work on actuators/operators.
- Only use clean, new lubricant when parts are being assembled.

5.0.0 ◆ TROUBLESHOOTING ELECTRIC VALVE ACTUATORS/OPERATORS

Successfully troubleshooting valve actuators/operators relies heavily on knowledge of the equipment, skills developed through training, and experience gained through working on them. Before attempting to troubleshoot an actuator/operator, you must be trained, qualified, and have the manufacturer's repair manual for the actuator/operator being examined on site. Before beginning to troubleshoot an electric actuator/operator, you should complete all the procedures listed earlier in this module.

> **WARNING!**
> Because troubleshooting an electric actuator/operator involves operating the valve, you should notify the control center and all affected personnel before beginning a troubleshooting task.

Task-specific procedures must be followed when applicable.

5.1.0 Preliminary Diagnoses

Typically, problems in valve actuators/operators are initially misdiagnosed as problems with valves because the operation of the valves is the demonstration of the problem. Field personnel first notice the problem when a valve fails to operate properly. You might be told that the valve is:

- *Hung* – Actuator/operator not functioning
- *Seized* – Actuator/operator not functioning
- *Out of control* – Actuator/operator overactive
- *Vibrating* – Could be in pipeline
- *Making strange noises* – Squeaking, hammering, or banging
- *Not powerful enough for the job* – Suggests the unit is slow or erratic in operation

Classroom

Ask a trainee to explain why out-of-service units may require maintenance. Discuss the general practices for maintaining units that are not being used.

Teaching Tip

See the Teaching Tip for Sectiona 4.0.0–4.5.0 at the end of this module.

Performance Testing

Have each trainee perform Performance Task 2 to your satisfaction. Fill out a Performance Profile Sheet for each trainee.

Homework

Have trainees review Sections 5.0.0–5.4.0 for the next session.

Materials

Ensure that you have everything required for testing during this session.

Safety

Discuss the consequences of failing to notify the control center before beginning a troubleshooting task.

Classroom

Explain how to prevent common misdiagnoses.

Review non-actuator/operator problems that affect actuator/operator function.

Audiovisual

Show Transparency 9 (Figure 14). Review the accessories that can cause electric actuators/operators to malfunction. Go over any questions trainees have.

Classroom

Review the factors that influence troubleshooting procedures for electric actuators/operators. Review the general steps for diagnosing problems.

Safety

Ask a trainee to explain why company safety policies must be followed when troubleshooting electric valve actuators/operators.

One or more such conditions might be reported when the only observable problem is that the valve is not operating properly. It is up to you to pinpoint the actual problem and repair it safely and efficiently. That is not to say that you should ignore the input from others because other personnel can be a valuable resource. At the same time, such input should be weighed with other solid evidence collected before deciding how to proceed. For example, if a valve is actually hung, you can determine this by simply attempting to turn the handwheel. Most likely, the problem is in the actuator/operator.

5.2.0 Non-Actuator/Operator Problems

Common problems that affect the actuator/operator often involve the accessories and additional equipment that are included to aid in the proper operation of the actuator/operator and valve. *Figure 14* shows an actuator/operator with a number of operating accessories.

A malfunction in any of the various accessory units could have a notable influence over the proper action of the electric valve actuator/operator and, thereby, the valve. If a problem can be traced to one of these accessories and corrected, the electric valve actuator/operator and the valve will operate properly. Some of the problems may require referral to E&I technicians. A list of such units includes but is not limited to:

- Valve positioning devices
- Transducers
- Solenoid valves
- Limit switches
- Handwheels
- Travel stops
- Valve position feedback devices
- Problems with a programmable logic controller (PLC) or a remote terminal unit (RTU)

5.3.0 Basic Procedures for Troubleshooting

Troubleshooting procedures depend upon the type of actuator/operator, the brand of the equipment, and sometimes the location of the valve. Company policy must be followed for any work that is done in the examination of a valve actuator/operator. Task-specific procedures and manufacturer's recommendations should be followed. The manufacturer's manual should be referred to for specifications and instructions.

When dispatched to troubleshoot an electric actuator/operator, you will need to operate the valve. Therefore, the preparation procedures listed earlier in this module must be completed. After those procedures are completed, you should follow these general steps to diagnose the problem:

> **WARNING!**
> The control center must be notified before any work is begun. Company safety policy must be followed when troubleshooting an electric valve actuator/operator.

Step 1 Understand how the valve and its actuator/operator function within the system.

Step 2 Learn about the problem by investigating the symptoms. Include input from local operations personnel, the unit's historical information, and personal assessment.

Step 3 If it is a repeat problem, contact the manufacturer's technical assistance department for information and suggestions about the particular actuator/operator.

Step 4 Take note of all attached equipment influencing the operation of the unit. The operation of limiting devices and other related units are especially important in diagnosing problems.

Step 5 Notify the control center, upstream and downstream pumping stations, and any other affected personnel on the pipeline.

Step 6 Switch the actuator/operator to manual control and run the unit through its cycle completely several times to observe the actual problem.

Step 7 Form a personal theory, supported by facts, about the cause of the problem.

Step 8 Form a plan of action for the most efficient solution to the problem and the return of the actuator/operator to normal operation.

Step 9 Initiate all applicable lockout/tagout procedures and follow company policy for PPE and safety procedures.

Step 10 Perform repairs/adjustments according to company policy and manufacturer's recommendations.

Step 11 Inspect the unit for worn or damaged parts. They should be repaired or replaced while the unit is disassembled.

Instructor's Notes:

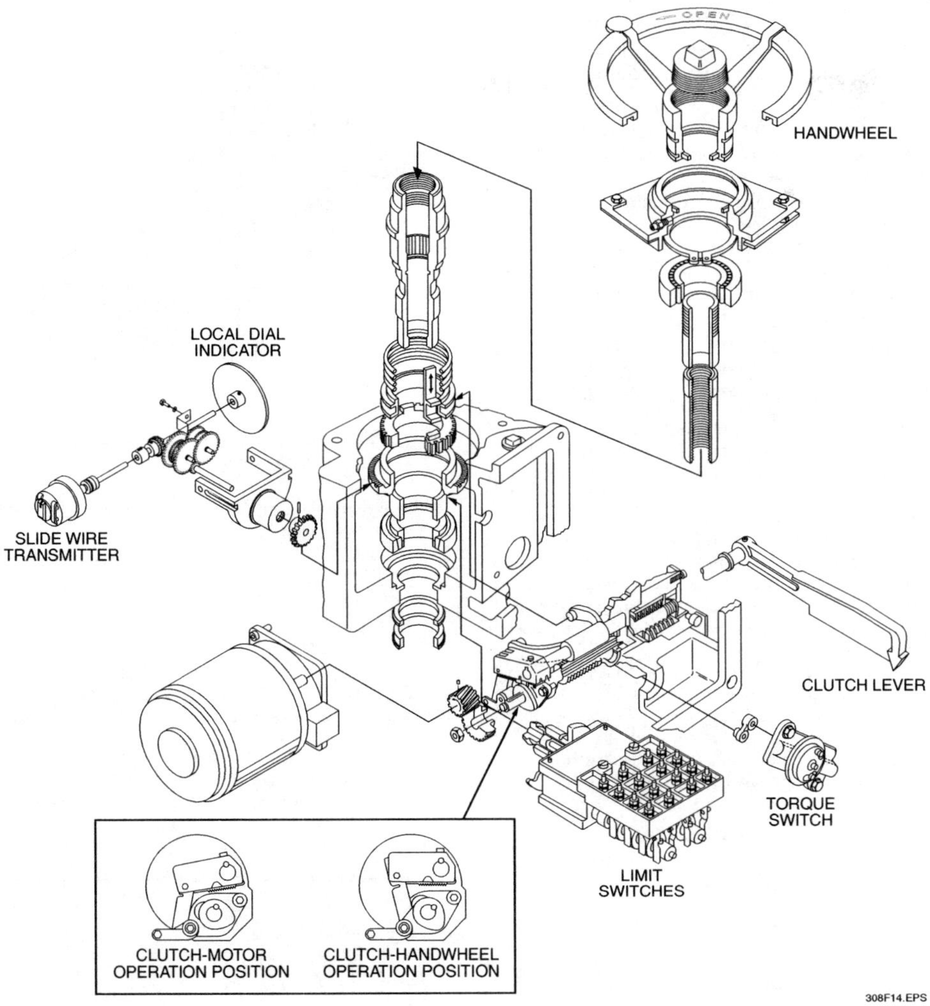

Figure 14 ◆ Actuator/operator with a number of operating accessories.

CAUTION

When disassembling an electric actuator/operator for inspection and repairs, clean the area, keep it clean, and keep the actuator/operator parts clean.

Step 12 Test the valve for proper operation.

Step 13 Return the unit to normal use.

Step 14 Notify the control center and other affected personnel. Complete all required documentation per company policy.

Safety

Explain problems that can result if the actuator/operator is not kept clean.

Teaching Tip

See the Teaching Tip for Sections 5.0.0–5.4.0 at the end of this module.

Performance Testing

Have each trainee perform Performance Task 3 to your satisfaction. Fill out a Performance Profile Sheet for each trainee.

Homework

Have trainees review Sections 6.0.0–6.4.4 for the next session.

Materials

Ensure that you have everything required for the demonstrations and laboratories during this session.

Classroom

Explain that the spur reduction gear is a separate attachment that should be repaired individually.

Safety

Emphasize the importance of isolating the valve and relieving all of the pressure from the pipeline before removing an electric worm gear actuator/operator.

Discuss the consequences of failing to secure the hoisting and lifting equipment before removing the actuator/operator.

Audiovisual

Show Transparency 10 (Figure 16). Review the layouts for machine-fitted and bolt-on spur gear attachments.

5.4.0 Repair Options

Common options for repairing an electric actuator/operator include but are not limited to:

- Repair the problem on site using manufacturer's repair manuals and recommendations.
- Replace the unit and send the failed unit to the company repair shop.
- Replace the unit and return the failed unit to the manufacturer. With some types of sealed units, this is the only recourse.
- Refer the problem to E&I technicians for electrical or communications problems.

6.0.0 ◆ ADJUSTING AND REPAIRING ELECTRIC ACTUATOR/OPERATOR COMPONENTS

Only trained and qualified mechanical technicians should attempt to adjust or repair an electric valve actuator/operator. When adjusting and repairing internal components, the manufacturer's manual and instructions for the exact unit should be secured and followed closely. Company policy must also be followed. You must be able to understand the manufacturer's manuals, technical bulletins, and the relationship of the unit in question to the pipeline and other systems associated with it.

The number of specific tasks that are involved in adjusting and repairing an electric valve actuator/operator and components is extensive. For this reason, this module only covers some of the more common activities. These descriptions of the tasks are very general in nature.

6.1.0 Removing an Electric Actuator/Operator from the Valve Stem

If an electric actuator/operator appears to have an internal problem, it will have to be removed from the valve stem and disassembled. A worm drive actuator/operator with a spur reduction gear will be used as an example of this procedure. Note that in this type of assembly, the spur reduction gear is a separate attachment and could be serviced or repaired separately if needed (see *Figure 15*). However, some worm gear electric actuators/operators with spur reduction gears are designed using an integral housing for both sets of gears, and the units cannot be separated for service.

308F15.EPS

Figure 15 ◆ Worm gear actuator/operator with a spur reduction gear.

> **WARNING!**
> Do not attempt to remove an electric worm gear actuator/operator from a valve until the valve has been isolated and all pressure from the pipeline has been relieved. Electric worm gear actuators/operators have a self-locking characteristic that, if removed without proper procedure, may release control of the valve and the pipeline.

You should first complete the preparation procedures listed earlier in this module. After all company safety policies and procedures are in place, follow the manufacturer's instructions for removal of the unit from the stem of the valve.

> **CAUTION**
> Appropriate hoisting or lifting equipment for an electric actuator/operator must be secured and in place before any attempt to remove the actuator/operator is begun.

6.2.0 Removing the Spur Gear Attachment

Some primary spur gear attachments are machined to fit the main worm gear housing. *Figure 16* shows layouts of a machine-fitted and a bolt-on-type

Instructor's Notes:

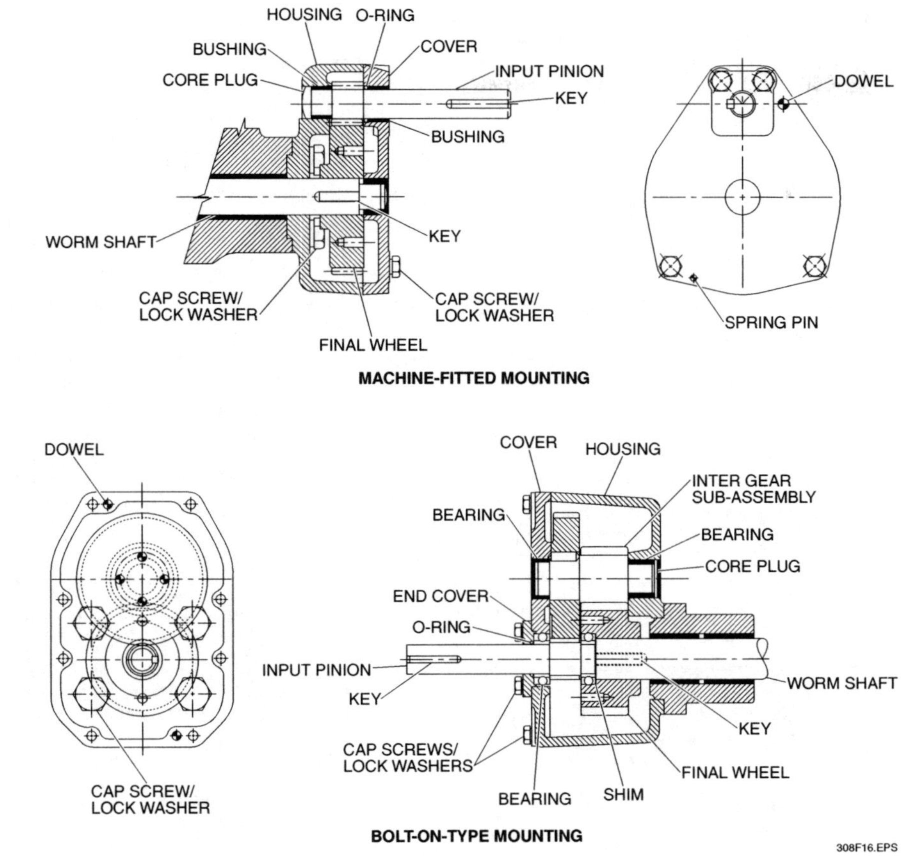

Figure 16 ◆ Machine-fitted and bolt-on type mountings.

mounting. In machine-fitted assemblies, it is necessary to disassemble the spur gear unit first and then proceed to disassemble the worm gear actuator/operator because the bolts holding the machine-fitted unit are located inside the housing, behind the gear train. Before beginning disassembly of the unit, be prepared to record the number and location of shims and parts as they are taken out. Follow these general steps to disassemble the spur gear attachment:

Step 1 Remove the bolts holding the cover to the housing.

> **NOTE**
> Record the number and position of shims, bearings, and any other small parts as they are removed.

Step 2 Carefully pull the cover off using jackscrews or another appropriate tool.

Step 3 Carefully pull out the internal spur gears and shaft.

Classroom

Review the procedures for disassembling the spur gear attachment. Ask whether trainees have any questions.

Demonstration

Display bolt-on and machine-fitted attachments. Demonstrate how to pull out the spur gears from both models.

Classroom

Explain how to remove the worm gear assembly after the spur gear attachments have been removed.

Laboratory

Using the demonstration models, have trainees practice removing the worm gear assemblies. Ask trainees to take turns completing the steps until the task is finished. Review trainees' work.

Step 4 For a machine-fitted attachment (mounting bolts inside housing), remove the inner bolts holding the housing to the worm gear actuator/operator and carefully separate the two units, taking care not to damage the worm shaft.

Step 5 For a bolt-on attachment (mounting bolts outside housing), remove the bolts holding the end cover and carefully and evenly lift and remove the end cover using jackscrews or another appropriate tool.

Step 6 Observe and note the location of any O-rings and other assembly sequences.

Step 7 Carefully pull out the spur gears, leaving the worm shaft in position and taking care not to damage the worm shaft and bearing areas.

Step 8 Clean all parts and remove all old grease so that gears, bearings, and other parts can be examined for wear and damage.

Step 9 Select replacements for all shafts, parts, bearings, and seals of questionable condition. Pay close attention to signs of wear on the worm and quadrant gears.

Step 10 Pack all bearings and gears with appropriate new grease.

> **NOTE**
> Before reinstalling gears and parts on a machine-fitted spur gear attachment, the unit must be reinstalled on the worm gear. Mounting bolts must be installed and tightened before installing bearings and gears.

Step 11 Carefully reassemble all parts in the reverse of the order in which they were taken out. Refer to your notes and the manufacturer's manual to confirm the assembly sequence and the number and placement of small parts and shims.

6.3.0 Removing the Worm Gear Assembly

Follow these steps to remove the worm gear assembly after the spur gear attachment has been removed:

Step 1 Take out the housing cover bolts from the actuator/operator so that the carrier (end caps) and accessory mounting equipment can be detached (refer to *Figure 6*).

Step 2 Take off the top cap, housing cover, and any motor mounting accessories, as applicable.

Step 3 Remove one or both end covers, as applicable to the model, by using jackscrews or another appropriate device to separate them from the housing.

Step 4 Remove the worm shaft using a careful twisting motion to avoid damaging the bearings, shims, shaft, or other parts.

Step 5 Lift the worm quadrant from the housing.

Step 6 Clean and inspect all parts for excessive wear.

Step 7 If worn, replace the worm quadrant and any bearings, bushings, and shims.

Step 8 Install bearings and shims on the worm shaft in the proper order. Refer to your notes made as the unit was being disassembled to confirm the proper installation sequence for bearings and the number of shims.

Step 9 Pack the bearings, shaft, and shims with appropriate grease during assembly.

Step 10 Pack the worm gear and shaft with appropriate grease before assembly.

Step 11 Reinstall the worm shaft back into the housing using a twisting motion in the direction of the worm threads.

Step 12 Pack the new worm quadrant in appropriate grease.

Step 13 Install the new worm quadrant into the housing in the approximate position it was in when it was removed.

Step 14 If a blanking plug is included, it must be pressed into place during the assembly process.

Step 15 Pack the housing with grease as instructed by the manufacturer's manual.

Step 16 Apply gasket compound or an appropriate gasket, as applicable, to the sealing surfaces of the case. Some manufacturers recommend installing new gaskets and O-rings any time a unit is disassembled to prevent future maintenance problems.

Step 17 Mate the carriers (end caps), cover plates, and other accessories as they were before disassembly.

Step 18 Insert and tighten bolts to the manufacturer's torque specifications.

Instructor's Notes:

6.4.0 Adjusting the Valve Travel

Electric actuators/operators on quarter-turn valves, such as ball valves, are designed with a quadrant, as opposed to a circular, gear. Most quarter-turn valves are equipped with adjustable travel stops to limit valve travel. These travel stops are usually in the form of threaded studs (stop screws) projecting through the gear housing with locking nuts, set screws, or other devices to prevent the setting from changing (see *Figure 17*). Always follow the manufacturer's recommendations and company policy when adjusting travel stops.

6.4.1 Adjusting the Stops

Because making electric actuator/operator stop adjustments will require cycling the valve, the control center must be notified before proceeding. You must complete all the preparation procedures included earlier in this module. Then you should follow these general steps to adjust the valve stops:

Step 1 Back off the locking nuts just enough to loosen the stop screws.

Step 2 Back out the stop screws slightly to eliminate any interference with moving the valve to its extreme positions

Step 3 Move the valve and actuator/operator to the fully closed position.

Step 4 Turn the travel stop screw on the closed side of the actuator/operator until it gently contacts the quadrant stop.

Step 5 Hold the stop screw with the appropriate tool to prevent the adjustment setting from changing and tighten the locking nut securely.

Step 6 Move the valve and electric actuator/operator to the fully open position.

Step 7 Turn the travel stop screw on the open side of the electric actuator/operator until it gently contacts the quadrant stop.

Step 8 Hold the stop screw with the appropriate tool to prevent the adjustment setting from changing and tighten the locking nut securely.

Step 9 Cycle the valve to verify that the stop settings are correct.

Step 10 Notify the control center and affected personnel that the valve has been returned to service.

Step 11 Complete all required documentation per company policy.

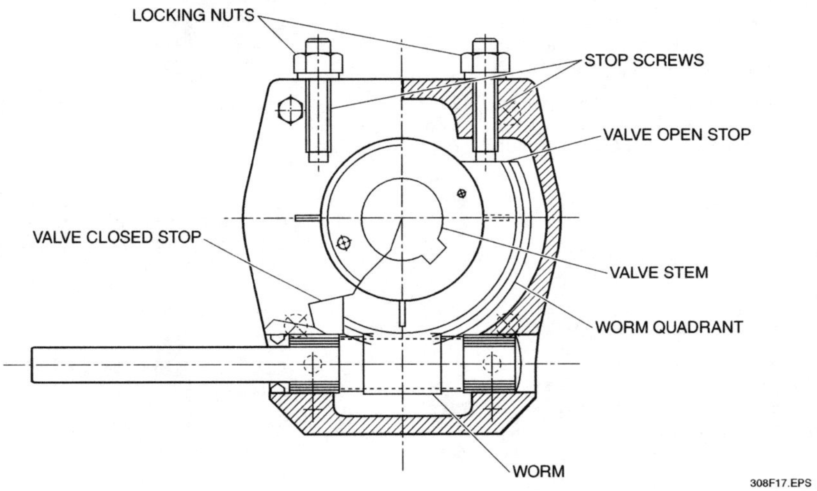

Figure 17 ◈ Worm gear actuator/operator with adjustable travel stops.

Audiovisual

Show Transparency 11 (Figure 17). Explain the role of the adjustable travel stops in an electric actuator/operator.

Classroom

Review the general steps to adjust the travel stops.

Safety

Stress that extra precautions must be taken when setting limits with the electrical power on.

Laboratory

Distribute copies of manufacturer's manuals for limit switches. Ask trainees to locate the information for setting limit switches and to identify the proper valve position and limit switch settings.

Classroom

Refer to Figure 20. Explain that resetting the torque switch allows it to operate more efficiently.

6.4.2 Setting the Limit Switches

Electric actuator/operator limit switches (see *Figure 18*) are used on powered units and must be adjusted to ensure proper valve positioning. Torque switches are also used and must be adjusted and set to properly protect the valve from over-torque situations. Since making limit switch adjustments will require cycling the valve, the control center must be notified before proceeding. After completing all the preparation procedures listed earlier in this module, you should follow these general steps to adjust the limit and torque switches:

308F18.EPS

Figure 18 ◆ Electric actuator/operator limit switch.

> **WARNING!**
> Special precautions must be taken because these procedures must be done with the electrical power on.

Step 1 Follow the associated task-specific procedures for work permits, safety policies, and procedures.

Step 2 Perform maintenance as instructed in the manufacturer's instruction manual and company policy and procedures.

Step 3 Verify proper valve position to determine the proper limit switch setting according to the manufacturer's instructions.

Step 4 Set the limit switches to ensure that the contacts operate at the proper valve positions according to the manufacturer's instruction manual.

Step 5 Set the torque switches (see *Figure 19*) to ensure that the switch disconnects the electric motor at the set mechanical torque value.

Step 6 Notify the control center and affected personnel when work is complete.

Step 7 Document the work accomplished and results for maintenance records.

308F19.EPS

Figure 19 ◆ Double torque limit switch.

6.4.3 Adjusting the Torque Switch

Torque switches (see *Figure 20*) are important for stopping the movement of an actuator/operator at the proper position to give the valve a tight seal but before the unit overloads and risks possible damage to the electric motor or valve. Many torque switches are set at the factory; however, due to normal operational changes including wear and repairs, it does become necessary to reset the torque switch for more efficient operation.

Because making the torque switch adjustments require cycling the valve, the control center must be notified before proceeding. You must complete all the preparation procedures listed earlier in this module. Then you should follow these general steps to adjust the torque switches:

Step 1 Lock out and tag all energy sources to the valve and actuator/operator.

Step 2 Release possible buildup of torque by shifting the clutch lever to the handwheel position.

Step 3 Remove the torque switch cover.

Instructor's Notes:

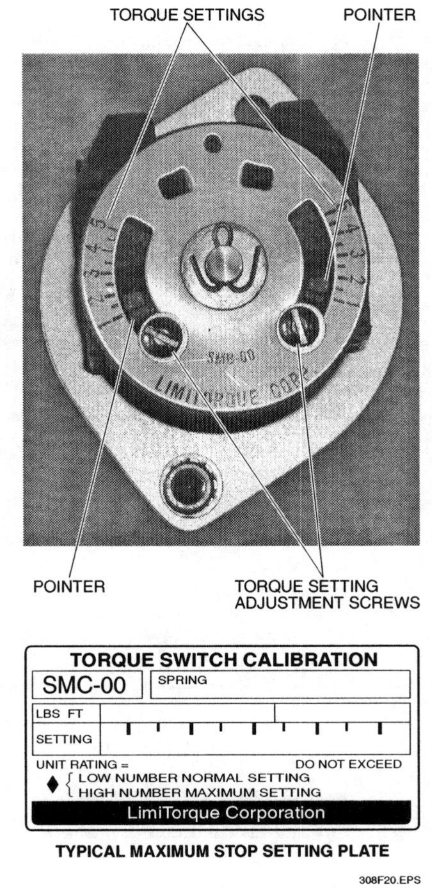

Figure 20 ◆ Torque switch.

> **NOTE**
> The procedure for setting the torque switch is the same for both directions of travel.

Step 4 Select the appropriate side of the torque switch for the setting required to be changed (open side or closed side).

Step 5 Loosen screw A.

Step 6 Slide pointer B to the number setting for torque required.

> **NOTE**
> The higher the number on the switch, the higher the torque being supplied.

Step 7 Retighten screw A without allowing the pointer to move.

Step 8 Turn on the actuator/operator power supply.

Step 9 Cycle the valve several times to confirm that it is seating properly and achieving a tight seal.

> **CAUTION**
> Do not remove the factory maximum stop setting plate on the side of the unit. Torque settings must not exceed the setting stated on this plate without first contacting the manufacturer for recommendations.

6.4.4 Setting the Limiting Valves

Follow these general steps to install an electric actuator/operator with limit switches on the worm gear actuator/operator:

> **WARNING!**
> Before beginning the replacement of any electric actuator/operator, verify that the power supply has been locked out and tagged.

Step 1 Remove the former actuator/operator according to manufacturer's recommendations, observing the various parts and fittings.

Step 2 Examine the new electric actuator/operator to verify that the flanges and shaft match the worm gear actuator/operator's flange and adapter, as applicable.

Step 3 Slide the actuator/operator onto the worm gear adapter flange and shaft.

Step 4 Tighten the bolts according to the manufacturer's torque specifications.

Step 5 Set the closed position stop as follows:

- Turn the valve to the closed position against the valve stop by hand.
- Adjust the actuator/operator limit switch setting to coincide with the closed valve stop.

Demonstration

Bring in a torque switch. Demonstrate how to make adjustments to the settings. Answer any questions trainees have.

Safety

Explain why the stop setting plate should not be removed from the side of the torque switch.

Discuss the consequences of failing to verify that the power supply has been locked out and tagged.

Classroom

Explain how to set the closed and open position stops on a limit switch.

Step 6 Set the open position stop as follows:

- Turn the valve to the open position against the valve stop by hand.
- Adjust the actuator/operator limit switch settings to coincide with the open valve stop.

Step 7 Consult the manufacturer's manual for additional information.

6.5.0 Replacing the Stem Nut

Some electric valve actuators/operators utilize a stem nut assembly, which is a splined drive sleeve on the valve stem connecting valve actuator/operator and the valve stem (see *Figure 21*). A stem nut may need to be replaced to maintain efficient operation of the valve. This is because some stem nuts tend to wear and hammer certain areas of the assembly, changing the shape and dimensions of the components and possibly causing components to bind or fail during usage.

Some stem nuts and drive sleeves are designed with lost motion built into the action in order to give the stem a hard jolt to help start it turning.

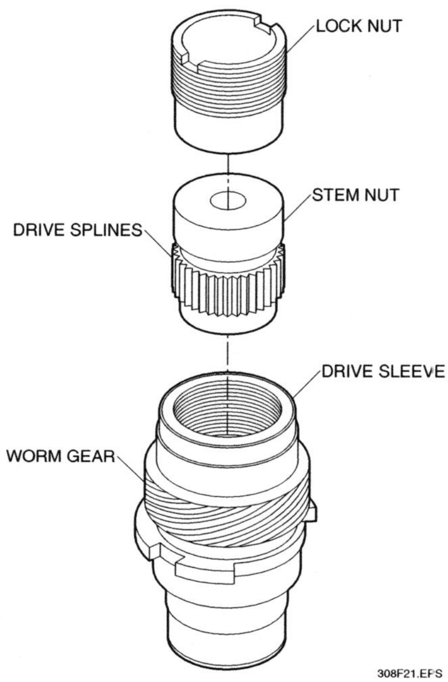

Figure 21 ◆ Stem nut assembly.

This characteristic is especially helpful when a large valve is sealed. Such stem nuts occasionally require replacement because of the hammer effect. On many of these units, the stem nut can be changed without removing the actuator/operator.

Because replacing a stem nut will require cycling the valve, the control center must be notified before proceeding. After completing all the preparation procedures listed earlier in this module, you should follow these general steps to replace the stem nut without removing the electric actuator/operator or valve from the installation:

> **NOTE**
> This general procedure is for a valve that opens with a rising stem. For other orientations, you should consult the manufacturer for instructions.

Step 1 Lockout and tag all power sources to the actuator/operator and its components.

Step 2 Move the valve to the open position.

> **WARNING!**
> Removing the stem nut without the valve in the open position may result in the stem rising suddenly when the lock nut is released, resulting in injury.

Step 3 Disassemble the protective stem coverings and plugs that may be in place.

Step 4 Drill out any staked areas of lock nut threads to release it.

Step 5 Turn the lock nut counterclockwise to remove it.

Step 6 Turn the valve by hand toward the closed position to cause the stem nut to follow the threads upward to a point where its splines disengage from the drive sleeve splines.

Step 7 Rotate the stem nut by hand, threading it off the stem completely.

Step 8 Lubricate the stem and the new stem nut assembly by hand according to the manufacturer's recommendations.

Step 9 Thread the new stem nut onto and down the stem by hand.

Step 10 Turn the valve toward the closed position by hand, making certain that the splines on the new stem nut and the drive sleeve engage properly.

Homework

Have trainees review Sections 6.5.0–6.8.3 and answer the Review Questions for the next session.

Materials

Ensure that you have everything required for the laboratories and testing during this session.

Audiovisual

Show Transparency 12 (Figure 21). Explain why it may be necessary to replace a stem nut assembly.

Laboratory

Bring in a stem nut assembly. Ask a trainee to demonstrate the procedures to replace a stem nut. Have trainees identify what was done correctly and/or incorrectly.

Safety

Explain problems that can result if the stem nut is removed without the valve in the open position.

Instructor's Notes:

Step 11 Continue turning the valve toward the closed position until the stem nut splines seat in the drive sleeve.

Step 12 Thread the locking nut onto the stem and tighten it. The locking nut should be flush with the top of the drive sleeve.

Step 13 Stake the locking nut at two points opposite each other, comparable to original staking points.

> **CAUTION**
> Incorrectly staking the locking nut may result in improper operation of the unit.

Step 14 Follow the manufacturer's instructions to reset the limit switch.

Step 15 Complete the reassembly of the electric valve actuator/operator, covers, and accessories.

Step 16 Return the valve to service and test operation.

Step 17 Notify the control center that the valve is back in service.

Step 18 Complete all required documentation per company policy.

6.6.0 Replacing the Motor

A failed electric actuator/operator motor requires replacement. Remember that the procedures vary according to the type of valve, actuator/operator, brand, and location.

Because replacing a motor on an electric actuator/operator will require cycling the valve, the control center must be notified before proceeding. After completing all the preparation procedures listed earlier in this module, you should follow these general steps to install an electric motor on the worm gear actuator/operator (see *Figure 22*):

> **WARNING!**
> Before beginning the replacement of an electric motor on an actuator/operator, verify that the power supply has been locked out and tagged.

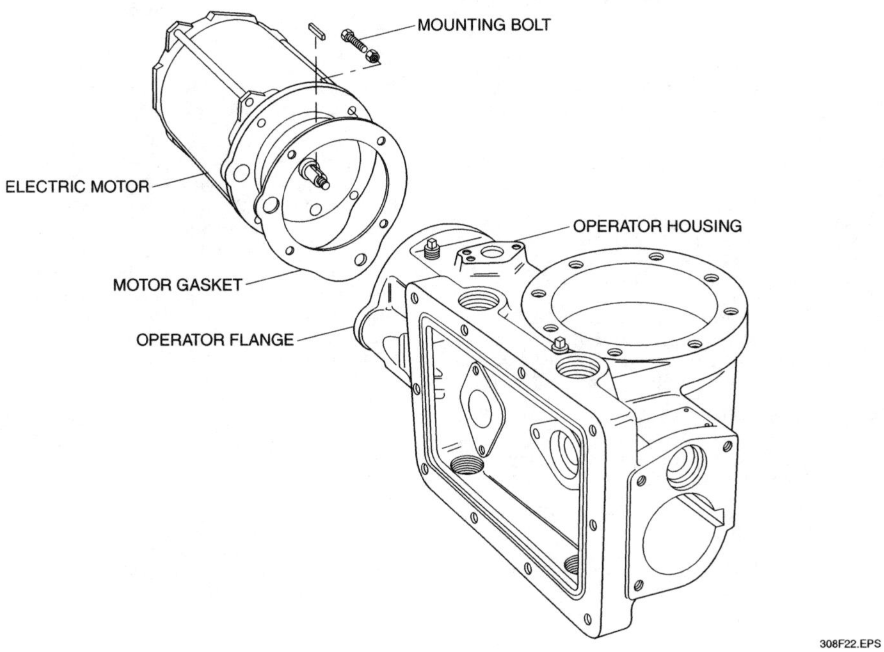

Figure 22 ◆ Electric motor mounted on worm gear flange.

Safety

Stress the importance of correctly staking the locking nut.

Audiovisual

Show Transparency 13 (Figure 22). Explain that the procedures to replace an electric actuator/operator motor vary according to type, brand, and location.

Safety

Ask a trainee to explain why the power supply must be locked out and tagged before replacing an electric motor.

Classroom

Emphasize that an improperly operating valve can cause a disaster on a pipeline.

Laboratory

Refer to company policies. Ask trainees to identify their company policies for installing electric actuators/operators. Compare each company's policies and discuss any significant differences.

Classroom

Explain how to install an electric actuator/operator that is mounted directly to a quarter-turn valve stem. Go over any questions trainees have.

Step 1 Remove the failed electric motor according to manufacturer's recommendations and company policy and procedures.

Step 2 Support and slide the replacement electric motor into place on the actuator/operator's flange and adapter, if applicable, by completing the following steps:

- Turn the motor shaft to align the key slots, splines, or adapter, if applicable, with the actuator/operator.
- Slide the electric motor onto the actuator/operator flange and shaft.
- Rotate the motor frame to align mounting-bolt holes.
- Insert and tighten bolts to the manufacturer's recommended torque specification.

Step 3 Verify that the control has been switched to local or manual control.

Step 4 Turn off all power, and lock out the unit.

Step 5 Turn the valve to the central position by hand.

Step 6 Connect the power circuit and the control circuit to the appropriate terminals. Refer to the manufacturer's schematic.

Step 7 Turn on the actuator/operator power.

Step 8 Observe the direction of motor rotation to confirm that the open switch opens the valve.

Step 9 Observe the motor rotation to confirm that the close switch closes the valve.

Step 10 Operate the opening cycle and verify that the open limit switch and the open torque switch stop the valve control when the valve is in the fully open position.

Step 11 Operate the closing cycle and verify that the close limit switch and the close torque switch stop the valve control when the valve is in the fully closed position.

Step 12 Refer to the manufacturer's instruction manual and company procedures for guidance and instructions for completing the procedure.

Step 13 Notify the control center that the valve is back on line.

Step 14 Complete all required documentation per company policy.

6.7.0 Installing Electric Actuators/Operators

Electric valve actuators/operators must be installed correctly because an improperly operating valve can result in a pipeline disaster. Each pipeline operator is required by federal law to prepare and follow a manual of written procedures for regular maintenance of valve actuators/operators. A valve maintenance log is used to record information about work done on actuator/operators.

6.7.1 Company Policy and Procedures

Installation procedures vary depending on the type of valve, type of actuator/operator, manufacturer, and the location of the installation. The following sections provide some broad examples of how such electric actuators/operators might be installed. All procedures here are very general. When installing actuators/operators or any other equipment, company policy and procedures must be followed. You should complete all preparation procedures listed earlier in this module. Manufacturer's recommendations and instructions for the exact units involved should be in hand, studied carefully, and followed.

6.7.2 Installing an Actuator/Operator on a Quarter-Turn Valve

Common industry design dictates that when the quarter-turn valve is fully opened, the stem keyway should line up with the run of the valve and pipeline. When the quarter-turn valve is fully closed, the keyway should be at an angle of 90 degrees to the run of the valve and pipeline.

After completing all the preparation procedures listed earlier in this module, you should follow these general steps to install an electric actuator/operator mounted directly to the quarter-turn valve stem (see *Figure 23*):

Step 1 Follow the manufacturer's recommendations for removing the former actuator/operator and associated equipment.

Step 2 Confirm that the valve stem and flange are in good condition.

Step 3 Verify that the valve stem and actuator/operator assembly are in the same relative position, either fully opened or fully closed.

Instructor's Notes:

308F23.EPS

Figure 23 ◆ Electric actuator/operator installed on a quarter-turn valve.

Step 4 Confirm that the keyway in the valve stem aligns with the keyway in the actuator/operator assembly by adjusting the position of one or the other slightly and installing a properly sized key.

Step 5 Place the actuator/operator in position on the valve stem flange.

Step 6 Notice the alignment of the bolt holes in the valve flange with those in the actuator/operator.

Step 7 Carefully rotate the actuator/operator on the valve stem flange until the bolt holes in the actuator/operator align themselves with the bolt holes in the valve stem flange.

Step 8 Bolt the actuator/operator to the flange using appropriate bolts and tighten to the manufacturer's torque specifications.

Step 9 Set the actuator/operator's limiting screws to coincide with the valve stops for the closed position by following these steps:

- Loosen both stop screw locking nuts or other devices, and back them off so that they do not interfere with the operation for the new settings.
- Operate the valve by hand until the closed position contacts the valve stop.
- Turn the actuator/operator closed-position stop screw until it contacts the actuator/operator stop.
- Holding the stop screw so that the setting cannot change, tighten the locking nut or screw.

Step 10 Set the electric actuator/operator's limiting screws to coincide with the valve stops for the open position by following these steps:

- Operate the valve by hand until the open position contacts the valve stop.
- Turn the open-position stop screw until it contacts the stop.
- Holding the stop screw so that the setting cannot change, tighten the locking nut or screw.

NOTE

If the actuator/operator has limit switches, follow the manufacturer's instructions to adjust the limit switch settings to coincide with the valve stops.

Step 11 Consult the manufacturer's manual for additional information.

6.8.0 Performing Adjustments and Repairs

The work of adjusting and repairing electric actuators/operators is critically important to the safety and operation of the pipeline. The safety and operation of the system depends upon quality work. In order to carry out the tasks of adjusting or repairing electric actuators/operators, you must be able to accomplish the work quickly and efficiently. However, you should not rush through a job just to finish quickly. Rather, each task should be approached in a planned, orderly manner with attention given to detail. Safe, quality work should be performed. Before beginning this task, you should complete all the preparation procedures listed earlier in this module. Then you should follow the general steps listed in the following sections.

6.8.1 Adjusting the Actuators/Operators

Before beginning to adjust an electric valve actuator/operator, you should complete the procedures listed in the section earlier in this module. All cautions, warnings, and company safety policy and

Classroom

Ensure that trainees understand how to set the limiting screws to coincide with the valve stops for the open and the closed positions.

Emphasize the importance of carefully planning all adjustments and repairs.

Safety

Have trainee discuss the consequences of working without maintaining an ongoing awareness of the valve and actuator/operator activities.

Explain problems that can result if the manufacturer's instructions are not followed.

Classroom

Explain how mechanical stop screws ensure proper operation of the valve. Ask whether trainees have any questions.

Safety

Stress the importance of identifying and responding to abnormal operating conditions.

procedures should be meticulously observed. You should follow these general steps to adjust an electric valve actuator/operator.

Step 1 Verify the location of valve with the actuator/operator to be adjusted.

> **NOTE**
> Ensuring that the correct valve and electric actuator/operator is located, especially before operating the valve, precludes any upsets.

Step 2 Verify the accessibility of the actuator/operator to be adjusted.

Step 3 Verify the number and nameplate data of the actuator/operator to be adjusted.

Step 4 Verify the actuator/operator type and manufacturer in order to verify documentation.

> **NOTE**
> If the nameplate is missing, it must be replaced according to company policy and procedures.

Step 5 Ensure that there are no obstructions blocking accessibility to the electric actuator/operator.

Step 6 Notify the control center and any other affected personnel that the work is beginning.

> **WARNING!**
> You should maintain an ongoing awareness of the valve and actuator/operator activities being performed to ensure safety during the performance of the task.

Step 7 All applicable task-specific procedures must be followed. Following a task-specific procedure helps to ensure job safety, efficient task performance, standardization, and appropriate documentation. Procedures should also include work permit information and notification requirements (control center, both the upstream and downstream pump stations for main line valves, and any other personnel possibly being affected).

> **CAUTION**
> You must take care to follow the manufacturer's instructions and industry recommendations in order to ensure personal safety, prevent mechanical damage to the electric actuator/operator being adjusted, avoid internal damage to the valve, and preserve the service life of the equipment.

Step 8 Verify the correct valve position to determine the appropriate limit switch setting.

Step 9 Set the **mechanical stop screws**, as applicable, according to the manufacturer's instructions to ensure the proper operation of the valve.

Step 10 Set the limit switches, as applicable, according to the manufacturer's instructions and recommendations to ensure that the limit switch contacts operate at the correct valve positions.

Step 11 Set the torque switches, as applicable, according to the manufacturer's instructions to ensure that the actuator/operator stops before the valve or the actuator/operator is damaged.

Step 12 Notify the control center and/or affected personnel to verify the integrity of the system and return it to normal operation.

Step 13 Complete all required documentation and results of the adjustment.

> **NOTE**
> Documenting data for compliance, historical review, and trending is an important element of valve maintenance.

> **CAUTION**
> You must identify any potential abnormal operating conditions and initiate appropriate response.

6.8.2 Repairing Actuators/Operators

Before attempting to repair an electric valve actuator/operator, you should complete the preparation procedures listed earlier in this module. All warnings, cautions, and company safety policy and procedures must be meticulously observed.

Instructor's Notes:

You should follow these general steps to perform a general overhaul of an electric valve actuator/operator:

Step 1 Verify the location of the actuator/operator to be repaired.

NOTE
Ensuring that the correct valve and actuator/operator is located, especially before operating the valve, precludes any upsets.

Step 2 Verify the accessibility of the actuator/operator to be repaired.

NOTE
Verification of accessibility is important, especially where large actuators/operators may necessitate the use of hoisting equipment (see *Figure 24*).

Step 3 Verify the actuator/operator type and manufacturer.

Step 4 Verify the number and the nameplate data of the actuator/operator to be repaired in order to verify documentation.

NOTE
If the nameplate is missing, it must be replaced according to company policy and procedures.

Step 5 Notify the control center and any other affected personnel that work is beginning.

WARNING!
You should maintain an ongoing awareness of the actuator/operator repair activities being performed to ensure safety during the performance of the task.

Step 6 All applicable task-specific procedures must be followed, which helps to ensure job safety, efficient task performance, standardization, and appropriate documentation. Procedures should also include work permit information and notification requirements (control center, both the upstream and downstream pump stations for main line valves, and any other personnel possibly being affected).

308F24.EPS

Figure 24 ◆ Using a hoist to lift an electric actuator/operator.

CAUTION
You must take care to follow manufacturer's instructions and industry recommendations in order to ensure personal safety, prevent mechanical damage to the actuator/operator, avoid internal damage to the valve being repaired, and preserve the service life of the equipment.

Step 7 Using all of the precautionary steps required by company policy and procedures and those recommended by the manufacturer's instructions, as applicable, to prevent damage and ensure personal safety, remove the actuator/operator from the system.

Step 8 Following the manufacturer's instruction manual to ensure proper procedures and safety, disassemble the actuator/operator.

WARNING!
Some gasket materials may contain asbestos. Company safety policy and procedures for working with asbestos must be followed.

Classroom

Discuss the procedures for performing a general overhaul of an electric valve actuator/operator.

Safety

Stress the importance of being vigilant while working on valves and actuators/operators.

Ask a trainee to explain why it is important to always follow industry recommendations.

Discuss the consequences of working with gasket materials that contain asbestos.

Safety

Explain how to safely remove a spring-loaded clutch shaft.

Classroom

Review the abnormal operating conditions that trainees may encounter while performing maintenance on electric actuators/operators.

Teaching Tip

See the Teaching Tip for Sections 6.0.0–6.8.3 at the end of this module.

Step 9 Inspect, repair, and/or replace worn or damaged parts according to manufacturer's instructions, recommendations, and specifications for the unit.

Step 10 Reassemble the actuator/operator according to the manufacturer's instructions and specifications for safe operation.

Step 11 Set the limit switches, as applicable, according to the manufacturer's instructions for safe and efficient operation of the valve.

Step 12 Disassemble the torque switch if applicable.

> **NOTE**
> Index marks must be made on the soft clutch components, and their positions should be recorded or roughly illustrated on paper in order to have a guide for proper reassembly.

Step 13 Disassemble the soft clutches (if applicable), inspect, and replace them as necessary.

> **WARNING!**
> Care must be exercised when removing the spring-loaded clutch shaft to prevent it from slipping and causing a serious injury.

Step 14 Reassemble the torque switch mechanism, using the manufacturer's instructions, personal notes, and index marks as guides.

Step 15 Set the torque switch, if applicable, according to the manufacturer's specifications for safe and efficient operation.

Step 16 Assemble the gear section of the actuator/operator, if applicable, according to the manufacturer's specifications.

Step 17 Notify control center and any affected personnel to verify the integrity of the system and return the system to normal operation.

Step 18 Complete all required documentation of the repair and results.

> **NOTE**
> Documenting data for compliance, historical review, and trending is an important element of valve maintenance.

6.8.3 Abnormal Operating Conditions

When performing maintenance on electric actuator/operators, situations may be encountered that require you to immediately respond in order to maintain proper operation of pipeline equipment. The following abnormal operating conditions could be encountered.

If a component failure or automated electric valve system failure allows uncontrolled valve travel, de-energize the valve at the power source and/or operate the valve manually.

Should you be unexpectedly confronted with a hazardous liquid or gas such as a stem leak or piping connection leak, attempt to isolate the release. If the release cannot be isolated safely at the site, the control center should be notified to isolate the site or to shut down the line as quickly as possible. You must evacuate the area and only take defensive action using the proper personal protective equipment.

If an unexpected status change occurs, attempt to stop valve travel and return the valve to the proper position. If you are unable to immediately return the valve to the proper position, take action to avoid an emergency condition. Notify the control center as quickly as possible.

In case of fire and/or explosion or ignition of hazardous, flammable liquids or gases, you should evacuate the area and only take defensive action using proper personal protective equipment. Contact the control center to isolate or to shut down the line as quickly as possible.

Instructor's Notes:

Summary

Valve actuators/operators are a key component in the operation and control of the pipeline system. The purpose of the actuator/operator is to convert power into valve movement, opening, closing, or setting a valve. Actuators/operators may be designed to operate by manual, electric, hydraulic, pneumatic, gas power, or a combination of these. In order to maintain safe and efficient operation, actuators/operators must be maintained well.

To perform mechanical work on actuator/operators, mechanical technicians must be trained and qualified and have a strong mechanical aptitude, as well as a certain amount of problem-solving ability. Troubleshooting a valve actuator/operator can be a long, tedious process requiring much patience. Like any other machine, an actuator/operator can be complex. The ability to stick to the task enables a technician to make a well thought-out diagnosis. The technician can then complete a repair or adjustment efficiently and return the pipeline to normal. Discipline is required to follow the many complex procedures required by manufacturer's instructions and company policies and procedures.

The valve actuators/operators govern the pipeline, maintaining safe and efficient operation of valves. In order to do that, actuators/operators must be maintained in good operating condition. It is up to you to learn, study, and train to become qualified to maintain such important equipment.

Classroom

Go over the Review Questions for this module, and answer any questions trainees have.

Review Questions

1. Federal law governing pipeline operations applies to _____.
   a. bench work
   b. inspections
   c. activities dangerous to personnel
   d. diagnostics

2. The major preventative maintenance task performed on electric actuators/operators is _____.
   a. cleaning
   b. lubrication
   c. documentation
   d. painting

3. _____ lubrication applies more to those electric actuators/operators using gear drives or automatic drives.
   a. Internal
   b. External
   c. Remote
   d. Manual

4. Many manufacturers recommend that electric actuators/operators seldom used should be cycled on a _____ basis.
   a. daily
   b. weekly
   c. monthly
   d. yearly

5. When applying grease to an electric actuator/operator, _____ should be consulted for lubricants suited to the climate, products, and type of installation involved.
   a. safety policy
   b. the One-Call system
   c. the E&I department
   d. manufacturer's recommendations

6. Manufacturers install _____ in the tops or specially drilled holes in the housings of some electric actuators/operators for lifting.
   a. lifting lugs
   b. flanges
   c. couplings
   d. hoists

7. When greasing a multiple reduction electric actuator/operator, missing a _____ that may be in an obscure location on several occasions may lead to a failure of that part.
   a. bearing
   b. joint
   c. shaft
   d. grease fitting

8. Some spur gear reduction units have multiple gear trains involving many gears, bearings, and shafts that may have to be disassembled and _____ packed.
   a. grease gun
   b. hand
   c. brush
   d. pressure

Examination

Administer the Module Examination. Be sure to record the results of the Exam on Craft Training Report Form 200 and submit the results to the Training Program Sponsor.

Performance Testing

Have each trainee perform Performance Tasks 4 and 5 to your satisfaction. Fill out a Performance Profile Sheet for each trainee.

Ensure that all Performance Tests have been completed and Performance Profile Sheets for each trainee are filled out. Be sure to record the results of the testing on Craft Training Report Form 200 and submit the results to the Training Program Sponsor.

9. For electric actuators/operators with many parts and complicated gear trains, the most acceptable place to disassemble them is in a well-equipped _____.

a. terminal
b. station
c. shop
d. control center

10. Because _____ an electric actuator/operator involves operating the valve, you should notify the control center and all affected personnel before beginning the task.

a. troubleshooting
b. lubricating
c. disassembling
d. hoisting

11. A safety concern with the electric worm gear actuator/operator is its self-_____ characteristic.

a. cleaning
b. contained
c. regulating
d. locking

12. For electric actuators/operators the appropriate _____ equipment should be secured and in place before any attempt to remove the actuator/operator is begun.

a. hoisting/lifting
b. hydraulic valve
c. repair shop
d. spur gear attachment

13. When reassembling an electric actuator/operator, bolts should be tightened to the manufacturer's _____ specifications.

a. seal
b. gasket
c. torque
d. case

14. _____ are important for stopping the electric actuator/operator before the motor overloads, risking possible damage to the actuator/operator and valve.

a. Limit switches
b. Torque switches
c. Lockout switches
d. Breakers

15. Common industry design dictates that when the quarter-turn valve is fully opened, the stem _____ should line up with the run of the valve.

a. handwheel
b. gear box
c. housing
d. keyway

Instructor's Notes:

GLOSSARY

Trade Terms Introduced in This Module

Mechanical stop screw: An adjustable screw that provides a way of limiting or adjusting the travel of the valve, preventing damage to the valve and valve seats caused by excessive travel.

Torque switch: A switch designed to sense the amount of torque being applied to a machine by an electric motor and to cut off power if torque exceeds a preset limit, preventing damage to the motor.

ACKNOWLEDGMENTS

Figure Credits

John Copeland, Industrial Development Corporation	308F23
Limitorque	308F06, 308F11, 308F12, 308F14, 308F15, 308F16, 308F17, 308F18, 308F19, 308F20
Marathon-Ashland Pipeline Company	308F13
Mokveld Valves	308F07
Steve Sanders	308F24

Instructor's Notes:

MODULE 63308-02 — TEACHING TIPS

The following are suggested activities or instructional methods to help you teach the material in this AIG.

General

When you call on someone to answer a question, the rest of the class relaxes or even tunes out because they expect that the question and answer will take place only between you and the trainee you called on. Instead, use this technique to involve more trainees in answering questions and to keep them on their toes.

1. Ask trainees to define a term or explain a concept.
2. After one trainee has answered, ask a trainee seated nearby if the answer is right. Then ask whether a trainee in the back of the room agrees.
3. Ask trainees to explain why they think an answer is right or wrong.
4. Use the session to clear up incorrect ideas and encourage trainees to learn from their mistakes.

Sections 2.0.0–2.5.0

Preventive Maintenance

This presentation will provide trainees with additional insight to preventive maintenance. For this presentation, invite an operations technician to speak to the class about preventive maintenance procedures for electric valve actuators/operators. Trainees will need their Trainee Guides and pencils and paper to take notes for discussion. Allow between 45 and 60 minutes for this exercise.

1. Tell the trainees that a guest speaker will be presenting information on maintenance procedures for electric valve actuators/operators.
2. Have trainees brainstorm questions for the guest speaker before the speaker arrives. Suggested questions could address the following:
   - Preventive maintenance methods
   - Scheduling
   - Manufacturer's warranties
   - Training
   - Documentation
3. Introduce the guest speaker. Remember to include the speaker's name, title, company, and relevant credentials.
4. Allow 30 to 45 minutes for the speaker and 15 minutes for questions.
5. After the presentation, have trainees share questions with the guest speaker.

Sections 3.0.0–3.3.4

Lubricating Electric Actuators/Operators

This exercise will familiarize trainees with techniques to lubricate electric actuators/operators. For this exercise, you will need tools and materials to apply external and internal lubricants. Trainees will need their Trainee Guides and pencils and paper to take notes for discussion. Allow 45 minutes for this exercise.

1. Set up several stations with the tools and materials for applying external and internal lubricants.
2. Ask trainees to visit each station and practice the following tasks:
   - Applying grease to a valve stem with a brush
   - Applying grease to a fitting using a grease gun
   - Disassembling an actuator/operator bearing housing
   - Applying grease to spur gear reduction units
   - Manually disassembling a bevel gear actuator/operator
   - Applying hand lubrication to a bevel gear actuator/operator
3. Have trainees complete as many of the procedures as time permits. Review and assess trainees' work and provide suggestions for efficient lubrication.
4. Answer any questions trainees have about lubricating electric actuators/operators.

Sections 4.0.0–4.5.0

Inspecting Electric Actuator/Operator Components

This exercise will give trainees the opportunity to practice inspecting electric actuator/operator components. For this exercise, you will need sample P&IDs and inspection checklists. Trainees will need their Trainee Guides and pencils and paper to take notes for discussion. Allow 30 to 45 minutes for this exercise.

1. Set up several stations with P&IDs, inspection checklists, and a description of a hypothetical electric actuator/operator.
2. Divide the class into small groups according to the number of stations.
3. Ask each group to conduct a mock inspection of the hypothetical electric actuator/operator at its station.
4. Discuss techniques for conducting efficient inspections.
5. Answer any questions trainees have about inspecting electric actuator/operator components.

Sections 5.0.0–5.4.0

Troubleshooting Electric Valve Actuators/Operators

This exercise will help trainees diagnose problems with electric actuators/operators. For this exercise, trainees will need their Trainee Guides and pencils and paper to take notes for discussion. Allow 30 minutes for this exercise.

1. Divide the class into small groups. Assign each group a hypothetical problem that could cause an electric valve actuator/operator to malfunction.
2. Have each group diagnose the problem and determine an appropriate solution.
3. Review and discuss trainees' answers. Ask whether trainees have alternative solutions to each group's problem.
4. Provide suggestions from your own experience with troubleshooting electric valve actuators/operators. Answer any questions trainees have.

Sections 6.0.0–6.8.3

Adjusting/Repairing Electric Actuator/Operator Components

This exercise will allow trainees to review the procedures for adjusting and repairing electric actuator/operator components. For this exercise, trainees will need their Trainee Guides and pencils and paper to take notes for discussion. Allow 45 minutes for this exercise.

1. Review the procedures for adjusting and repairing electric actuator/operator components.
2. On the whiteboard/chalkboard, write the first step for each of the following tasks:
   - Removing an electric actuator/operator from the valve stem
   - Removing the spur gear attachment
   - Removing the worm gear assembly
   - Adjusting the valve travel
   - Replacing the stem nut
   - Replacing the motor
   - Installing electric actuators/operators
   - Adjusting electric actuators/operators
   - Repairing electric actuators/operators
3. Ask trainees to identify the subsequent steps until all of the steps are listed in order. Encourage trainees to use their Trainee Guides only when necessary.
4. Ensure that trainees understand how to perform each of the procedures. Emphasize that these procedures are guidelines and that manufacturers' instructions and company policies should always be referenced.
5. Answer any questions trainees have about adjusting and repairing electric actuator/operator components.

MODULE 63308-02 — ANSWERS TO REVIEW QUESTIONS

Answer	Section
1. c	2.0.0
2. b	3.0.0
3. a	3.1.0
4. c	3.0.0
5. d	3.0.0
6. a	3.3.0
7. d	3.3.3
8. b	3.3.1
9. c	4.0.0
10. a	5.0.0
11. d	6.1.0
12. a	6.1.0
13. c	6.3.0
14. b	6.4.3
15. d	6.7.2

CONTREN™ LEARNING SERIES — USER UPDATES

The NCCER makes every effort to keep these textbooks up-to-date and free of technical errors. We appreciate your help in this process. If you have an idea for improving this textbook, or if you find an error, a typographical mistake, or an inaccuracy in NCCER's Contren™ textbooks, please write us, using this form or a photocopy. Be sure to include the exact module number, page number, a detailed description, and the correction, if applicable. Your input will be brought to the attention of the Technical Review Committee. Thank you for your assistance.

Instructors – If you found that additional materials were necessary in order to teach this module effectively, please let us know so that we may include them in the Equipment/Materials list in the Instructor's Guide.

Write: Curriculum Revision and Development Department
National Center for Construction Education and Research
P.O. Box 141104, Gainesville, FL 32614-1104

Fax: 352-334-0932

E-mail: curriculum@nccer.org

Craft ______ Module Name ______

Copyright Date ______ Module Number ______ Page Number(s) ______

Description

______

______

______

(Optional) Correction

______

______

(Optional) Your Name and Address

______

______

Module 63309-02

Maintain, Troubleshoot, and Repair Metering Devices and Provers

Maintain, Troubleshoot, and Repair Metering Devices and Provers Annotated Instructor's Guide

Module 63309-02

MODULE OVERVIEW

This module reviews the techniques used to inspect, maintain, and repair metering devices and provers. Trainees will learn how to repair and calibrate prover systems.

PREREQUISITES

Please refer to the Course Map in the Trainee Module. Prior to training with this module, it is recommended that the trainee shall have successfully completed the following:

Core Curriculum; Pipeline Core, Modules 66101-02 and 66102-02; Pipeline Mechanical Levels One and Two; Pipeline Mechanical Level Three, Modules 63301-02 through 63308-02

OBJECTIVES

Upon completion of this module, the trainee will be able to:

1. Inspect and maintain metering devices.
2. Repair metering devices.
3. Inspect and maintain prover systems.
4. Repair prover systems.
5. Calibrate prover systems.

PERFORMANCE TASKS

Under the supervision of the instructor, the trainee should be able to:

1. Inspect, maintain, and repair metering devices.
2. Inspect, maintain, and repair prover systems.
3. Calibrate prover systems.

NCCER STANDARDIZED CRAFT TRAINING PROGRAM

The National Center for Construction Education and Research (NCCER) provides a standardized national program of accredited craft training. Key features of the program include instructor certification, competency-based training, and performance testing. The program provides trainees, instructors, and companies with a standard form of recognition through a National Craft Training Registry. The program is described in full in the *Guidelines for Accreditation*, published by the NCCER. For more information on standardized craft training, contact the NCCER by writing us at P.O. Box 141104, Gainesville, FL 32614-1104; calling 352-334-0911; or e-mailing info@nccer.org. More information may be found at our Web site, www.nccer.org.

HOW TO USE THIS ANNOTATED INSTRUCTOR'S GUIDE

Each page presents two sections of information. The larger section displays each page exactly as it appears in the Trainee Module. The narrow column ties suggested trainee and instructor actions to each page and provides icons to call your attention to material, safety, audiovisual, or testing requirements. The bottom of each page includes space for your notes.

Teaching Tip

If you see the Teaching Tip icon, that means there is a teaching tip associated with this section. Also refer to any suggested teaching tips at the end of the module.

SAFETY CONSIDERATIONS

Ensure that the trainees are equipped with appropriate personal protective equipment. Emphasize the importance of following all safety precautions and procedures when working with power tools.

PREPARATION

Before teaching this module, you should review the Module Outline, Objectives, Performance Tasks, and the Materials and Equipment List. Be sure to allow ample time to prepare your own training or lesson plan and gather all required equipment and materials.

MATERIALS AND EQUIPMENT LIST

Transparencies

Markers/chalk

Blank acetate sheets

Transparency pens

Pencils and scratch paper

Module Examinations*

Performance Profile Sheets*

Overhead projector and screen

Whiteboard/chalkboard

Appropriate personal protective equipment

Copies of *49 CFR Part 192* (Gas) and/or *49 CFR Part 195* (Liquid)

Copies of your company policy and procedures manual

API Manual of Petroleum Measurement Standards

Typical strainer assembly

Tools and materials to clean a typical strainer assembly and to disassemble a turbine meter

Meter clearance records

Practice meters

Appropriate measuring gauges

Components of a disassembled positive displacement meter

Materials to clean positive displacement meter components

Proving spheres

Calipers

Sizing rings

Steel tape

* Located in the Test Booklet.
**Located at the end of this module.

ADDITIONAL RESOURCE

This module is intended to present thorough resources for task training. The following reference work is suggested for both instructors and motivated trainees interested in further study. This is optional material for continued education rather than for task training.

Manual of Petroleum Measurement Standards. Washington, DC: American Petroleum Institute.

TEACHING TIME FOR THIS MODULE

An outline for use in developing your lesson plan is presented below. Note that each Roman numeral in the outline equates to one session of instruction. Each session has a suggested time period of 2½ hours. This includes 10 minutes at the beginning of each session for administrative tasks and one 10-minute break during the session. Approximately 20 hours are suggested to cover *Maintain, Troubleshoot, and Repair Metering Devices and Provers.* You will need to adjust the time required for hands-on activity and testing based on your class size and resources.

Topic	Planned Time
Session I. Maintaining Metering Devices	
A. Straightening Elements	________
1. Laboratory	________
B. Lubrication and Maintenance of a Mechanical Stack	________
C. Cleaning Strainers	________
1. Laboratory	________
Session II. Repairing Metering Devices, Part One	
A. Repairing Turbine Meters	________
B. Disassembling Turbine Meters	________
1. Laboratory	________
Session III. Repairing Metering Devices, Part Two	
A. Repairing Positive Displacement Meters	________
1. Laboratories	________
B. Troubleshooting Metering Devices	________
1. Performance Testing	________
Session IV. Maintaining Prover Systems	
A. Inspecting Unidirectional Provers	________
B. Inspecting Bidirectional Provers	________
C. Inspecting Tank-Type Provers	________
D. Inspecting Small-Volume Provers	________
Session V. Repairing Prover Systems, Part One	
A. Verifying Valve Integrity	________
1. Laboratory	________
B. Inspecting and Sizing Proving Spheres	________
Session VI. Repairing Prover Systems, Part Two	
A. Verifying Detector Switches	________
B. Verifying Internal Coatings	________
1. Laboratory	________
C. Repairing Backflow Barriers	________
Session VII. Calibrating Prover Systems, Part One	
A. Calibrating Waterdraw	________
1. Laboratory	________
B. Performing Waterdraw Procedures	________
C. Verifying Reliability	________
Session VIII. Calibrating Prover Systems, Part Two	
A. Summary	
1. Summarize module	________
2. Answer questions	________

B. Review Questions ____________

C. Module Examination ____________

1. Trainees must score 70% or higher to receive recognition from the NCCER.
2. Record the testing results on Craft Training Report Form 200 and submit the results to the Training Program Sponsor.

D. Performance Testing ____________

1. Trainees must perform each task to the satisfaction of the instructor to receive recognition from the NCCER.
2. Record the testing results on Craft Training Report Form 200 and submit the results to the Training Program Sponsor.

Module 63309-02

Maintain, Troubleshoot, and Repair Metering Devices and Provers

Instructor's Notes:

Instructor's Notes:

Introduction to the Pipeline Operator Qualification Rule, API, and NCCER

It could be said that petroleum products keep America running. Petroleum products like oil and natural gas are the principal sources of energy in the United States and are used in many manufacturing and chemical processes. Petroleum products are primarily transported by pipeline. There are more than a million miles of liquid and natural gas pipeline in this country, ranging in diameter from two inches to four feet. These pipelines bring crude oil and natural gas from production fields to processing plants and refineries—and ultimately to the consumer.

Pipeline Operator Qualification

Petroleum products are considered hazardous, so pipeline personnel must take extra precautions and follow specific procedures when working with them. To help ensure the safe handling and transportation of petroleum products, the United States Department of Transportation (DOT) mandates safe practices as federal regulations.

One such regulation, the Operator Qualification (OQ) rule, requires pipeline operators to develop a written qualification plan to evaluate their personnel and their contract personnel (see *49 CFR Part 192 Subpart N—Gas,* and *49 CFR Part 195 Subpart G—Liquid*). The evaluation hinges on the operator's ability to perform certain covered tasks and to be able to recognize and respond to abnormal operating conditions that may be encountered while performing these covered tasks.

Covered Tasks

A pipeline performance task is deemed *covered* if the activity passes the following four-part test:

1. Is it performed on a pipeline facility?
2. Is it an operations or maintenance task?
3. Is it performed as a requirement of *49 CFR Part 195 (Liquid)* or *49 CFR Part 192 (Gas)*?
4. Does it affect the operation or integrity of the pipeline facility?

If the answer to all four questions is *yes,* then the activity is considered *covered* and must figure into the evaluation process of pipeline personnel.

American Petroleum Institute

The OQ rule and its related covered tasks provided the catalyst to develop a unified pipeline training and assessment program—so that pipeline operators and owners have ready access to the assessment and training tools they need. This effort was the direct result of the vision and leadership provided by The American Petroleum Institute (API) and API members.

Efforts to unify the industry are nothing new to API. Indeed, the impetus for forming the Institute in 1919 was the need to standardize engineering specifications for drilling and production equipment. Today, the most pressing issues revolve around public perceptions and government policies toward the industry. Speaking with one voice on these issues has become as essential as having interchangeable parts in the field. API's members determine the petroleum industry's positions on public policy as well as the standards that govern its day-to-day operations worldwide.

Consortium on Operator Qualification

API assembled a focus group whose primary purpose was to identify covered tasks and provide compliance recommendations to pipeline operators.

The Consortium on Operator Qualification (COOQ) became the guiding force for the identification and development of covered tasks for the liquid pipeline industry. Their work product, API publication 1161, *Guidance Document for the Qualification of Liquid Pipeline Personnel,* formed the basis for the resulting Pipeliner Training and Assessment Program.

The covered task list created in API publication 1161 has now evolved to include gas covered tasks and is maintained by NCCER. The most current covered task list can be downloaded at www.nccer.org.

National Center for Construction Education and Research

With the guidance and support of API, its members, and the Consortium on Operator Qualification, the National Center for Construction Education and Research (NCCER) was able to design and implement the Pipeliner Training and Assessment Program.

NCCER is a not-for-profit 501(c)(3) education foundation established in 1995 by the world's largest and most progressive construction and maintenance contractors and national contractor associations. NCCER was founded to address the

severe workforce shortage facing the construction and maintenance industries and to develop a standardized training process and curricula. Today, NCCER is supported by hundreds of leading construction and maintenance companies, manufacturers, and national associations and developed more than 35 construction and maintenance titles.

NCCER also maintains a National Registry, which contains the records of trainees and instructors who have received NCCER training. All completion information for every trainee is sent to the NCCER and filed within the Registry. The Registry can then confirm training and skills for workers as they move from company to company, state to state, or even within their own company.

Standardized Curriculum (NCCER's Contren™ Learning Series)

NCCER curricula are developed using the expertise of industry subject matter experts (SMEs). The use of SMEs as an integral part of every new development or revision ensures that the curriculum is current, industry-driven, and indicative of industry practices. The development and/or revision process is facilitated from start to finish by NCCER project managers, who work closely with the SMEs and a select team of technical writer(s) on the development and presentation of the curriculum.

Pipeline: Five Training Paths

The pipeline curriculum focuses on essential skills and competencies within five skill areas:

- Corrosion Control
- Electrical and Instrumentation
- Maintenance
- Mechanical
- Operations (Field and Control Center)

SMEs from the pipeline industry have determined the skills and competencies needed based on API's covered tasks and the consensus of the group/industry. Specific company information is generally not incorporated into the curriculum unless the SMEs determine that the information is the industry standard.

Training Modules and Covered Tasks

The smallest unit of instruction in the curriculum is a *module*. Each curriculum is composed of modules (the equivalent of stand-alone chapters) that are logically sequenced into complete *levels* of instruction. Every module is task-oriented and requires the successful completion of a written (knowledge-based) examination and a performance (hands-on) test. Modules may be individually selected and/or combined for efficient task training. Modules in the pipeline curricula, to the extent possible, will align with API-developed covered tasks. While not all modules will relate to covered tasks, all covered tasks will relate to at least one module.

Benefits of Training to the Pipeline Marketplace

Applying NCCER's standardized training process allows pipeline owners and contractors in the United States to establish the benchmark for acceptable skill and competency levels within the pipeline industry. This process will assist training organizations in developing and/or improving standardized pipeline training programs that effectively match their resources and operating conditions while responding to regulatory demands and to the needs of the industry. The process will also recognize training organizations that meet the minimum standards of competence, warranting regulatory, public, and professional confidence. This in turn will foster national unity within the pipeline industry for the mutual benefit of individual technicians, owners, contractors, and ultimately pipeline-users and society.

Additional Resources

To review the Pipeliner Training and Assessment Program Implementation and Resource Guide and the latest version of pipeline covered tasks, visit NCCER's Web site at www.nccer.org, or contact:

National Center for Construction Education and Research
PO Box 141104
Gainesville, FL 32614-1104
(888) 622-3720
Web site: www.nccer.org

For more information on API, please contact:

American Petroleum Institute
1220 L Street, Northwest
Washington, D.C. 20005
(202) 682-8000
Web site: www.api.org

To order *API Pub 1161*, contact Global Engineering Documents by calling 1-800-854-7149 or visiting their Web site at www.global.his.com.

Additional sources of information on the pipeline operator qualification regulation include:

DOT – Office of Pipeline Safety
www.ops.dot.gov

Transportation Safety Institute
www.tsi.dot.gov

Instructor's Notes:

COURSE MAP

This course map shows all of the modules in the third level of the Pipeline Mechanical curriculum. The suggested training order begins at the bottom and proceeds up. Skill levels increase as you advance on the course map. The local Training Program Sponsor may adjust the training order.

PIPELINE MECHANICAL LEVEL THREE

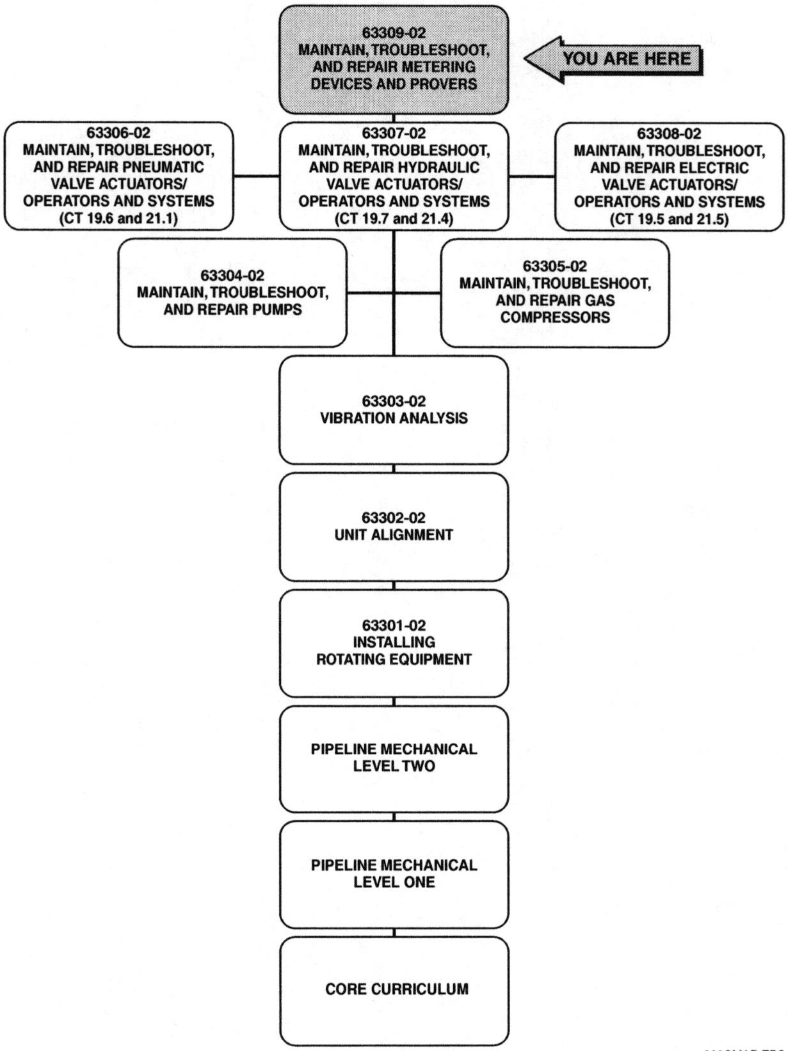

309CMAP.EPS

Homework

Assign reading of Module 63309-02.

Instructor's Notes:

MODULE 63309-02 CONTENTS

Figures

Instructor's Notes:

MODULE 63309-02

Maintain, Troubleshoot, and Repair Metering Devices and Provers

Objectives

When you have completed this module, you will be able to do the following:

1. Inspect and maintain metering devices.
2. Repair metering devices.
3. Inspect and maintain prover systems.
4. Repair prover systems.
5. Calibrate prover systems.

Prerequisites

Before you begin this module, it is recommended that you successfully complete the following: Core Curriculum; Pipeline Core, Modules 66101-02 and 66102-02; Pipeline Mechanical Level One; Pipeline Mechanical Level Two; Pipeline Mechanical Level Three, Modules 63301-02 through 63308-02.

Required Trainee Materials

1. Pencil and paper
2. Appropriate personal protective equipment

1.0.0 ◆ INTRODUCTION

Meters used in the pipeline industry must be closely calibrated and monitored. Accuracy of meters used for custody transfer must be certified and documented. Accurate measurement is the key to keeping losses to a minimum for both company and customer. Mechanical technicians are responsible for keeping meters and associated equipment in good condition to maintain the necessary accuracy. To ensure accuracy of measurement and protect the company s reputation for integrity, meters are subjected to regular proving operations. Proving establishes the level of accuracy of individual meters.

It is the responsibility of the mechanical technician to inspect, troubleshoot, and repair meters and provers that experience wear, changes in accuracy, or failure. Mechanical technicians must be trained and qualified to perform the work on meters and provers. Ongoing training and experience must be combined with mechanical skills, problem-solving ability, and the discipline to read and follow manufacturer s technical manuals and company procedures.

This module provides general information on the repair procedures involved in maintaining meters and provers.

2.0.0 ◆ MAINTAINING METERING DEVICES

Inspecting and maintaining metering equipment is an ongoing process that should be done meticulously. Operations personnel should be encouraged to notice the condition of the equipment as they are working and to report problems promptly. Mechanical technicians should visually inspect the metering banks and associated equipment whenever they are called to the site for repairs and other maintenance tasks. Many problems can be discovered by personnel during their routine work simply through alertness. Personnel should take note of any condition that might affect the operation and accuracy of the metering and proving equipment and report it in accordance with company policy. Although many of the important parts of meters and provers cannot be inspected unless the units

Materials

Ensure that you have everything required to teach the course. Check the Materials and Equipment List at the front of this module.

Classroom

Explain that terms shown in bold are defined in the Glossary at the back of this module.

Audiovisual

Show Transparency 1, Course Objectives.

Show Transparency 2, Performance Tasks.

Teaching Tip

See the general Teaching Tip at the end of this module.

Classroom

Explain how provers are used to ensure the accuracy of metering devices.

Discuss the role of mechanical technicians in inspecting metering banks and maintaining metering equipment.

Classroom

Review the conditions to observe when inspecting metering equipment.

Audiovisual

Show Transparency 3 (Figure 1). Review the components in a flow-straightening element.

Classroom

Explain how to avoid swirl generation in the pipeline using fluid-straightening elements.

Laboratory

Refer to the *API Manual of Petroleum Measurement Standards.* Ask trainees to read and discuss the standards in Chapter 5. Answer any questions trainees have.

are disassembled, some conditions that should be noticed include:

- Liquid product or vapor being released to the environment
- Strong or unusual odors
- Loose, damaged, or broken wiring and connections
- Damaged or bent gauges or gauges with broken glasses
- Damaged thermometers
- Visible damage to cases, housings, or piping
- Loose or broken parts on meters, counters, or other accessories
- Unusual noises

NOTE
Metering and proving installations require a zero tolerance for leakage, both internally and externally.

Company policy should establish a clear division of maintenance tasks according to the complexity of the task, allowing less experienced personnel to perform daily maintenance, such as replacing a gauge, and leaving more complex tasks for trained and qualified mechanical technicians.

2.1.0 Straightening Elements

To achieve accurate measurement, the fluid entering a meter should be in a calm, evenly flowing condition. In turbine meter installations, straightening elements are installed in the pipe to control the flow by reducing the swirling motion in the fluid before it reaches the meter. Standards for installing straightening devices appear in the *API Manual of Petroleum Measurement Standards, Chapter 5, Metering, Section 3, "Measurement of Liquid Hydrocarbons by Turbine Meters."*

Flow-straightening elements (see *Figure 1*) normally consist of a cluster of tubes or long metal vanes that are installed lengthwise in the portion of pipe immediately preceding the meter. Straightening elements can also be made from a group of perforated discs or wire mesh screens installed in a series. However, disc and screen types are not as common, because of the larger pressure drop resulting from their design.

Fluid-straightening elements must be long enough to relieve swirling conditions in the pipe, to prevent the same conditions from reforming, and strong enough to withstand the highest flow rates. Elements must be designed so that they will not create more swirl as the fluid passes on toward the meter. In order to avoid the generation of swirl, follow these guidelines:

- Cross sections of the straightening elements in the pipe should be as smooth and evenly proportioned as possible.
- Elements should be strong enough to resist any kind of deformation or change resulting from high flow rates.
- Construction should be free of any type of metal projections, such as rough welding or other objects that could interfere with flow.
- Meter runs should always be located far enough from equipment or piping that would cause swirl or any kind of rough velocity profile.
- The flow-straightening section should be at least the minimum length required by API standards for that installation.
- Pipe sizes for the meter run and the meter bore should be concentric and properly aligned.
- Flanges and gaskets must be aligned with the bore of the pipe and have no edges protruding inside the circumference of the bore.

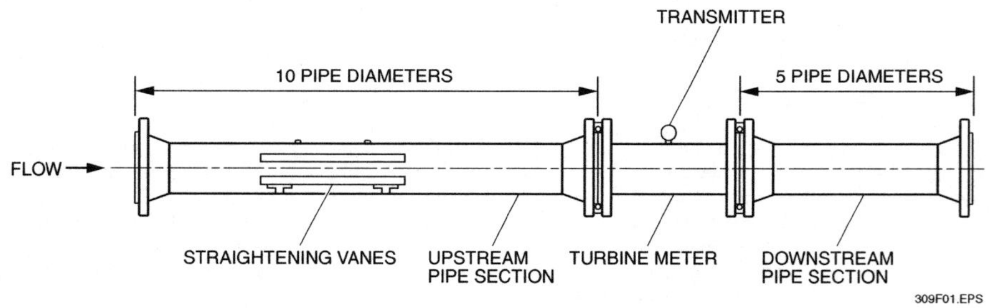

Figure 1 ◆ Flow-straightening elements.

Instructor's Notes:

2.2.0 Lubrication and Maintenance of a Mechanical Stack

Mechanical metering stacks are installed above meters. The stack displays the total amount of liquid measured by the meter as it turns. The mechanical stack is driven by a shaft that penetrates the housing and engages the rotor of the meter. The counter drivetrain, calibrator, and mechanical stack are connected by small gears on top of the meter (see *Figure 2*).

There are three basic divisions of the counter: the set of gears and shaft, the packing gland, and the calibrator. The packing gland is the pressure seal for the shaft that penetrates the meter housing. It requires the occasional addition of the appropriate lubricant to enable it to maintain a good seal.

The calibrator is a complicated mechanism containing a clutching assembly that allows the shaft to slip slightly as the meter rotates. While the meter rotates smoothly, the clutch slippage at every revolution of the meter causes the shaft to the stack to move somewhat in a cycle, adjusting the meter reading. It should be remembered that when proving this type of meter setup, this movement makes it harder to achieve repeatability.

The mechanical stack (see *Figure 3*) normally includes one or more of the following units:

- Compensators
- Right-angle drives
- Counters
- Pulse generators
- Mechanical register

These additional units on the stack and the bearings, shafts, and gears associated with them increase the friction load, or drag, on the meter. Lubrication and maintenance of these small parts and bearings is necessary to overcome the effects of the extra load on the meter. With an instrument of such accuracy, manufacturer's instructions and recommendations should be followed closely.

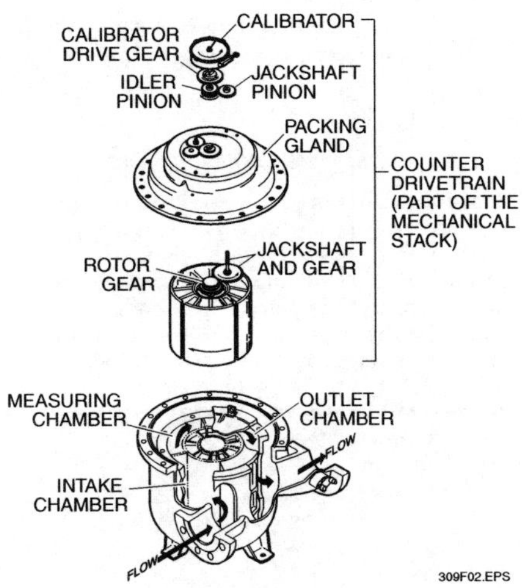

Figure 2 ◆ Layout of a typical PD meter and drivetrain.

2.3.0 Cleaning Strainers

Strainers are installed in the pipeline upstream of meters. The purpose of the strainers is to protect the meters from debris, sand, dirt, rust, welding slag, or any other contaminants that might pass through the pipeline and damage the meters. Strainers are especially important for the protection of turbine metering installations.

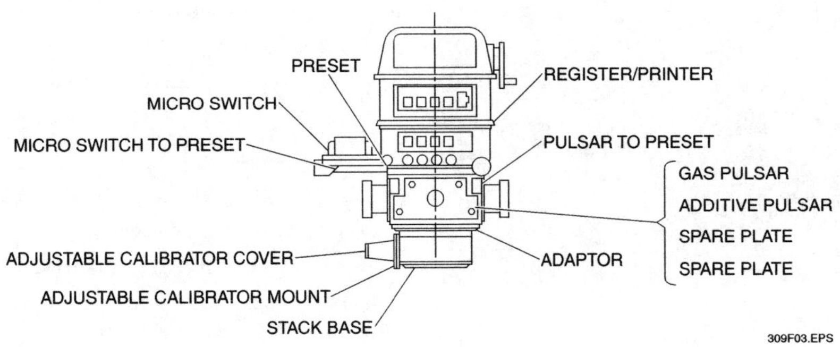

Figure 3 ◆ Mechanical stack.

Audiovisual

Show Transparency 4 (Figure 2). Explain how mechanical stacks are installed above meters.

Classroom

Review the basic divisions of a meter's counter.

Refer to Figure 3. Discuss the units that may increase friction load or drag on the meter.

![Classroom]

Explain the role of the strainer in the pipeline. Discuss the effects of clogs on pressure in the pipeline.

Review the essential preparations and general procedures for cleaning a strainer.

![Safety]

Explain why vehicles with internal combustion engines should not be allowed near a strainer cleaning procedure.

Strainers become clogged and must be cleaned periodically. One of the main indicators that the strainer is clogged is an excessive pressure differential from one side of the strainer to the other, as indicated by the gauges. If the flow of liquid cannot be shut off to clean the strainer, two strainers should be installed so that one can be bypassed while the other is being cleaned. Some strainers can be backwashed to clean them, and the backwashed material is then diverted to a sump.

Because of the size of some operations, the strainers must be large and may require overhead cranes or lifting devices to change them. The larger strainers may require a heavy impact wrench and a chain hoist to open them. Mechanical technicians are responsible for this maintenance task. *Figure 4* shows typical strainers installed upstream of a strainer installation with permanently installed overhead cranes for cleaning the strainers.

2.3.1 Preliminary Preparations

Before attempting to clean out any strainer, you should make proper preparation. Company policy and procedure should be followed. Some of the procedures that should be completed before attempting to clean a strainer include:

Figure 4 ◆ Layout of a typical strainer.

- Securing the appropriate work orders and permits
- Enlisting enough personnel to accomplish the job (at least two)
- Obtaining the proper PPE, gas meters, and other equipment
- Gathering a supply of cleanup supplies
- Collecting the appropriate wrenches and tools for the job
- Securing appropriate hoisting and rigging equipment
- Requisitioning the needed components, screens, and filters
- Obtaining the appropriate manufacturer's instruction manual
- Posting any warning signs that are applicable
- Completing any other required company procedures

2.3.2 Strainer Cleaning Procedure

Because strainer models vary, it is not possible to give one set of procedures that would apply accurately to every model. You should follow the manufacturer's instructions and recommendations for the exact model, along with company policy. However, these general steps might be followed to clean a strainer and restore it to serviceability.

> **WARNING!**
> Hot work must be prohibited near a strainer cleaning procedure. Vehicles with internal combustion engines should not be allowed in the proximity of this procedure, whether in the field or in a building with vents and windows.

Step 1 Notify the control center that the operation is beginning.

Step 2 Follow control center instructions.

Step 3 Isolate all energy sources associated with the strainers.

Step 4 Close, chain, and lock all valve handwheels.

Step 5 Open, lock out, and tag the related breakers.

Step 6 Verify that all valves are closed.

Step 7 Verify that associated breakers are open.

Step 8 Engage local stop button safety mechanisms (if applicable).

Step 9 Deactivate associated alarms (if applicable).

Instructor's Notes:

Step 10 Gradually open each drain valve at the meter.

> **CAUTION**
> Don't leave the strainer drain valve closed as it will clog and damage the equipment.

Step 11 Verify that the strainer pressure gauge is at zero.

Step 12 Gradually open the strainer cover vent.

> **CAUTION**
> If liquid flows from the vent, close the vent and hand-seat the metering installation valves to stop all flow.

Step 13 Verify that the system is draining by detecting air being drawn into the vent.

Step 14 Continue draining until no air is being drawn into the strainer vent and the sump level stops increasing.

> **WARNING!**
> Full-face respirators must be worn in the following steps of the procedure to protect your eyes and face from splashing fluid.

Step 15 Slowly loosen the cover-positioning mechanism.

> **WARNING!**
> If a whistling noise is heard or if liquid leaks from the screws, allow the pressure to bleed off before continuing. Otherwise, a sudden release of pressure may cause fluid to flash and injure personnel.

Step 16 Raise the positioning mechanism clear of the fasteners.

Step 17 Open the clamping mechanism to allow the cover to be lifted past it.

Step 18 Raise the cover clear of the clamping mechanism, but hold the position just above the strainer, allowing excess liquid to drain back into the strainer housing.

Step 19 Prepare absorbent material near the strainer to catch the residual drainage when the cover is completely removed.

Step 20 Hoist the cover to the side, and lower it onto the absorbent material.

> **CAUTION**
> Take care to avoid damaging small fittings and pipes when using hoisting equipment to lift the cover.

> **WARNING!**
> Never allow any body part or extremity to pass under any part of a load being lifted or moved with hoisting equipment. The load could drop or fall, causing serious injury to personnel below.

Step 21 Take out the strainer components and place them in the appropriate container (see *Figure 5*).

Step 22 Clean out any residue inside the strainer housing.

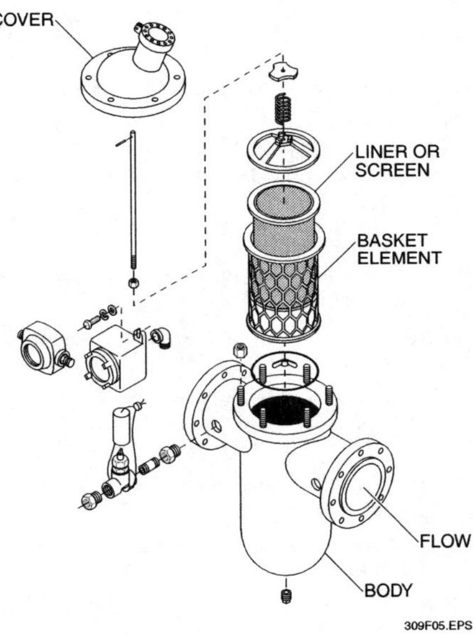

Figure 5 ◆ Layout of a typical strainer assembly.

Safety

Discuss the consequences of leaving the strainer drain valve closed.

Stress the importance of stopping all flow of liquid from the vent.

Emphasize that full-face respirators must be worn during the strainer cleaning process.

Ask a trainee to explain what to do if a whistling noise is heard or if liquid leaks from the screws.

Explain problems that can result when using hoisting equipment.

Stress that trainees must not pass under any part of a load being lifted or moved.

Audiovisual

Show Transparency 5 (Figure 5). Explain how to properly remove the strainer components.

Classroom

Explain how to avoid stripping when clamping the mechanism bolts.

Laboratory

Set up a station with a typical strainer assembly. Ask a trainee to demonstrate the cleaning procedures. Have trainees observe the procedure and identify what was done correctly and/or incorrectly.

Classroom

Explain how to return the strainer to service. Go over any questions trainees have.

Homework

Have trainees review Sections 3.0.0–3.1.2 for the next session.

Materials

Ensure that you have everything required for the laboratory during this session.

Step 23 Install the new strainer components.

Step 24 Clean the strainer housing.

Step 25 Clean and dry all the closure components and the cover.

Step 26 Examine the cover O-ring seal and replace it if necessary.

Step 27 Hoist the cover back over the strainer housing.

Step 28 Carefully lower the cover into position on the housing.

Step 29 Maneuver the cover into its exact position using the hoist.

Step 30 Close the clamping mechanism, tightening the bolts equally according to the manufacturer's recommended torque.

> **CAUTION**
> To avoid stripping, do not overtighten the clamping mechanism bolts.

Step 31 Make certain that the cover and assembly components are correctly positioned.

Step 32 Complete all the documentation.

2.3.3 Returning the Strainer to Service

The cleaned strainer must be returned to service by the mechanical technicians. These general steps might be followed to return the strainer to service.

Step 1 Notify the control center that startup is beginning.

Step 2 Follow the control center's instructions to proceed.

Step 3 Close all drain valves and the strainer vent.

Step 4 Very slowly open the inlet valve, and allow the strainer and metering installation to fill with liquid.

Step 5 Observe the strainer closure for signs of release.

Step 6 Release all lockout and isolation protection.

Step 7 Open valves in the appropriate sequence to bring the unit on-line.

Step 8 Notify the control center that the unit is now on-line.

Step 9 Transport the clogged strainer to the appropriate location for steam cleaning.

Step 10 Complete all the required documentation.

3.0.0 ◆ REPAIRING METERING DEVICES

As mentioned previously, metering devices are usually trouble-free for long periods. However, when a problem does occur, it is the responsibility of the mechanical technician to go to the site, troubleshoot the system to determine what the problem is, and devise a plan of action to solve the problem as efficiently as possible. To do that, you must have a strong understanding of meters and all the associated equipment and must rely heavily on training, experience, and the manufacturer's manuals.

The first indication that something is wrong with a meter is normally a change or a trend toward change as displayed by the meter factor when proving the meter. Company policy designates how much change in the meter factor is allowable before intervention is required. By calculation, it is determined whether the meter has reached the action level or not. The trend is demonstrated by the proving record on the meter factor log sheet (see *Figure 6*).

3.1.0 Repairing Turbine Meters

If a turbine meter requires internal cleaning and repair, it is generally removed from the installation and carried to a shop. Some meters are returned to the factory for service, especially if still under warranty.

Although most meters are carried to a shop or returned to the factory, some may be repaired in the field. Field repair to the meters could apply if the debris in the meter can be easily cleaned by disassembling the meter or if you are in a location that is far from a station and you do not have a replacement.

A turbine meter contains few parts to maintain, and a simple inspection may reveal the problem. For example, all meters should turn freely; if not, the problem is in the shaft mounting or bearings, and the meter's accuracy will decrease. You can test for this condition by blowing into the meter or using low-pressure compressed air. The application of gentle air should cause the rotor to spin and come slowly to a stop. Some rotors may then rock back and forth a few times as the magnet in the pickup aligns with the rotor. If the rotor stops abruptly, it suggests the bearings need cleaning. This can often be accomplished by using a compatible solvent to flush the bearings. Many manufacturers recommend this procedure as opposed to disassembly.

Classroom

Refer to Figure 6. Explain how to review a meter factor log sheet and how to identify a change or trend that may indicate a malfunction.

Explain when a turbine meter should be repaired in the field and when it should be returned to the factory for service.

Instructor's Notes:

XYZ Company

METER FACTOR LOGGING SHEET

Location: Anywhere | **Meter Make/Style:** Generic PD | **Flow Range: Min:** 85 **Max:** 850 BPH

Serial #: G123456 | **Temp Comp: Yes/No** ATG | **Action Limit:** ±0.50%

Product: Crude | **ProverType:** Bi-Dir Eall | **Prover Serial #:** C12345

Sequence No.	Date	Meter Factor	Proving Report #	Meter Temp.	Meter Press.	Gravity API@60°	Flow Rate	% Deviation Last Proving	% Deviation Baseline	Remarks
1	1-2-87	1.0005	101	78.0	60	35.5	180	INITIAL	INITIAL	Baseline factor
2	2-3-87	1.0008	102	76.5	58	35.4	178	+0.03	+0.03	
3	3-5-87	1.0010	103	80.5	61	35.8	181	+0.02	+0.05	
4	4-3-87	1.0015	104	81.2	60	36.0	180	+0.05	+0.10	
5	5-5-87	1.0021	105	84.0	61	35.8	179	+0.06	+0.16	
6	6-4-87	1.0019	106	83.5	60	35.5	182	-0.02	+0.14	
7	7-7-87	1.0028	107	87.0	62	35.8	180	+0.09	+0.23	
8	8-4-87	1.0037	108	88.0	60	35.4	180	+0.09	+0.32	
9	9-6-87	1.0048	109	89.0	60	35.8	181	+0.11	+0.43	
10	10-3-87	1.0042	110	85.0	60	35.2	185	+0.06	+0.37	
11	11-4-87	1.0061	111	81.0	60	35.4	180	+0.19	+0.55	Repaired meter
12	11-5-87	1.0002	112	81.5	59	35.6	176	–	–	First factor
13	11-6-87	1.0010	113	83.5	61	36.1	181	–	–	Baseline factor
14	12-8-87	1.0002	114	78.5	62	35.7	175	-0.08	-0.08	
15	1-4-88	1.0009	115	77.0	61	36.1	172	+0.07	-0.01	
16	2-6-88	1.0018	116	74.5	63	36.6	170	+0.09	+0.08	
17	3-4-88	1.0015	117	76.0	61	36.1	175	-0.03	+0.05	
18	4-6-88	1.0028	118	81.5	60	36.0	180	+0.13	+0.18	
19	5-4-88	1.0020	119	82.5	59	35.9	181	-0.08	+0.10	

309F06.EPS

Figure 6 ◈ Meter factor logging sheet.

Safety

Explain why unregulated, high-pressure compressed air should not be used to test meters.

Classroom

Review the procedures to remove a turbine meter for repair. Ask whether trainees have any questions.

Safety

Emphasize that lifting and hoisting equipment must be properly secured before removing large meters.

Ask a trainee to explain how to properly support a meter.

Audiovisual

Show Transparency 6 (Figure 7). Review the layout of a disassembled meter.

> **CAUTION**
> Do not use unregulated, high-pressure compressed air to test the meter as this may overspin the meter, causing serious damage.

If the bearings in a meter are made of graphite, caution should be exercised when handling them. Graphite bearings are very hard and brittle and break easily. Operators of such meters should keep spare sets on hand. Often graphite bearings break if dropped or if the meter receives a hard jolt.

3.1.1 Removing Turbine Meters for Repair

Meters can be expected to operate with minimal problems for extended periods unless they are damaged. Haphazard mechanical turning and disassembly of meters is neither necessary nor recommended. Removal of an operational meter is not recommended unless necessary. The manufacturer's recommendations and company policy should be followed.

When technicians determine that a meter requires removal, company policy must be followed. Once a work order has been issued and all required permits have been secured, the mechanical technician can proceed to the site and begin preparation for the repair.

Before beginning the repair, you should complete the following steps:

Step 1 Secure the appropriate manufacturer's manuals.

Step 2 Notify the control center.

Step 3 Observe all company-required safety procedures, including the wearing of PPE.

Step 4 Isolate the metering installation.

Step 5 Drain off all pressure from the system.

Step 6 Lock out and tag all energy sources.

After completing the preliminary steps, follow the manufacturer's recommendations and company procedure for removing the meter from the installation. Some of the parts of the meter are only accessible when disassembled. Work on a meter should only be attempted by trained and qualified mechanical technicians, preferably in a shop, or by the manufacturer. If it is necessary to disassemble the unit, follow the manufacturer's instructions and recommendations. Keep written notes of the procedure. Make drawings of assemblies for reference when reassembling the unit.

> **CAUTION**
> For large meters, the appropriate lifting or hoisting equipment must be secured and in place before beginning removal.

Step 1 Make certain that company policy and procedure for repairs to metering equipment are followed.

Step 2 Support the meter so that it cannot fall.

> **CAUTION**
> Take care to prevent the meter from falling or being dropped.

Step 3 Disconnect any accessories and wiring that may be involved.

Step 4 Remove the bolts from the flanges on both ends of the meter.

Step 5 Carefully lift the meter from its place, either by hand or by hoist, as required by size.

Step 6 Place the appropriate gaskets or sealing devices in place on the flanges.

Step 7 Carefully position the replacement meter into the proper location.

Step 8 Install the bolts and tighten them according to factory recommendations.

Step 9 Reattach electrical connections to the meter.

Step 10 Restore the metering system to operation.

Step 11 Make the appropriate notifications, and document the repair.

3.1.2 Disassembling a Turbine Meter

Meters come in many brands and designs, and procedures for disassembly vary widely. The procedure supplied here is general in nature. Mechanical technicians should consult the manufacturer's instructions for the exact model before proceeding. However, these general steps might be followed to disassemble a turbine meter (see *Figure 7*):

Step 1 Use the appropriate tool to remove the sending unit and associated wiring.

Step 2 Take the rotor and support mechanism out of the meter, handling them carefully to avoid damaging the rotor.

Instructor's Notes:

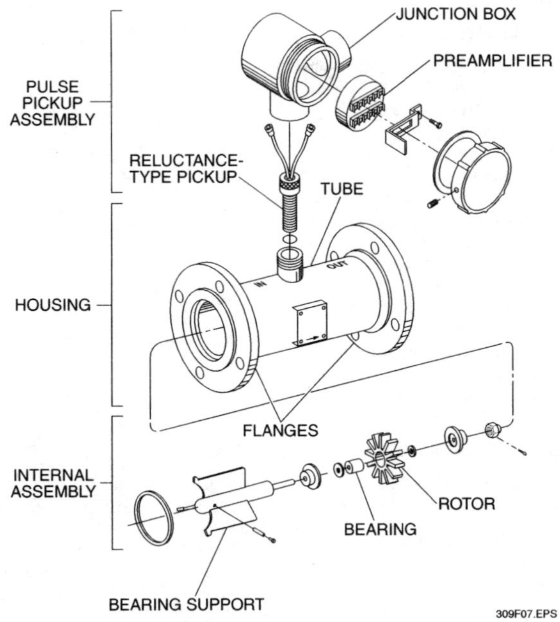

Figure 7 ◆ Layout of a disassembled turbine meter.

Step 3 Wash the rotor and support assembly in a recommended solvent or alcohol.

Step 4 If bearing wear or damage is observed, company policy and procedure should be followed to repair the unit or return it to a shop.

> **NOTE**
> Some meters are designed with a replaceable cartridge containing the rotor and bearing assembly as a unit.

> **CAUTION**
> If bearings, rotors, or other components are to be replaced in the field, you must verify that the part to be used is the exact replacement. Incorrect parts can cause severe damage or failure.

Step 5 Reassembly requires following the reverse of the disassembly.

Step 6 Test the meter for free operation by gently blowing air through the unit or spinning it by hand.

> **CAUTION**
> To avoid damage to the mechanism, do not spin the meter at high speed.

Step 7 Flush the lines of the metering assembly.

Step 8 Reinstall the meter in the metering installation.

Step 9 Make appropriate notifications.

Step 10 Return the metering installation to operation.

Step 11 Prove the meter immediately.

Step 12 Make appropriate notifications and prepare all required documentation.

Laboratory

Bring in a turbine meter. Ask a trainee to demonstrate how to disassemble the meter. Have trainees identify any bearing wear or damage.

Safety

Discuss the consequences of using incorrect replacement parts.

Explain how to prevent problems caused by spinning the meter at high speeds.

Homework

Have trainees review Sections 3.2.0–4.0.0 for the next session.

Materials

Ensure that you have everything required for the laboratories and testing during this session.

Classroom

Review preparations for repairing a positive displacement meter.

Laboratory

Distribute copies of meter clearance records. Ensure that trainees understand the importance of referencing these records. Ask trainees to identify measurement information that you specify.

Safety

Explain how to relieve the pressure from an operational meter before beginning disassembly.

Discuss problems that can result if rotor components are damaged during disassembly.

3.2.0 Repairing Positive Displacement Meters

Most positive displacement (PD) meters can be disassembled and repaired in the field while they are still physically connected to the pipeline. If a PD meter has to be removed, refer to the general steps explained earlier for removing a turbine meter.

Before beginning disassembly or removal of a PD meter, mechanical technicians should make certain to secure the manufacturer's instruction manual for the exact model of PD meter to be disassembled. Manuals should also be secured for the accessories that may be included in the installation. Company policy and procedure must be followed in all cases. The procedures presented here include disassembly of the PD meter, assembly of the meter, adjustments that must be made following assembly, and replacing components.

3.2.1 Making Preliminary Preparations

In preparing to repair a PD meter, technicians should accomplish the following preliminary tasks:

- Obtain the appropriate work permits.
- Make proper notifications.
- Obtain the required PPE.
- Secure any specialty or manufacturer-supplied tools required.
- Acquire a set of feeler gauges and a supply of emery cloth.
- Requisition the commonly needed parts and components.
- Prepare a wooden support block for the rotor assembly (see *Figure 8*).
- Obtain a meter clearance record (see *Figure 9*).

> **NOTE**
> Use a meter clearance record to document all the facts about the meter's condition. Some companies supply meter clearance records so that all important measurement information can be easily recorded in a well-organized method. If a meter clearance record is not available, technicians should carefully record the clearances of each component on a sheet of paper or in a notebook.

3.2.2 Disassembling a PD Meter

All accessories, including the calibrator, metering stack, air eliminator, and strainer should be examined closely and eliminated as the cause of a problem before disassembly of the meter is considered. Minor adjustments on the meter should also be evaluated as a possible cause of the problem.

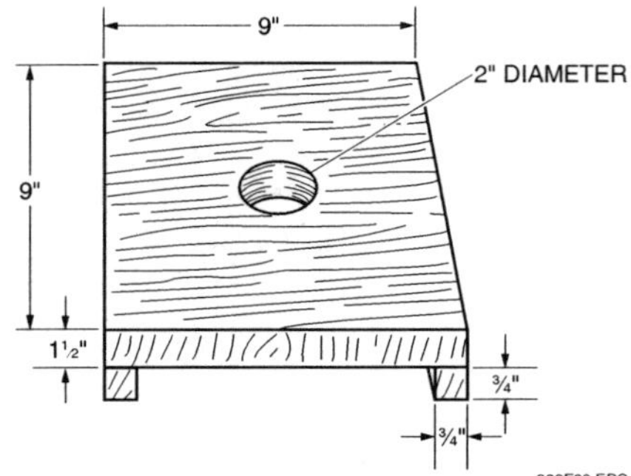

Figure 8 ◆ Prepared rotor support block.

Once normal operation of all accessories and adjustments has been established, the disassembly of the meter is the next logical step.

The disassembly of a meter should follow the sequence of steps and procedures that are specified in the manufacturer's instruction manual for the exact model. Although this module cannot supply a single set of procedures that will apply to all brands and models of PD meters, the technician may follow these general steps to disassemble a PD meter.

> **WARNING!**
> All pressure must be relieved from an operational meter before beginning disassembly. High-pressure leaks can cause serious injury to eyes and skin.

Step 1 Notify the control center that the procedure is beginning.

Step 2 Isolate the meter from the system.

Step 3 Lock out and tag all energy sources.

Step 4 Drain off all pressure and liquid.

Step 5 Dismantle the top accessories.

Step 6 Remove meter top bolts.

Step 7 Pry the meter top loose, and lift it off.

Step 8 Carefully raise the rotor out of the housing.

Step 9 Invert the rotor and place it on a prepared support block.

> **CAUTION**
> Use care. Rotors are assembled with perfectly matched components at the factory and must be replaced as a unit if any part is damaged.

Instructor's Notes:

Meter Clearance Record

Meter & Model________________ Serial Number________________

Item	Date	As Found	As Assembled	Date	As Found	As Assembled
Rotor to Block						
Blade Adj. Total End Clearance						
Blade Slot						
Blade end to Rotor						
Blade Roller Over Radius Portion of Cam						
Blade Tip Toward Housing						
Item	**Date**	**As Found**	**As Assembled**	**Date**	**As Found**	**As Assembled**
Rotor to Block						
Blade Adj. Total End Clearance						
Blade Slot						
Blade end to Rotor						
Blade Roller Over Radius Portion of Cam						
Blade Tip Toward Housing						

309F09.EPS

Figure 9 ◆ Meter clearance record.

3.2.3 Disassembling a PD Rotor

Rotor assemblies should be taken apart for inspection and cleaning using great care. To disassemble the rotor, technicians should follow these general steps:

Step 1 Loosen and remove the locating arm.

Step 2 Take off the retaining ring and key.

Step 3 Take off the blade blocks and clamps.

Step 4 Simultaneously pry opposite sides of the rotor cover, using two prying tools.

> **CAUTION**
> Pry upward only, so as not to damage, mark, or create burrs on the sides of the rotor.

> **NOTE**
> Observe and mark the positions of the locating pins for future reference at reassembly.

Step 5 Disassemble the rotor cover assembly.

Step 6 Examine the bearing for wear or damage and replace according to company procedure and the manufacturer's instructions.

> **NOTE**
> If mechanical technicians do not have the proper equipment to align, reset, and glue tungsten carbide bearings in place, this assembly may have to be replaced with a completely new assembly from the manufacturer.

Step 7 Mark the back edge of each rotor blade and its respective slot in the rotor with a magic marker so that the factory-matched blades can be reassembled in their place.

Step 8 Lift and work the lower blade assembly out of the rotor, using care not to damage the blade edges.

> **NOTE**
> Although the terms "upper" and "lower" seem reversed, it must be remembered that the rotor unit is inverted on the support.

> **CAUTION**
> Use care not to damage the blade edge, as damage to any part of the rotor assembly will require the replacement of the entire unit.

Classroom

Review the procedures to disassemble a positive displacement rotor.

Safety

Explain how to properly pry the opposite sides of the rotor cover.

Stress that damage to any part of the rotor assembly requires that the entire unit be replaced.

Laboratory

Set up stations with the components of a disassembled positive displacement meter and the materials to clean them. Ask trainees to clean and inspect the components.

Classroom

Discuss the special considerations for assembling a positive displacement meter.

Review the procedures to make end clearance and blade tip adjustments. Answer any questions trainees have.

Step 9 Examine and replace blade rollers as necessary, according to the manufacturer's instructions.

Step 10 Remove the blade cam and shaft.

Step 11 Pull the blade cam and key off the shaft.

Step 12 Index the upper blade assembly and rotor with a magic marker.

Step 13 Take out the upper blade assembly, taking care not to lose the bearing key.

Step 14 Take out the clamping ring.

Step 15 Remove the snap ring and lift out the thrust bearing set.

> **NOTE**
> Two types of thrust bearings are used in various applications, standard and tungsten carbide.

Step 16 Invert the rotor and take off the gear plate.

The PD meter has now been completely disassembled. Technicians should clean all parts and closely examine each unit for wear, damage, or any sign of problems. Parts that are in question should be replaced as applicable. Company procedure and manufacturer's instructions and recommendations should be followed.

3.2.4 Assembling a PD Meter

Assembly of the PD Meter is essentially the reverse of the steps taken to disassemble the unit. However, some steps must be performed with care. Manufacturer's instructions and recommendations should be followed. All clearances and settings should be adjusted as specified by the manufacturer. Manual schematics and line drawings should be studied closely. As the technician proceeds through the assembly of the meter, the new settings and clearances should be recorded on the meter clearance record in the as-assembled column.

Certain precautions should be taken as the technician assembles the PD meter. Some parts are easily left out. Others, if not properly positioned, can damage the unit as it is assembled. The following items should be considered during assembly:

- When installing the meter cover, the adjusting screw must be retracted so that the rotor does not interfere with tightening the cover, causing internal damage or cracking the cover.
- The manufacturer's recommended gasket sealant or the equivalent should be applied to the mating surfaces before they are assembled and tightened.
- Technicians should make certain that the internal gears are meshed before tightening the cover.
- Technicians should verify that the locating arm slot and the locating pin in the bottom housing are properly aligned with each other.
- Ensure that the rotor adjusting spring is installed.
- Bolts and fittings that may have been removed from the side of the housing during disassembly should have the manufacturer's recommended sealant applied before being installed.
- All index marks on blades and slots should be carefully repositioned during assembly.

3.2.5 Making End Clearance Adjustments

End clearance must be adjusted on all PD meters that are assembled after maintenance and repairs; and clearance is sometimes required on in-service PD meters. These general steps may be followed to adjust the end clearance on meters that do not require the removal of the calibrator assembly from the top to adjust the rotor.

> **NOTE**
> Follow Steps 1 through 3 for PD meters in service.

Step 1 Isolate the meter.

Step 2 Lock out and tag all energy sources.

Step 3 Drain off all pressure and liquid.

Step 4 Locate the access plug on the bottom of the meter.

Step 5 Take out the access plug.

Step 6 Insert the appropriate tool into the access hole.

Step 7 Rotate the meter manually while turning the tool to the left until the rotor begins to bind, signaling the full stop position. (If the rotor is already hard to turn, move the tool to the right until the rotor can be turned.)

> **NOTE**
> Index the position of the tool and the hub of the housing for reference.

Instructor's Notes:

Step 8 Start the rotor turning and begin turning the tool to the right until the rotor just begins to bind; mark this position (full down position).

Step 9 Turn the tool to the left until the index marks on the tool and those on the housing are aligned.

Step 10 Turn the tool clockwise until the mark on the tool is 120 degrees from the mark on the housing.

> **NOTE**
> At the completion of these steps, the rotor should turn freely. If it does not, the procedure should be followed again, making certain all stops are included.

Step 11 When the end clearance is properly set, install the plug with a new gasket.

Step 12 Complete all the required documentation.

Step 13 Complete all the required notifications.

3.2.6 Making Blade Tip Adjustments

PD meters may require an adjustment of the clearance between the blade tips or edges and the housing. In order to adjust the clearance between the blade tips and the housing, technicians should follow the manufacturer's instructions and company procedure. These general steps may be followed as a guide.

> **NOTE**
> Using a tool called a *spider* to center the rotor will make it easier to turn in the housing.

Step 1 Turn the rotor to move a blade into the measuring chamber.

Step 2 Force the blade against the wall of the measuring chamber, and with the appropriate gauge, measure the clearance along the entire blade edge.

Step 3 Record the reading on the meter clearance record column for *as-found readings*.

Step 4 Repeat Steps 2 and 3 for all other blades in the meter.

Step 5 Refer to the clearance chart from the manufacturer's manual to determine if a blade is within the range of tolerance.

> **NOTE**
> If readings are greater than the manufacturer's allowable specifications, the blade assembly must be replaced.

Step 6 If the clearance is too small, the blades affected must be dressed down as directed in the manufacturer's manual.

Step 7 When proper clearance is achieved, record the new clearance in the *as-assembled* column on the meter clearance record.

3.2.7 Adjusting Rotor Clearance

PD Meters may require an adjustment of the clearance between the rotor and the block. In order to do so, you should follow the manufacturer's instructions. These general steps should be followed to adjust the clearance between the rotor and the block:

> **NOTE**
> Clearance should be constant along the entire length of the rotor.

Step 1 Using the appropriate gauge, measure the clearance along the length of the rotor.

Step 2 Record the measurement in the *as-found column* of the meter clearance record.

Step 3 Compare the measurement to the clearance chart in the manufacturer's manual.

> **NOTE**
> Clearance that is more than the manufacturer's specification requires that the rotor be replaced.

Step 4 If the clearance if less than the manufacturer's specifications, technicians should follow the manual's instructions to reduce the block facing, using emery cloth.

Step 5 When proper clearance is achieved, the measurement should be recorded in the *as-assembled column* on the meter clearance record.

Step 7 Assemble the meter.

Step 8 Complete the required documentation

Step 9 Make the required notifications.

Classroom

Emphasize the importance of completing the required documentation.

Explain how to adjust the clearance between the rotor and the block.

Classroom

Review additional components for positive displacement meters that may require replacement.

Teaching Tip

See the Teaching Tip for Sections 3.0.0–3.2.9 at the end of this module.

Classroom

Ask trainees to discuss troubleshooting procedures for metering devices. Drawing on troubleshooting techniques trainees have learned elsewhere, ask trainees to provide possible solutions for common meter problems.

Performance Testing

Have each trainee perform Performance Task 1 to your satisfaction. Fill out a Performance Profile Sheet for each trainee.

Homework

Have trainees review Sections 5.0.0–5.4.0 for the next session.

Materials

Ensure that you have everything required to teach this session.

Audiovisual

Show Transparency 7 (Figure 10). Review the layout of a traditional bidirectional pipe prover.

3.2.8 Adjusting Blade Slot Clearance

PD meters may experience wear on the blade slots, increasing clearance to an unacceptable point. You should consult the manufacturer's instruction manual clearance charts to evaluate the amount of clearance. These general steps may be followed to adjust the blade slot clearance:

Step 1 Insert the appropriate gauge between the slot side and the side of the blade.

Step 2 Record the reading on the meter clearance record in the *as-found* column.

Step 3 Compare the reading to the manufacturer's clearance chart.

> **NOTE**
> If the clearance is too great (slots too wide), the rotor must be replaced.

Step 4 Record the measurement of those slots within tolerance in the *as-assembled* column of the meter clearance record.

Step 5 Complete the required documentation.

Step 6 Make the required notifications.

3.2.9 Replacing Components

PD meters use many parts that are necessary to the operation and accuracy of the meter. Many require a simple mechanical procedure to replace them and are not covered in this module. Some of the procedures that would be involved in this group would include the replacement of:

- Stem assembly adjustment
- Packing gland
- Blade rollers
- Roller cam

4.0.0 ◆ TROUBLESHOOTING METERING DEVICES

When carrying out a troubleshooting procedure, you may find one or more of these common meter problems:

- Meter blocked
- Rotor damaged
- Electrical short in the pickup coil
- Preamplifier failure
- Improper wiring
- Power supply problem
- Bearing failure
- Mechanical stack problem
- Insufficient backpressure

Most of the problems can be repaired in the field, although company policy may require the meter to be taken to a shop. Manufacturer's manuals and recommendations for the exact model should be followed. Mechanical technicians should be careful to verify exact replacements.

5.0.0 ◆ MAINTAINING PROVER SYSTEMS

The *API Manual of Petroleum Measurement Standards, Chapter 4, Section 2, Conventional Pipe Provers, Paragraph 4.2.2* states: "All types of metering devices operate on the common principle of the repeatable displacement of a known volume of liquid from a calibrated section of pipe being displaced between two signaling detectors. Displacement is achieved by means of a slightly oversized sphere or piston that is driven along the pipe by the stream that is being metered. The corresponding metered volume is simultaneously determined." Such is the process of proving. *Figure 10* shows a layout of a traditional bidirectional pipe prover.

In order to carry out proving operations accurately, provers must be kept in good operating condition. Operations personnel should inspect the prover closely before beginning a proving operation. The prover should be inspected periodically by mechanical technicians. Company policy and procedure should be followed along with the manufacturer's instructions and recommendations.

5.1.0 Inspecting Unidirectional Provers

All portable provers should have the grounding system inspected on a regular basis and every time the prover is used.

- Inspect for external leaks.
  - Inspect flanges and fittings.
  - Check hydraulic system.
  - Verify hydraulic ram packing.

Instructor's Notes:

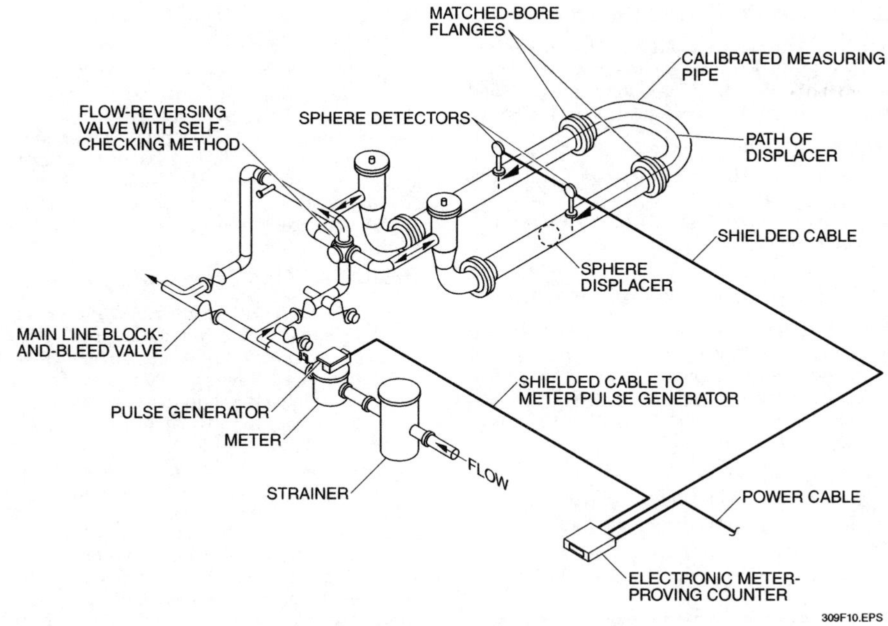

Figure 10 ◆ Layout of a traditional bidirectional pipe prover.

- Inspect the prover sphere. (Prover must be isolated and drained.)
  - Verify the proper sphere texture (color of the ball).
  - Verify agreement with the manufacturer's specification.
  - Verify the proper inflation size (use tape or caliper).
  - Inspect for roundness (in at least two directions).
  - Examine the surface condition for scratches, gouges, and consistency by manually feeling.
- Inspect the prover seals. (The interchange section must be opened.)
  - Visually inspect the seals for scratches, cuts, or gouges.
  - Feel for surface consistency .
- Inspect the backflow barrier.
  - Check for physical damage.
  - Look for missing or broken segments.
- Inspect the detector switches.
  - Clean the switch mechanisms.
  - Verify proper operation.
- Inspect the hydraulic system.
  - Verify the fluid level.
  - Verify the fluid cleanliness.
  - Drain off any water in the reservoir.
  - Verify the specified hydraulic pressures.
  - Verify the appropriate operation of the system.
  - Change the hydraulic fluid according to company procedure.
- Inspect the prover valves.
  - Inspect the inlet, outlet, and bypass valves.
  - Verify valve integrity of block-and-bleed-type valves.
- Inspect the prover piping (if practical).
  - Examine the internal coating.
  - Look for any buildup of foreign material on the surfaces.
  - Look for buildup of product on the surfaces.

Classroom

Discuss the inspection elements for a unidirectional pipe prover.

Classroom

Review the guidelines for inspecting bidirectional, tank-type, and small-volume provers.

Safety

Stress that manufacturer's instructions and recommendations should be followed when working with vapor recovery equipment.

Emphasize the need to inspect the grounding system for all portable provers on a regular basis.

Classroom

Ask trainees to list the different types of prover systems. Have trainees explain how to properly inspect each type of prover. Go over any questions trainees have.

Homework

Have trainees review Sections 6.0.0–6.2.2 for the next session.

Materials

Ensure that you have everything required for the laboratory during this session.

Laboratory

Refer to the *API Manual of Petroleum Measurement Standards, Chapter 4*. Ask trainees to review the information on valve integrity. Discuss the specifications.

5.2.0 Inspecting Bidirectional Provers

All portable provers should have the grounding system inspected on a regular basis and every time the prover is used. Inspection items not included on the list for unidirectional provers includes:

- Inspect the four-way valve.
  - Verify the valve integrity (block-and-bleed valve).
  - Lubricate according to manufacturer's recommendations.

5.3.0 Inspecting Tank-Type Provers

Tank-type provers have certain types of associated equipment that is not found on a permanently mounted pipe prover. A list of items that should be inspected and repaired as necessary, and that are characteristic of a tank-type prover, includes but is not limited to:

- Grounding system
- Leveling equipment (adjusters and level glasses)
- Gauge glasses (broken, cracked, dirty)
- Vapor recovery equipment
- Pressure/vacuum vents

> **CAUTION**
> Manufacturer's instructions and recommendations should be followed for vapor recovery equipment and pressure/vacuum vents.

> **WARNING!**
> All portable provers should have the grounding system inspected on a regular basis and every time before the prover is used. Ground faults can cause serious injury or electrocution.

> **NOTE**
> Any maintenance or repair performed on a prover should be well documented.

5.4.0 Inspecting Small-Volume Provers

Small-volume provers are more technically involved than unidirectional provers are. You should follow the manufacturer's instructions for inspecting and maintaining a small-volume prover. You should have a strong understanding of the operating principles of the small-volume prover.

6.0.0 ◆ REPAIRING PROVER SYSTEMS

Repairing prover systems consists of repairing the valves, inspecting and sizing the spheres, repairing the detector switches, applying internal coatings, and repairing backflow barriers. Only trained and qualified personnel are allowed to make repairs on prover systems and components. Always follow the manufacturer's guidelines and your company policy when performing these activities.

6.1.0 Verifying Valve Integrity

The API *Manual of Petroleum Measurement Standards, Chapter 4, Section 2, Conventional Pipe Provers, Paragraph 4.2.4.6* states "All valves used in pipe prover systems that can provide or contribute to a bypass of liquid around the prover or meter or to leakage between the prover and meter shall be of the block-and-bleed type or an equivalent with a provision for seal verification."

All valves that could contribute to the bypass of any amount of liquid around the prover or meter or to leakage between the prover and meter must be tested and their seals verified according to company policy. If a problem is detected, operations personnel should report the problem to the control center immediately so that the repair can be scheduled. The repair of valves involved in proving operations should only be done by trained and qualified mechanical technicians.

Flow-reversing valves are used on bidirectional provers. Interchange valves are used on unidirectional provers. The purpose of both types of valves is to control the flow through the prover.

All prover valves must demonstrate a verifiable tight seal during the proving run. There can be no leakage of fluid past the valve seal while it is in the closed position. You should inspect the valve by closing it and opening the small leak-detection valve on the side or bottom of the prover valve. If any fluid escapes, the valve is not sealing. Technicians must examine and repair the valve before proving proceeds. The manufacturer's instructions and recommendations for the exact valve model should be followed.

6.1.1 Verifying Valve Cycle Speed

You should time the cycling speed of the valves to make certain that the units are within the manufacturer's specifications. If a diverter valve is not cycling at the proper rate, the power unit should be examined. If the valve operator is turning within the manufacturer's specifications, an operator with a different gear ratio should be selected

Instructor's Notes:

to increase or decrease the valve speed as required. The required speed can be determined by calculating the pre-run time and comparing it to the cycle time of the valve. An operator can be selected who will cycle the valve within the allotted time frame of the pre-run part of the proving run.

6.1.2 Flow Velocity

The velocity of the flow through the prover should be calculated, and if the velocity is higher than is recommended by the manufacturer's specifications for the meter being proved, the control valves should be adjusted to maintain the recommended flow rate. The technician should consult the prover manufacturer's instructions and recommendations for the velocity for the model and the liquid being proved.

Generally, the recommended flow velocity for a bidirectional prover is about 4.5 feet per second (fps) with a maximum of 5 fps. If the flow reaches a higher level, water hammer will become a problem unless there are other compensations built into the prover.

Because a unidirectional prover does not change direction of the flow, velocities up to 10 fps can be tolerated. However, flow velocities above 10 fps may cause turbulence that will not allow the sphere to drop into the chamber and will result in launching problems.

6.1.3 Verifying Prover Velocity

The pre-run section of the prover is the distance of the pipe from the launching chamber barrel to the first detector switch. The prover velocity must be interactive with the speed of the interchange valve or the four-way valve so that the valves have an opportunity to form a complete seal before the displacer reaches the first detector switch. If there is a discrepancy in their timing or the flow velocity, the prover will be inaccurate. Technicians should test the valve cycling speed and the velocity of the flow passing through the prover.

6.2.0 Inspecting and Sizing Proving Spheres

Traditional pipe provers use a sphere or ball as the displacer in the measuring pipe (see *Figure 11*). Spheres are commonly constructed of polyurethane, nitrile, or neoprene. Prover spheres should be removed, inspected, and their size measured on a regular schedule according to manufacturer's recommendations and company policy. Note the composition of the sphere's outer shell to make certain that it is compatible with the product commonly being used in the prover, especially if there has been a recent change in product.

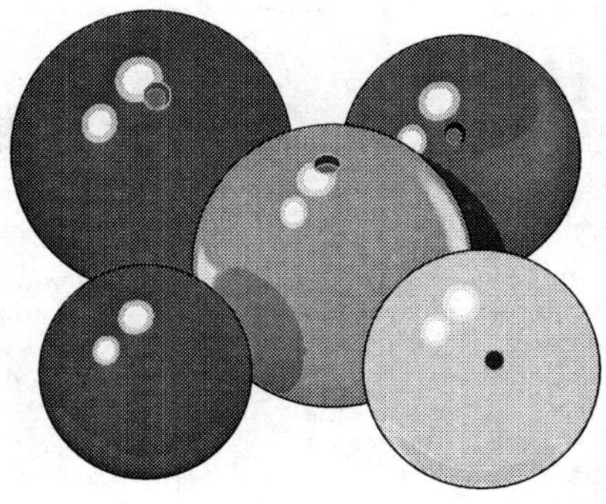

309F11.EPS

Figure 11 ◆ Prover spheres.

The entire circumference of the sphere should be examined closely for any signs of wear, cuts, scrapes, and abrasions that could interfere with efficiency or indicate that failure could occur. Spheres should also be examined for flat spots, out-of-round shapes, or any other phenomenon related to shape. If the sphere shows signs of damage, wear, or being out of shape, it should be replaced with a new one or one that is verified to be in good condition.

> **WARNING!**
> Before attempting to remove a sphere from the prover, the prover must be isolated and decommissioned to avoid injury to personnel or damage to equipment.

6.2.1 Taking Measurements

Measurement of a sphere is accomplished with a caliper made for the purpose, a sizing ring, or a steel tape. Serviceable measurement should be between 1% and 4% larger than the inside diameter of the prover pipe to create a good seal (see manufacturer's recommendations and company policy and procedure for the particular sphere and application). The amount of size increase or decrease needed, if required, can be calculated by multiplying the inside diameter of the prover pipe by 0.01 to 0.04, as recommended, and adding the result to the diameter. Then the ball can be inflated or deflated to achieve that size.

Classroom

Emphasize the importance of consulting manufacturer's specifications when verifying valve cycle speed, flow velocity, and prover velocity.

Safety

Explain how to isolate and decommission a prover before removing the sphere.

Classroom

Review the equipment used to accomplish sphere measurement.

Audiovisual

Show Transparency 8 (Figure 12). Review the typical fittings used for sphere inflation.

Safety

Emphasize the importance of ensuring that the sphere's reservoir is never empty.

Classroom

Review the procedures to fill and size a sphere. Ask whether trainees have any questions.

NOTE
Displacer spheres should never be stored on a flat surface in inflated condition; they should be suspended in a net or placed in a hollowed-out bed of sand to prevent the development of a flat spot.

Inflation or adjustment of sphere size should be accomplished by adding a mix of equal parts of ethylene glycol (anti-freeze) and water. A small funnel, a pump, and fittings are required to increase size (see *Figure 12*). The sphere can be deflated to the required size by releasing fill through the valve. Size can be verified by use of a sizing ring or caliper, or measurement with a steel tape, and then it should be documented.

6.2.2 Performing Sizing Procedure

Spheres vary somewhat from one manufacturer to another; however, the procedure for filling spheres is generally the same. The procedures listed here should be adjusted for the brand and application of the sphere being serviced. To fill and size a sphere, follow these general steps:

Step 1 Pry out the seal with a screwdriver or other appropriate tool.

Step 2 Screw out the valve core by turning it with a core extractor.

Step 3 Using a small funnel or other device, fill the sphere with appropriate liquid.

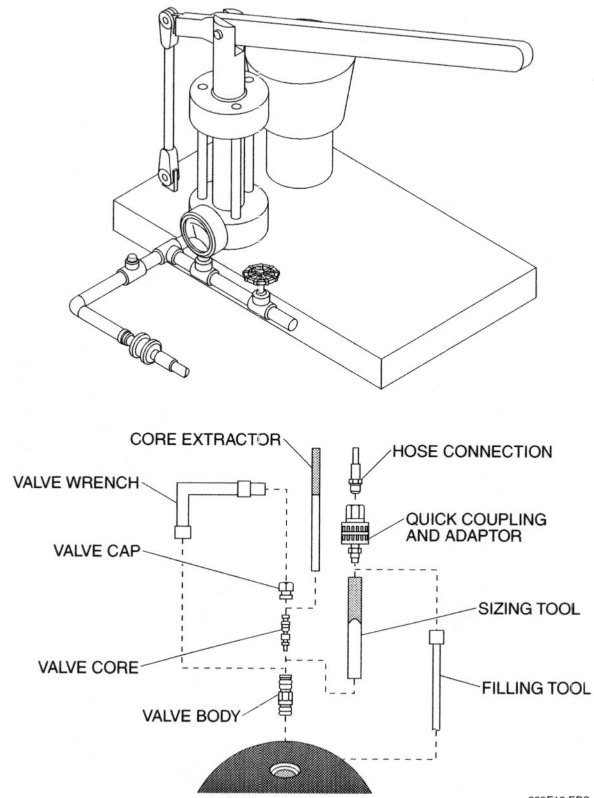

Figure 12 ◆ Typical fittings for sphere inflation.

Instructor's Notes:

> **NOTE**
> It may be necessary to tap or jolt the sphere while filling to make certain that all air is evacuated. On larger spheres, the pump may eliminate any air trapped inside. Only hand-tighten sphere sizing adapters and fittings to avoid damaging the O-rings.

Step 4 Install the valve core by turning it with the core extractor or other appropriate tool.

Step 5 Add the liquid to the reservoir on the pump.

> **CAUTION**
> Ensure that the reservoir is never empty.

Step 6 Attach the pressure hose from the pump to the sizing adapter.

Step 7 Pump until the air is eliminated from the hose and the pump.

Step 8 Attach the sphere to the sizing adapter, tightening only by hand.

Step 9 Pump the sphere to the required circumference.

> **CAUTION**
> When the sphere is filled, release the pressure from the pressure hose with the relief valve before removing the hose fittings to avoid damaging the fittings.

Step 10 Relieve the pressure from the pressure hose.

Step 11 Detach the sizing adapter from the sphere.

Step 12 Replace the valve cap using the appropriate tool.

Step 13 Observe the valve for possible leakage.

> **CAUTION**
> If the sphere valve leaks, replace the entire assembly.

6.3.0 Verifying Detector Switches

Detector switches (see *Figure 13*) signal the passing of the displacer at either end of the calibrated measuring pipe (see *Figure 14*). Each pipe prover requires at least two detector switches to be able to detect the start and finish of each run. Detectors are normally constructed with a ballpoint plunger assembly penetrating the measuring pipe wall. When the sphere passes under it, the plunger is forced upward, tripping the detector switch.

You should inspect and test the detector switches on the prover to verify that they are in good operating condition. Fittings, seals, and wiring should be tight with no leaks. Detectors that have failed or that give questionable signals must be replaced. The manufacturer's recommendations and company policy should be followed to replace the switch. If company policy include cleaning of the detector switch in the field, the manufacturer's instructions and recommendations should be followed for the exact unit.

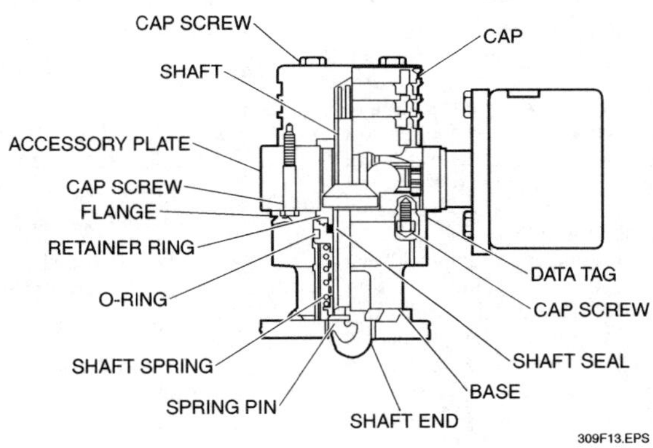

Figure 13 ◆ Layout of a sphere detector switch.

Safety

Discuss the consequences of failing to release the pressure from the hose before removing the hose fittings.

Ask a trainee to explain what to do if the sphere valve leaks.

Teaching Tip

See the Teaching Tip for Section 6.2.0 at the end of this module.

Homework

Have trainees review Sections 6.3.0–6.5.2 for the next session.

Materials

Ensure that you have everything required for the laboratory during this session.

Audiovisual

Show Transparency 9 (Figure 14). Discuss the role of the detector switches in a calibrated measuring pipe.

Classroom

Explain how the depth at which a detection unit is installed affects the point of detection.

Classroom

Review the procedures for replacing a detector switch, and answer any questions trainees have.

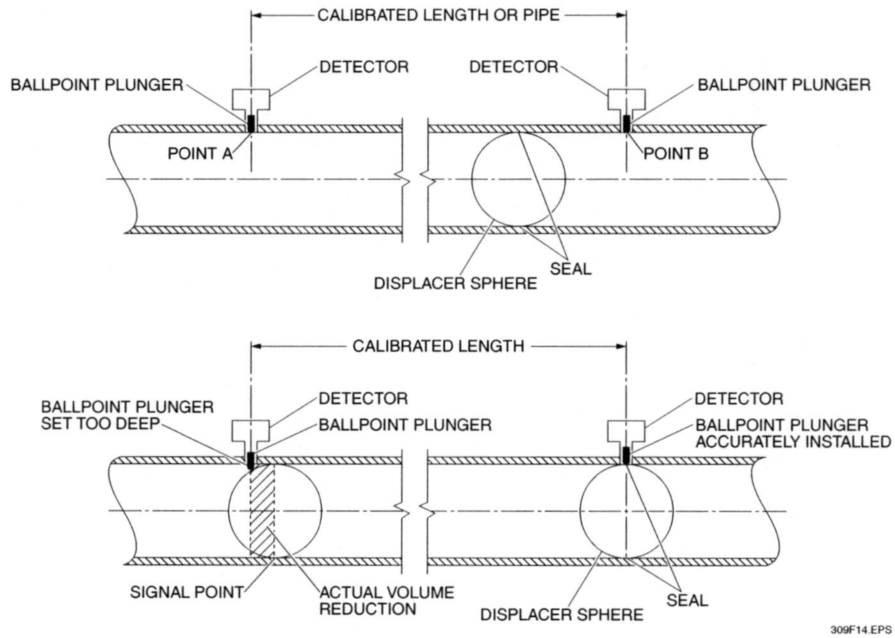

Figure 14 ◆ Calibrated section of a prover measuring pipe.

> **NOTE**
> Some manufacturers are now producing self-calibrating detector switches that need no recalibration after installation.

Installing a new detector switch normally changes the volume that the prover measures by some degree. The point of detection is greatly affected by the depth to which the unit is installed. Installing the switch contact point at too great a depth will cause the switch to trip early, reducing the calibrated volume. Installing the switch at too shallow a depth could have the affect of not causing the switch to trip at all (see *Figure 14*).

The distance between Point A and Point B must be maintained as closely as possible. When installing a new detector switch, you should closely calibrate the point at which the displacer is detected to within 0.02% of the calibrated volume between the switches. The detection point must be repeatable. The prover, especially a unidirectional prover, should be calibrated as soon as possible after a switch has been replaced. *Figure 15* shows sphere detector switches on a portable pipe prover.

> **NOTE**
> Some companies require detector switches to be replaced in sets if there is a problem with one.

In order to replace a detector switch, follow these general steps:

Step 1 Secure the applicable work order and permits.

Step 2 Obtain and follow the appropriate manufacturer's instruction manual.

Step 3 Notify the control center and other affected personnel.

Step 4 Isolate the prover system involved in the replacement of the switch.

Step 5 Lock out and tag all energy sources.

Step 6 Drain off all pressure and any necessary liquid.

Step 7 Using the appropriate tools, remove and replace the detector switch.

Step 8 Restore pressure and flow to the prover.

Instructor's Notes:

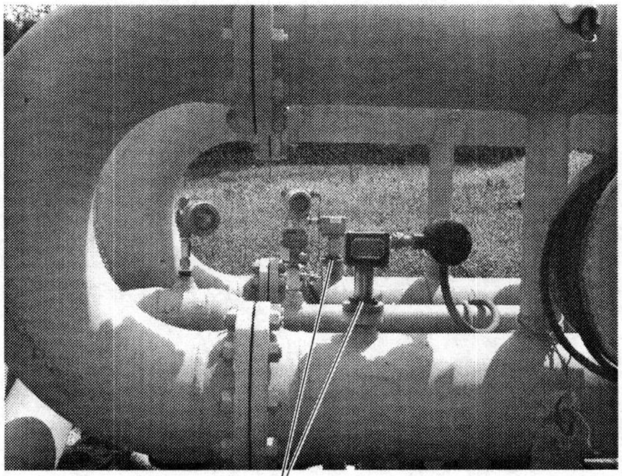

309F15.EPS

Figure 15 ◆ Sphere detectors on a portable prover.

Step 9 Bleed off collected air and vapor from the prover.

Step 10 Operate the prover to make certain the new detector switch is functioning properly.

Step 11 Notify the control center that the work is completed.

Step 12 Complete all documentation of the work done.

6.4.0 Verifying Internal Coatings

According to the API *Manual of Petroleum Measurement Standards*, Chapter 4—*Proving Systems, Section 8, "Operation of Proving Systems," Paragraph 4.8.3.3*, the internal surfaces of prover pipes are normally coated to protect them from corrosion that can result in changes in the pipe's calibrated volume. The coating also provides a smoother surface, reducing drag and wear on the sphere. Internal coatings must be inspected for coating failures, adhesions, and corrosion. Inspectors should pay special attention to the bends of the pipe where wear is often more prominent. In all cases, the manufacturer's manual and recommendations should be consulted and followed.

If the prover has experienced adhesions of foreign material building up, it may be possible to clean away the material. Manufacturer's recommendations and procedures should be followed. When cleaning adhesions, the work should be done carefully to avoid damaging the pipe coating underneath the adhesion. The condition should be investigated closely to determine what has caused the buildup or adhesions and to determine if it might be prevented or reduced in the future.

The development of corrosion creates rough places in the prover. Smoothing up small areas of corrosion might be possible if done with care. Corrosion begins in hard-to-reach areas such as an elbow or a long bend. In larger developments of corrosion, you might need to contact management to decide if the entire unit needs reconditioning. The factory may need to be consulted to determine the best procedure.

6.5.0 Repairing Backflow Barriers

Backflow barriers are installed as part of the interchange mechanism of unidirectional provers to maintain flow through the loop section of the prover during the reloading cycle. The backflow barrier is actually a very tough semi-check valve, not a sealing device (see *Figure 16*).

The function of the backflow barrier is to make certain that a continual flow through the prover loop is maintained while the sealing plug of the prover interchange mechanism is off the seat.

Continual flow through the prover is necessary for the sphere to be able to descend into the sealing ring area, ready for launch. With the backflow barrier in place and operating properly, the

Laboratory

Refer to the *API Manual of Petroleum Measurement Standards, Chapter 4, Section 8*. Ask trainees to review and discuss the information about verifying internal coatings.

Classroom

Ask a trainee to explain how to clean foreign material built up in a prover.

Audiovisual

Show Transparency 10 (Figure 16). Explain the role of backflow barriers in unidirectional provers.

Audiovisual

Show Transparency 11 (Figure 17). Review the operation principles of a backflow barrier.

Classroom

Explain how the sealing plug controls the flow of fluid through the launching mechanism.

Discuss problems that can result if seals leak. Explain how to replace the differential pressure switch.

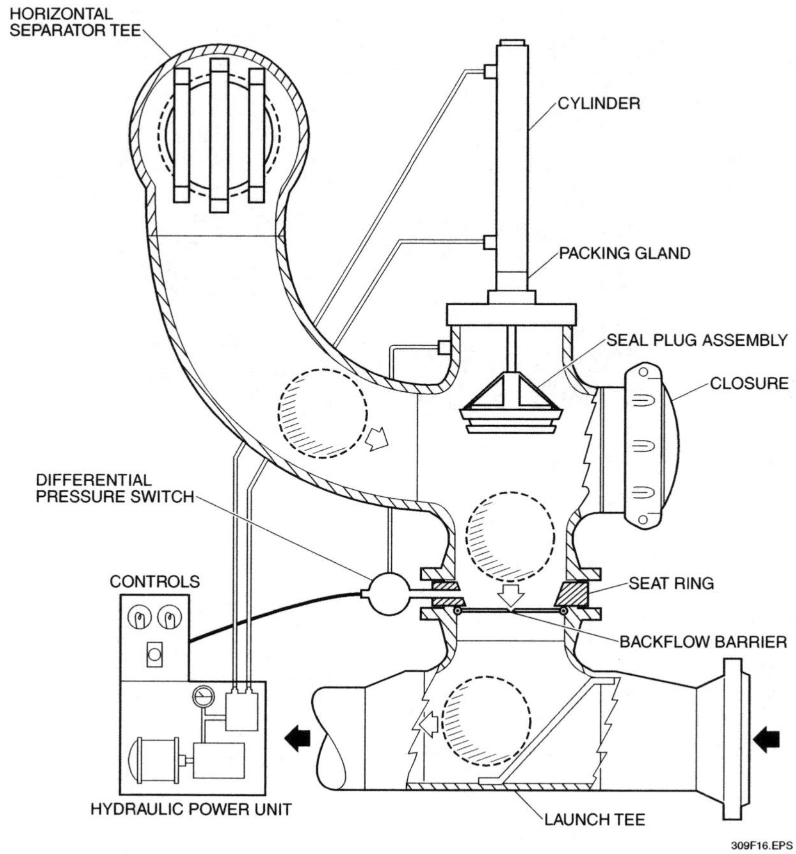

Figure 16 ◆ Unidirectional sphere launcher with backflow barrier.

descent of the next sphere cannot be disrupted by liquid rushing backward, suspending the sphere out of the launching position.

6.5.1 Verifying Operating Principles

When operations personnel signal the prover to begin the proving run, the cylinder lifts the sealing plug off the sealing ring, allowing the next sphere to descend into the launching chamber, coming to rest on the tapered seal ring. At this point, the backflow barrier minimizes backflow up into the interchange mechanism while the sealing plug is in the elevated position (see *Figure 17*).

When the plug seal descends, the sphere is pushed down through the sealing ring, forcing the backflow barrier to open, allowing the sphere to pass into the proving pipe flow. The motion of the sealing plug continues downward onto the tapered seat, automatically sealing the loading mechanism away from the flow of the prover. This sealing action must take place before the sphere travels into the calibrated area of the measuring prover pipe.

With the sealing plug firmly seated in the seal ring during the proving run, all liquid is forced to travel through the measuring pipe. No fluid is allowed to flow through the launching mechanism

Instructor's Notes:

during the proving run. When the proving cycle is over, the mechanical pressure on the sealing plug is reduced to relieve pressure on the sealing rings. However, the sealing plug remains in sealing position.

6.5.2 Verifying Differential Pressure Switch

Differential pressure switches are used on unidirectional provers to verify that there is no seal leakage through the seal plug during the proving run. The sealing plug is designed with a tapered seat that traps liquid and compresses it in the seal area. The pressure is transmitted through a tube and detected by a pressure differential switch (see *Figure 18*). If seal leakage occurs, pressure decreases in the sealing plug area, and the differential switch senses the reduction in pressure.

The differential pressure switch sends an alarm warning to personnel. Many provers are fitted with an automatic shut-off switch. The prover automatically stops the proving procedure if there is a sealing problem. Due to the extreme pressure of the sealing plug, the sealing ring and seat may have become worn or damaged, but they can easily be replaced on site. If the differential pressure switch fails, it can also be replaced easily.

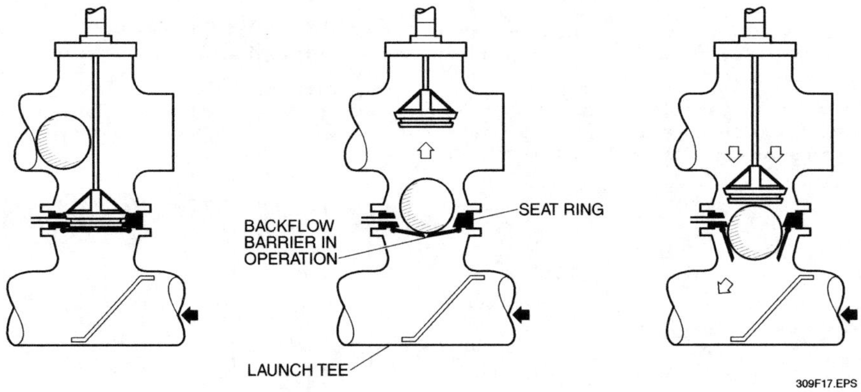

Figure 17 ◆ Backflow barrier in operation.

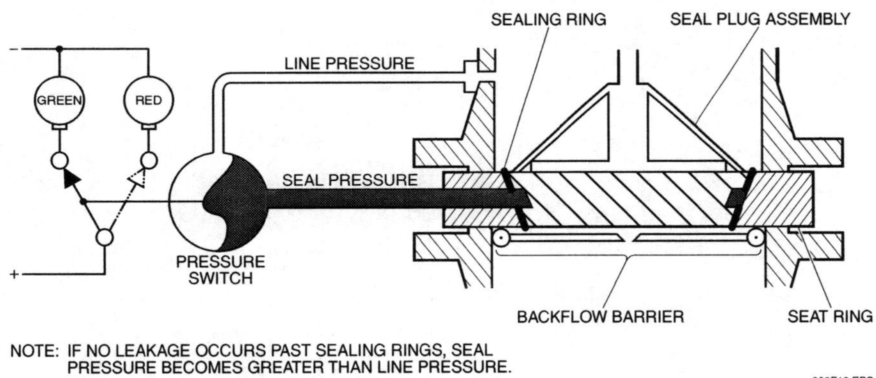

Figure 18 ◆ Backflow barrier pressure differential switch.

Homework

Have trainees review Sections 7.0.0–7.3.0 for the next session.

Materials

Ensure that you have everything required for the laboratory during this session.

Classroom

Review the applications for which water calibration can be used to calibrate provers. Discuss alternative methods.

Laboratory

Refer to the *API Manual of Petroleum Measurement Standards, Chapter 12.* Ask trainees to review and discuss the specifications for calibrating prover systems.

Classroom

Explain how to perform waterdraw procedures. Stress the importance of verifying the reliability of calibration runs. Go over any questions trainees have.

7.0.0 ◆ CALIBRATING PROVER SYSTEMS

The *API Manual of Petroleum Measurement Standards, Chapter 4, Proving Systems, Section 2, Conventional Pipe Provers, Paragraph 4.2.11* states:

> *A pipe prover must be calibrated before it is placed in service to determine its base volume at standard conditions [usually atmospheric pressure and 600F (150C)]. Periodic recalibration of the prover is also required. The base volume shall be documented on a calibration certificate....The base volume for a unidirectional prover is the calibrated volume corrected to standard temperature and pressure conditions and displaced between the detectors for a single pass. The base volume of a bidirectional prover is the sum of the volumes displaced between detectors for a round trip of the displacer corrected to standard temperature and pressure conditions.*

In order to be accepted for use in custody transfer and financial transactions, the volume corrections must be calculated according to the *API Manual of Petroleum Measurement Standards, Chapter 12, Calculation of Petroleum Quantities, Section 2, Calculation of Petroleum Quantities Using Dynamic Measurement Methods and Volumetric Correction Factors.*

7.1.0 Calibrating Waterdraw

Waterdraw calibration (see *Figure 19*) is considered a trustworthy method although master meter calibration is used in some cases where waterdraw methods are not convenient. Water is the standard liquid used in calibration because its properties are stable and well known.

7.2.0 Performing Waterdraw Procedures

Preliminary preparations should follow company policy including all safety procedures and notifications. The prover should be clean and isolated from the pipeline.

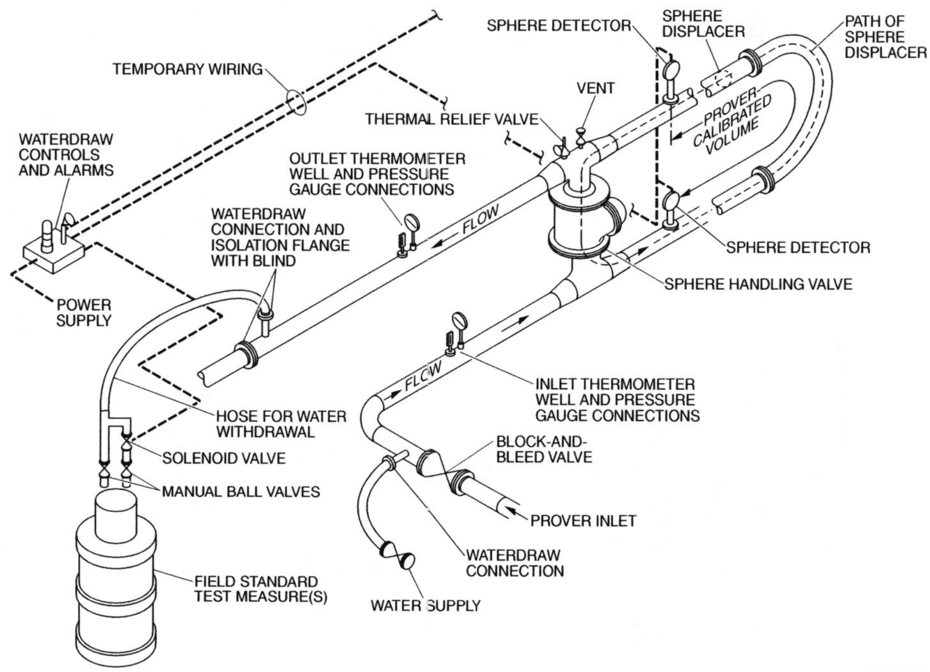

Figure 19 ◆ Waterdraw calibration of a unidirectional prover.

Instructor's Notes:

Step 1 Secure all applicable work orders and permits.

Step 2 Collect a supply of appropriate forms and manuals, as applicable to the installation.

Step 3 Arrange a substantial supply of clean water.

Step 4 Assemble a set of the largest volumetric field-standard test measures available.

Step 5 Connect lines and equipment for the water draw to the prover following company policy.

Step 6 Notify the control center that calibration is beginning.

It is recommended that one preliminary run be made so that the approximate volume of water can be measured. Selection of the appropriate number and sizes of volumetric field-standard test measures can be made and assembled for the operation. After all preliminary preparations have been made, follow these general steps to perform a waterdraw calibration:

Step 1 Observe the temperature and pressure of the preliminary discharge, and record those for the start of calibration.

Step 2 Continue with the series of calibration runs, carefully observing, recording, and calculating the volume from all volumetric field-standard test measures and other factors for each prover run.

Step 3 Make required corrections for temperature, pressure, and any other factors required to achieve an accurate base volume according to guidelines from *API Manual of Petroleum Measurement Standards, Chapter 12, Calculation of Petroleum Quantities, Section 2, Calculation of Petroleum Quantities Using Dynamic Measurement Methods and Volumetric Correction Factors.*

> **NOTE**
> The volume of each prover pass shall be corrected individually to obtain the base prover volume. It is implied that the volume is at a standard condition of 60°F (15°C).

Step 4 Examine the results to verify that two or more consecutive runs fall within 0.02% (plus or minus 0.01%) of the average of all runs.

Step 5 Complete all the required documentation.

7.3.0 Verifying Reliability

Since all the initial calibration runs are made at the same pressure and flow rate, it is possible that a consistent leak exists that has not been detected, and therefore, all calculations are in error. To verify that the initial calibrations are reliable, another calibration run should be executed with a flow rate that is changed by at least 25%. With this change in the flow rate, the resulting prover volume should be the same as the original (within 0.02%, plus or minus 0.01%). If it is not, the possibility of a leak exists, and it must be repaired before a reliable calibration can be performed.

Summary

Metering and proving systems are at the heart of pipeline business integrity. The operation of metering and proving systems controls the accuracy of financial exchanges between company and customer. The importance of their condition and accuracy cannot be stressed too much. It is the responsibility of the mechanical technician to maintain, troubleshoot, and repair these systems and related components. Such work requires technicians who are trained and qualified to do the work efficiently.

Metering and proving systems must be maintained in such a way that their installation and operation meet API standards. Mechanical technicians should be completely familiar with API standards relating to meters, provers, and their related components before attempting to perform work on a unit.

Mechanical technicians must also be completely familiar with company policy and procedures relating to the maintaining of metering and proving systems. Mechanical technicians must be trained and qualified in order to secure this area of company business. When you have finished this module, you will have gained important skills to enable you to become a trained and qualified mechanical technician.

Teaching Tip

See the Teaching Tip for Sections 7.0.0–7.3.0 at the end of this module.

Homework

Have trainees answer the Review Questions and prepare for testing during the next session.

Materials

Ensure that you have everything required for testing during this session.

Classroom

Go over the Review Questions for this module, and answer any questions trainees have.

Review Questions

1. _____ establishes the level of accuracy of each individual meter.
   a. Proving
   b. Certifying
   c. Inspection
   d. Monitoring

2. Many of the important parts of meters and provers cannot be inspected unless the units are _____.
   a. standard
   b. disassembled
   c. turned off
   d. operating

3. _____ elements are used in turbine meter installations to reduce the swirling motion in the fluid.
   a. Tempering
   b. Pressurizing
   c. Turbulizing
   d. Straightening

4. _____ are installed above meters.
   a. Mechanical stacks
   b. Strainers
   c. Straighteners
   d. Turbines

5. The counter drive train, calibrator, and mechanical stack are connected by _____.
   a. couplings
   b. small gears
   c. cables
   d. the shaft

6. The calibrator _____ makes it harder to achieve repeatability when the meter is being proved.
   a. clutch slippage
   b. meter rotation
   c. alignment
   d. gears

7. Strainers are installed in the pipeline _____ of meters.
   a. on both sides
   b. downstream
   c. upstream
   d. on top of

8. Some strainers are available that can be _____ to clean them, the material from which is then diverted to a sump.
   a. filtered
   b. flushed
   c. scraped
   d. backwashed

9. _____ must be prohibited near a strainer cleaning procedure.
   a. Draining
   b. Hot work
   c. All vehicles
   d. Maintenance

10. Company policy and procedure design states how much change in the meter _____ is allowable before intervention.
    a. accuracy
    b. motion
    c. factor
    d. sensitivity

11. Graphite bearings are easily _____.
    a. cleaned
    b. broken
    c. removed
    d. scratched

12. Most _____ meters can be disassembled and repaired in the field.
    a. PD
    b. turbine
    c. mechanical
    d. flow

13. PD meter _____ must be replaced as a unit if any part is damaged.
    a. gaskets
    b. flanges
    c. bearings
    d. rotors

14. _____ must be adjusted on all PD meters that are assembled after maintenance.
    a. Alignment
    b. Springs
    c. End clearance
    d. Internal gears

Instructor's Notes:

15. All _____ should have the grounding system inspected on a regular basis.

a. strainers
b. portable provers
c. straightening elements
d. mechanical stacks

16. _____-type provers have certain types of associated equipment that are not found on a permanently mounted prover.

a. Sphere
b. Small-volume
c. Bidirectional
d. Tank

17. _____ valves are used on bidirectional provers.

a. Flow-reversing
b. Interchange
c. Relief
d. Globe

18. Mechanical technicians should time the _____ of the valves.

a. output
b. cycling speed
c. operation
d. opening

19. Traditional pipe provers use a _____ as the displacer.

a. plug
b. bypass
c. sphere
d. piston

20. Installing a new detector switch normally changes the _____ that the prover measures.

a. mass
b. volume
c. point of detection
d. direction

21. The internal surfaces of prover pipes are normally _____.

a. roughened
b. irradiated
c. textured
d. coated

22. The function of the _____ is to make certain that a continual flow through the prover loop is maintained.

a. backflow barrier
b. detector switch
c. sphere valve
d. sealing plug

23. Differential pressure switches are used on _____ provers.

a. all
b. bidirectional
c. unidirectional
d. tank-type

24. _____ is the standard liquid used in calibration.

a. Oil
b. Water
c. Saline solution
d. Liquefied gas

25. To verify that the initial calibrations are reliable, another calibration run should be executed with a flow rate changed by at least _____%.

a. 15
b. 25
c. 35
d. 50

Examination

Administer the Module Examination. Be sure to record the results of the Exam on Craft Training Report Form 200 and submit the results to the Training Program Sponsor.

Performance Testing

Have each trainee perform Performance Tasks 2 and 3 to your satisfaction. Fill out a Performance Profile Sheet for each trainee.

Ensure that all Performance Tests have been completed and Performance Profile Sheets for each trainee are filled out. Be sure to record the results of the testing on Craft Training Report Form 200 and submit the results to the Training Program Sponsor.

Instructor's Notes:

REFERENCES & ACKNOWLEDGMENTS

Additional Resources

This module is intended to present thorough resources for task training. The following reference work is suggested for further study. This is optional material for continued education rather than for task training.

Manual of Petroleum Measurement Standards. Washington, DC: American Petroleum Institute.

Figure Credits

John Copeland, Industrial Development Corp. 309F04, 309F15, 309F17

MODULE 63309-02 — TEACHING TIPS

The following are suggested activities or instructional methods to help you teach the material in this AIG.

General

When you call on someone to answer a question, the rest of the class relaxes or even tunes out because they expect that the question and answer will take place only between you and the trainee you called on. Instead, use this technique to involve more trainees in answering questions and to keep them on their toes.

1. Ask trainees to define a term or explain a concept.
2. After one trainee has answered, ask a trainee seated nearby if the answer is right. Then ask whether a trainee in the back of the room agrees.
3. Ask trainees to explain why they think an answer is right or wrong.
4. Use the session to clear up incorrect ideas and encourage trainees to learn from their mistakes.

Sections 3.0.0–3.2.9 *Repairing Metering Devices*

Measuring and adjusting clearance is an important part of repairing a positive displacement meter. This exercise will familiarize trainees with the procedures to measure and adjust clearances. For this exercise, you will need practice meters, appropriate measuring gauges, and meter clearance records. Trainees will need their Trainee Guides and pencils and paper to take notes for discussion. Allow 30 minutes for this exercise.

1. Set up a variety of stations with practice meters and the appropriate measuring gauges for performing the following tasks:
   - Making end clearance adjustments
   - Making blade tip adjustments
   - Adjusting rotor clearance
   - Adjusting blade slot clearance
2. Ask trainees to visit each station, measure the appropriate clearance, and record the reading on the meter clearance record.
3. Review trainees' answers and discuss any problems trainees had performing the procedures.
4. Answer any questions trainees have about repairing metering devices.

Section 6.2.0 *Inspecting and Sizing Proving Spheres*

This exercise will give trainees the opportunity to practice the procedures to fill and size a sphere. For this exercise, you will need proving spheres, calipers, sizing rings, and steel tape. Trainees will need their Trainee Guides and pencils and paper to take notes for discussion. Allow 20 minutes for this exercise.

1. Set up stations with proving spheres, calipers, sizing rings, and steel tape.
2. Ask trainees to practice the procedures for inspecting and sizing proving spheres. Emphasize that the procedures vary according to the brand and application of the sphere.
3. Have trainees add liquid to the reservoir to test for leakage.
4. Review and assess trainees' work. Answer any questions trainees have about inspecting and sizing proving spheres.

Sections 7.0.0–7.3.0

Calibrating Prover Systems

This exercise will give trainees the opportunity to observe prover calibration procedures. For this exercise, you will need to locate a nearby site where a waterdraw calibration procedure is scheduled. Trainees will need their Trainee Guides and pencils and paper to take notes for discussion. Allow at least 60 minutes for this exercise.

1. Arrange a visit to a local site where a waterdraw calibration procedure is scheduled.
2. Ask the operations personnel to explain the scheduled procedure to the trainees.
3. Ask trainees to observe the procedures and assist as appropriate. Have trainees take note of any corrections or adjustments to temperature, pressure, and other factors.
4. Following the site visit, review trainees' observations.
5. Answer any questions trainees have about adjusting and repairing hydraulic valve actuators/operators.

MODULE 63309-02 — ANSWERS TO REVIEW QUESTIONS

Answer	Section
1. a	1.0.0
2. b	2.0.0
3. d	2.1.0
4. a	2.2.0
5. b	2.2.0
6. a	2.2.0
7. c	2.3.0
8. d	2.3.0
9. b	2.3.2
10. c	3.0.0
11. b	3.1.0
12. a	3.2.0
13. d	3.2.2
14. c	3.2.5
15. b	5.1.0
16. d	5.3.0
17. a	6.1.0
18. b	6.1.1
19. c	6.2.0
20. b	6.3.0
21. d	6.4.0
22. a	6.5.0
23. c	6.5.2
24. b	7.1.0
25. b	7.3.0

CONTREN™ LEARNING SERIES — USER UPDATES

The NCCER makes every effort to keep these textbooks up-to-date and free of technical errors. We appreciate your help in this process. If you have an idea for improving this textbook, or if you find an error, a typographical mistake, or an inaccuracy in NCCER's Contren™ textbooks, please write us, using this form or a photocopy. Be sure to include the exact module number, page number, a detailed description, and the correction, if applicable. Your input will be brought to the attention of the Technical Review Committee. Thank you for your assistance.

Instructors – If you found that additional materials were necessary in order to teach this module effectively, please let us know so that we may include them in the Equipment/Materials list in the Instructor's Guide.

Write: Curriculum Revision and Development Department
National Center for Construction Education and Research
P.O. Box 141104, Gainesville, FL 32614-1104

Fax: 352-334-0932

E-mail: curriculum@nccer.org

Craft ______________________ Module Name ______________________

Copyright Date ____________ Module Number ____________ Page Number(s) ____________

Description

____

____

____

____

(Optional) Correction

____

____

____

(Optional) Your Name and Address

____

____

____

Index

Index